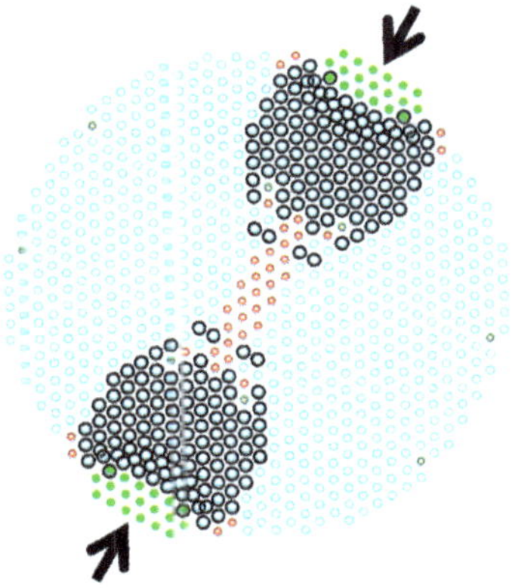

INTRODUCTION TO PRACTICAL PERIDYNAMICS

Computational Solid Mechanics Without Stress and Strain

Frontier Research in Computation and Mechanics of Materials and Biology

ISSN: 2315-4713

Series Editors: Shaofan Li *(University of California, Berkeley, USA)*
Wing Kam Liu *(Northwestern University, USA)*
Xanthippi Markenscoff *(University of California, San Diego, USA)*

Vol. 1 Introduction to Practical Peridynamics:
Computational Solid Mechanics Without Stress and Strain
by Walter Herbert Gerstle

Frontier Research in Computation and Mechanics of Materials and Biology – Vol. 1

INTRODUCTION TO PRACTICAL PERIDYNAMICS

Computational Solid Mechanics Without Stress and Strain

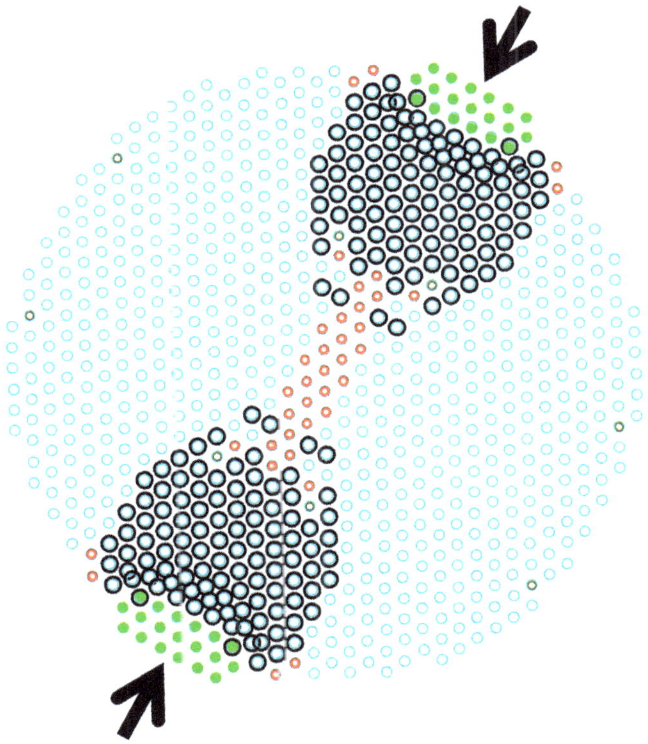

Walter Herbert Gerstle
University of New Mexico, USA

World Scientific
NEW JERSEY • LONDON • SINGAPORE • BEIJING • SHANGHAI • HONG KONG • TAIPEI • CHENNAI • TOKYO

Published by

World Scientific Publishing Co. Pte. Ltd.

5 Toh Tuck Link, Singapore 596224

USA office: 27 Warren Street, Suite 401-402, Hackensack, NJ 07601

UK office: 57 Shelton Street, Covent Garden, London WC2H 9HE

Library of Congress Cataloging-in-Publication Data
Gerstle, Walter.
Introduction to practical peridynamics : computational solid mechanics without stress and strain / Walter Herbert Gerstle, University of New Mexico, USA.
pages cm. -- (Frontier research in computation and mechanics of materials and biology)
ISBN 978-9814699549 (hardback : alk. paper)
1. Mechanics, Applied--Mathematics. 2. Materials--Mathematical models.
3. Solids--Mathematical models. I. Title.
TA342.G47 2015
620.1'054--dc23
2015020143

British Library Cataloguing-in-Publication Data
A catalogue record for this book is available from the British Library.

In-house Editor: Amanda Yun

Dedication

In memory of my father and mother
Kurt and Eva Gerstle

To my wife and son
Irene and David Gerstle

And, of course, to all of my students

Preface

This book proposes a departure from business-as-usual in the computational simulation of solids. Solid mechanics has, until now, usually been framed in terms of stress and strain. Indeed, the modern concept of "material" arises from, and is almost inseparable from, the concepts of strain and stress, which Augustin-Louis Cauchy invented in the early 1820s. The Navier-Cauchy linear partial differential equations of elasticity, as well as the Navier-Stokes equations of fluid mechanics, served as the archetype from which the more general discipline of continuum mechanics arose in the 1950s, primarily through the publications of Truesdell and Noll. Concurrently, to model cracking, which continuum mechanics fails adequately to address, Griffith, Irwin and many others invented the discipline of fracture mechanics.

With the advent of computers in the 1950s, structural engineers sought methods to solve realistic problems in structural mechanics, and the finite element method provided the necessary bridge that allowed engineers to solve the Navier-Cauchy equations of linear elasticity on a computer. The finite element method was wildly successful, and it continues to be very important in analyzing and designing countless modern technologies.

Following the success of the finite element method in solving problems of linear elasticity, engineers also began to solve nonlinear problems including such behaviors as plasticity, damage, creep, and fracture. However, a number of difficulties emerged; principally engineers found that the theory of continuum mechanics was inadequate to solve fracture problems, which are discontinuous at their core.

Most researchers, however, were reluctant to deviate from the theory of continuum mechanics, so they patched up the theory, with contrivances such as singularity elements, the crack band model, discrete fracture propagation models, and localization limiters. Researchers for the most part continued to toe the continuum mechanics line.

I have taught structural analysis, mechanics of materials, advanced mechanics of materials, finite elements, and fracture mechanics during the thirty years of my academic career. In recent years, computers have become more and more powerful. I, and many others, have become increasingly uncomfortable teaching engineering practices involving assumptions that ignore the power of computers – and which are therefore perhaps unlikely to survive the current computer revolution. We no longer teach once-important engineering practices like the moment distribution method and the conjugate beam method. We now teach matrix methods in every engineering curriculum. Engineering education continues to change in response to improvements in computer technology.

Digital computers are not capable of directly representing the basic ideas of continuum mechanics (like continuous analytic fields of strain and stress). Consequently, continuum mechanics theories must be discretized, in the form of a finite element (or other discrete) model, to be represented by a computer.

The situation is currently very peculiar. Why do we take a discrete physical model, and then through mathematical gymnastics (like taking limits) turn it into a set of partial differential equations, only to then later, through yet more mathematical manipulation, turn the problem back into a set of algebraic equations that can only then be represented and solved by a computer?

Why not just start from the outset, in describing our physical model, with a discrete algebraic computational model, directly solvable on the computer?

This book strives for the simplest possible computational model for solid bodies that both yields practical engineering results and that also connects in a meaningful way with the human mind.

I have written this book both as a treatise and as a textbook for a graduate-level class in computational solid mechanics, and therefore it includes chapters that introduce the classical theory of elasticity, continuum mechanics, fracture mechanics, peridynamics, solid modeling, plasticity, and damage mechanics. However, I introduce all of these theories with a critical eye, because our ultimate goal is to develop a theory that is at the same time sufficiently true to physical reality, sufficiently comfortable for the human mind, and sufficiently computer-friendly. While important physical behaviors must be included, our model both demands and provides no more precision than is warranted by the solid materials that we seek to represent. While this book focuses on solids, the methods described herein are quite general, and one can extend these methods to many other physical phenomena (like fluids, thermo-mechanics, electromagnetics, and soils).

Fundamental inspiration for this book comes from my father, Kurt H. Gerstle, of the University of Colorado, graduate advisor Anthony R. Ingraffea, of Cornell University, and Stewart Silling, the inventor of peridynamics, of Sandia National Laboratories. Professor Shaofan Li, of University of California, Berkeley, invited me to write this book. I appreciate the encouragement and guidance provided by my colleague and friend Timothy Ross.

Thanks go to my graduate students who have helped to develop the state based peridynamic lattice method described in this book. They are listed in the sequence that they worked with me: Nicolas Sau, Eduardo Aguilera, Navid Sahkavand, Kiran Tuniki, Asifur Rahman, Hossein Honarvar, Raybeau Richardson, Aziz Asadollahi, Seth McVey, and Shreya Vemuganti.

I also thank the University of New Mexico for allowing me the time to write this book during a sabbatical leave, and Susan Atlas, director of the Center of Advanced Research Computing at UNM, for collaborating with me in the development of the parallel particle simulation code, pdQ, and for hosting me during the writing of this book.

Walter Herbert Gerstle
University of New Mexico,
Albuquerque, New Mexico

June 2015

Contents

Chapter 1

Deformable Solids

Chapter objectives are to:

- explain the purpose of this book;
- review the historical development of solid mechanics;
- discuss current methods for solid modeling;
- explain why existing theories of solid mechanics are incomplete;
- introduce the peridynamic model;
- introduce discrete lattice-based geometric models of solids;
- describe the capabilities and limitations of digital computers;
- describe the scope of this book.

Engineering is the application of mathematics, science, and human ingenuity to devise the trappings of civilization. Today we call the early engineers and scientists of the Age of Reason "natural philosophers", and most of them were practical people who invented abstract engineering models at the same time as they were designing and constructing philosophical frameworks and technologies to serve humanity.

Emerging technology – ships, trains, airplanes, dams, high-rise buildings – required ever better understanding of the behavior of deformable solid bodies. Imagine trying to design a jumbo jet without an understanding of the strength and stiffness of deformable bodies! Natural philosophers started to develop quantitative models for the behavior of solid bodies about 350 years ago, resulting in what we now (too grandiosely) call the "theory of elasticity". All of these models require judicious simplifications of the real physical world. The "theory of elasticity" is really no theory at all, involving as it does many engineering judgments and simplifications. We should more appropriately call it a "model of elasticity". Other models for deformable solids are possible, and some of these are more appropriate to our time.

About sixty years ago, digital electronic computers came into being, rendering some of the early models of deformable solids obsolete. For example, engineers no longer use graphical methods to solve truss problems, or the moment-distribution method to solve frame problems.

The digital computer opens up new possibilities for computational methods to simulate solids, fluids, and other forms of matter. We focus upon solids in this book.

The purpose of this book is both critically to review the long-standing methods of solid mechanics and to develop new methods that are better suited to our modern computerized world.

We focus upon the development of practical new method of computational solid mechanics called the "state-based peridynamic lattice model – SPLM".

1.1 Difficulties with differential equations

In engineering practice, the discipline of continuum mechanics suffers from three principal difficulties:

1. It is difficult to comprehend;
2. It is difficult to implement on digital computers; and
3. It has difficulty in simulating observed behaviors of materials.

One purpose of this book is to provide a practical, yet rigorous, alternative to continuum mechanics.

Classical elasticity and its generalization, continuum mechanics, have become dominant over the years. We pose the classical theory of elasticity as a set of linear partial differential equations. However, prior to the advent of computers, the general theory of elasticity was much too impractical for direct engineering application, so to be practical for engineering design and analysis, simplifications of the general theory of elasticity, particularly Euler-Bernoulli beam theory, Timoshenko beam theory, and various theories for plates and shells, have been developed. Thus, the analysis of deformable solid bodies has really split into two disciplines: "mechanics of materials", and "continuum mechanics". In "mechanics of materials", reasonable approximations are made (like plane cross-sections of axial members remain plane after deformation) to enable closed-form analytical solutions to be developed for certain classes of structures (like axial members subject to tension, bending, shear, and torsion).

"Mechanics of materials", being posed as reasonably easily solved ordinary differential equations, is comfortable for hand calculations, and is easily understood by the human mind; really, it has become the bread-and-butter of engineering design. However, the mechanics of materials approach involves many approximations, and it is limited in application to axial members, and perhaps flat plates and pressure vessels. It is not a general approach, and involves many simplifications and perhaps oversimplifications. One must exercise much engineering judgment in applying the methods of mechanics of materials in the design of structures.

The "linear theory of elasticity" is more general than the mechanics of materials approach, but still is incapable of modeling many important features of the deformation of solids. The linear theory of elasticity assumes both small and spatially continuous deformation, as well as linear stress-strain behavior, and these assumptions are unrealistic except in very special conditions. Physically unrealistic behaviors such as singular strain and stress fields emerge in the solutions of the equations of the linear theory of elasticity.

On the other hand, with "continuum mechanics", we seek and achieve much more generality, with the consequence that the resulting partial differential equations are nonlinear and highly mathematical and are, with rare exceptions, much too difficult to solve analytically in closed form. Continuum mechanics provides a framework for solving problems in which deformations may be large and the relationship between strain and stress may be nonlinear.

Despite the generality achieved by continuum mechanics, it is still insufficiently general to model solids that might potentially deform in a discontinuous manner. In fact, continuum mechanics assumes that all fields associated with a material body are continuous. Consequently, strictly speaking, continuum mechanics cannot tolerate discontinuous features like corners and cracks, which necessitate discontinuous fields describing the surfaces of the body. In addition, the limiting processes required for defining the intrinsic objects of continuum mechanics, for example density, stress, and strain are not justified in many situations. In light of the power and emerging dominance of electronic digital computers, and because materials do not always deform continuously, the continuum mechanics model is not always best.

Continuum mechanics, cast in terms of partial differential equations, is not well suited for modern digital computers because exact solutions to partial differential equations with boundary conditions (boundary value problems) are, in general, impossible to obtain. Finite element, boundary element, and finite difference methods provide only approximations of these boundary value problems. In most problems, for example at re-entrant corners, the computational solution is radically, even infinitely, wrong. Although the computational solution may provide useful information to a skilled engineer, years of experience are required to develop adequate judgement, and even then, engineers often draw erroneous conclusions from the computational solution.

The situation is ironic. With continuum mechanics, we start with a material body, which is perhaps most easily thought of as a system of discrete objects (be they atoms, molecules, crystallites, or bits of sand and gravel), and then through the "smearing" concept of "mass density", we assume that this collection of discrete objects is approximated as a spatially continuous "continuum material body". In conceptually transforming the collection of discrete particles into a continuous medium, however, we lose any information we might have had regarding the size scale of the constituent particles making up the body. Because we now define the body within a real analytic space[a], it is amenable to the methods of real functional analysis (also called calculus). Because continuum mechanics assumes that all properties of the body possess spatial continuity, it is necessary to define the analytical boundaries of the body using piecewise-continuous spatial functions called surfaces

[a] "Real analysis" is a classical discipline, started by Descartes, and developed further by Newton and Leibniz, who invented calculus. Euler, Bolzano, and Cauchy later developed the subject. The essential idea of real analysis is the notion of continuity. Bolzano rigorously defined continuity in 1816.

and edges. These analytical bounding surfaces and edges are crisp – although the boundaries of the body are in reality fuzzy, because the concept of mass density breaks down near the boundaries of the body. Continuum mechanics assumes that the body consists of "continuum particles" – a considerably different concept than the concept of discrete particles. The material body is assumed to occupy a domain in a real three-dimensional Cartesian space, $\mathbb{R}^3$. The space of material particles, called the *material space*, is *not* a physical space – rather, it is an abstract space describing the material particles. Prior to deformation, we endow these continuum particles with a property called the "initial reference position", which, in continuum mechanics, is assumed to be a *continuous* mapping from the material space to the physical space.[b] After deformation due to some type of applied loading, the particles move, as a *continuous* mapping, to new positions in physical space, called the "spatial positions". To get a handle on deformation of a material particle, an entity called the "deformation gradient, $\boldsymbol{F}$" is developed, which, after being decomposed into a rigid-body component and a deformational component, provides a description of the spatial "change in shape" of the body at a material point.[c] Using the deformation gradient, an entity called "strain, $\boldsymbol{\varepsilon}$" is defined, along with an associated work-conjugate entity called "stress, $\boldsymbol{\sigma}$".[d] By means of a material constitutive law relating stress to strain (or perhaps stress to strain *rate*), the stress at a material point is determined. The divergence of the stress field at a material point gives the force per unit volume acting at each point within the material. Using Newton's second law, the acceleration of each of the infinite number of particles is thus determined. This entire model is analytical, and except for a few simple cases, one can find no closed-form mathematical solutions. Because the number of particles is infinite, one cannot evaluate the model numerically until one first discretizes it. Thus, engineers developed the finite difference method and the finite

---

[b] Material space has no physical dimensions; physical space has dimensions of physical length.

[c] The deformation gradient $\boldsymbol{F}$, a tensor quantity, requires a continuous mapping from the reference to the spatial configuration. Otherwise, the gradient would be undefined.

[d] Both strain and stress are tensors.

element method, with which they converted the analytical solid model back into a discrete model, with a finite number of grid points, or nodes, and finite elements, and degrees of freedom. They could then evaluate numerically the discretized computational model on a digital computer, although the resulting solution only approximately satisfied the original differential equations of motion.

The question arises: since computers deal with floating point numbers and arithmetic operations, not real numbers and the associated operations of calculus, like differentiation, integration, divergence, gradient, and curl, would it not be much simpler just to start off with a discrete, rather than an analytical, model of the solid body?

Of course, the early developers of continuum mechanics – specifically Navier and Cauchy in the 1820's – had very good reasons for wanting to consider solids as continua rather than as a finite number of discrete particles. The early developers simply did not have the computational power to model solid deformable structures as a collection of many interacting particles. However, with modern digital computers, it now makes sense, at least in some cases, to consider structures as collections of a large, but finite, number of interacting discrete particles.

Another problem with continuum mechanics is that of geometric size scale. The traditional elasticity and continuum mechanics models provide no lower limit on size scale. Even more fundamentally, analytical geometric models allow one to define features (like holes or inclusions of another material) of very small size. This is a severe problem, because the concepts of density, stress, and strain, and hence the concept of "material", fail as geometric features of the continuum model approach the size scale of the material fabric. The lower limit on the size scale of an engineering solid model must be large compared to the size scale of the material fabric – otherwise the limiting processes of continuum mechanics are invalid.

To summarize, the classical concept of "material" is fundamentally associated with concepts of density, stress, and strain. All of these concepts fail for modeling features smaller than the size scales specific to each type of material.

When Navier and Cauchy developed their namesake theory of elasticity, there existed only very primitive digital computers – the abacus, tables of functions, a very crude mechanical calculator invented by Blaise Pascal in 1642, and slide rules, which are mechanical analogue computers that had been developed over one hundred years earlier by John Napier (1550–1617) and William Oughtred (1575–1660). Thus, most of natural philosophy at the time involved seeking methods for describing nature using mathematical models that did not require a significant amount of numerical computation.[e] With the introduction of the modern digital electronic computer, in the mid-twentieth century, widespread numerical computation suddenly became feasible.[f]

With inexpensive numerical computations now available, the use of analytical methods (real calculus) in describing engineering models is no longer a necessity. Discrete physical models are perfectly fine, and are in fact preferable with regard to the three difficulties listed in the first paragraph of this section.

Therefore, the question remains. Can we develop engineering models more suitable for computers, based perhaps not upon real numbers and the associated calculus, but rather upon integer and floating-point numbers and their associated arithmetic operations? Until recently, almost all models for solid mechanics treated the Cauchy theory of strain and stress as "reality". True, the Cossarat brothers developed a "micropolar" theory that included particle rotations and micro-moment densities in addition to strain and stress components (Cosserat, E. and Cosserat, F., 1909). In addition, various particle and lattice models, for example (Krumhansl, 1968), (Kunin, 1983), and (Eringen, 2002) have been developed over the years, but most of these models still treated the stress-strain theory of Cauchy as the benchmark "reality" that was to be modeled.

Only recently was a truly coherent alternative to the Cauchy stress-strain theory developed, with the framework called "peridynamics" of Stewart Silling (Silling, 2000). ("Peridynamic" is derived from the Greek roots "peri", or "near' and "dynamic" or "force".) Yet even this

---

[e] Many modern pure mathematicians still prefer to avoid computational methods.

[f] As recently as 1943, the Manhattan Project used a room full of women – human computers – for computationally simulating the physical models used in the development of the atom bomb in Los Alamos, New Mexico.

framework, which was modified to become "continuum state-based peridynamics" (Silling, S.A., Epton, M., Weckner, O., Xu, J., and Askari, E., 2007), continued to assume a continuous $\mathbb{R}^3$ initial reference (and hence, implicitly, material) space. Nonetheless, the pioneering work of Silling, with peridynamic states, showed clearly that strain and stress are not the only way to rigorously model the mechanical behavior of solid bodies at the macroscale.

However, Silling continued to characterize peridynamics as a continuum theory of solid mechanics, continuing to employ, for example, the continuum concepts of mass density and material particle. In this book, we completely shed the continuum paradigm, and introduce a theory of solid mechanics called the state-based peridynamic lattice model (SPLM). The SPLM has the advantage that problems that have quantitative inputs are guaranteed to produce quantitative (none-infinite) outputs. Continuum mechanics does not guarantee quantitative (finite) outputs. Furthermore, both continuum mechanics and continuum peridynamics require some sort of further discretization in order to achieve numerical solutions using computers. Why not just set out with a discrete model of the solid material from the get go?

Although many researchers have developed particle models, most of them have been at the same time more complex than necessary, and insufficiently general to model the deformations of solids. For example, the discrete element method (DEM) (Radjaï, Farhang and Dubois, Frédéric, 2011), developed particularly in the discipline of soil mechanics, endows particles with geometry (spheres, ellipsoids, etc.) and rigid or elastic material behavior at the particle level, with tractions acting between particles thought of as normal and frictional contact tractions. In this sense, the DEM has not yet entirely shed the stress-strain paradigm.

The random particle methods, as for example developed by (Cusatis, 2001) and (J. E. Bolander and G. S. Hong, 2002), assume particles of random size with force interactions between particles depending in somewhat complicated and ad-hoc ways upon areas of adjacency between particles, which are placed randomly. They have developed these random particle methods with the aim of accurately modeling the meso-scale[g] material structure, and these models are more complicated than is necessary to model macroscale phenomena like elasticity, plasticity, and damage.

In contrast, in this book, we seek the simplest possible particle model that is capable of simulating observed material phenomena at the macroscale.

Lattice models developed to date, for example (Schlangen, E. and Van Mier, J.G.M., 1992), led the way, but these models are in certain ways more complex than necessary, and are in other ways also too limited, to efficiently and realistically model macro behavior or material bodies. In the work of Schlangen and van Mier, the fundamental objects of lattice models are axial solid members: truss and beam elements, each of which follow the principles of mechanics of materials, except that one removes these members from the model when the maximum stress or strain within the element exceeds a limiting value.

In this book, with the state-based peridynamics lattice model (SPLM), we develop a model that is as simple as we can imagine, starting with a regular array of particles – a lattice – defining the material body. The material particles P_i of a body, being finite and countable, may be named using an N-dimensional Cartesian *integer* reference space, called $\mathbb{Z}^N$. The material particles, labeled using integers, are then mapped to an initial reference configuration $\boldsymbol{X}$ in real N_R-dimensional physical space $\mathbb{R}^{N_R}$ – but not continuously, as continuity has no meaning for integers. Material behavior emerges through a constitutive function relating the *peridynamic discrete deformation state* and the *peridynamic discrete force state*, rather than through a constitutive relation relating strain and stress. These "discrete peridynamic states" are alternate ways of thinking about "strain" and "stress". Discrete peridynamic states are more natural for computers, and after some practice, connect with the human mind more naturally than the continuum mechanics definitions of strain and stress.

[g] The meso-scale is one size level down from the macro-scale. The macro-scale is the size scale used to define the geometry of the material body. Thus, for concrete, the meso-scale is the size of the largest particles of aggregate, while the macroscale is that used to define the dimensions of the structural element.

By providing an appropriate mathematical and computational framework, we believe that the SPLM will be sufficiently well defined and robust that engineers will use it confidently to solve a wide variety of practical engineering problems that are difficult to solve using the classical and continuum mechanics approaches.

1.2 Classical solid mechanics

The purpose of this section is not to introduce classical solid mechanics. We assume that the reader already has had an introductory class in mechanics of materials. Rather, the purpose of this section is to show that classical solid mechanics has some very fundamental conceptual difficulties.

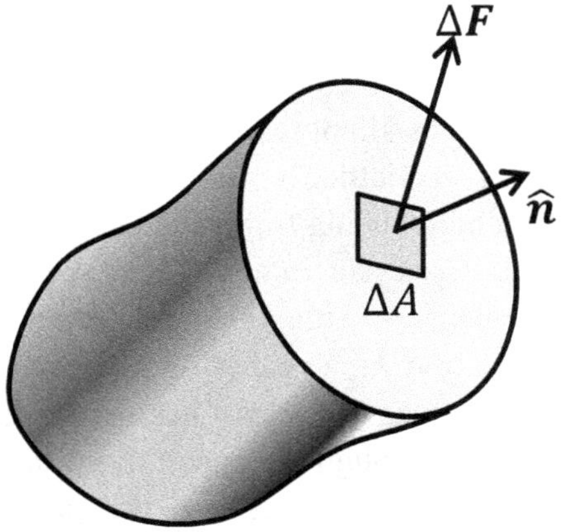

Figure 1.1. The traction vector is defined as the limit as ΔA goes to zero of $\Delta \boldsymbol{F}/\Delta A$.

Many books have been written on classical solid mechanics (also called mechanics of materials, mechanics of deformable solids, etc.), including as typical examples (Hibbeler, 2005) and (Boresi, A.P., Schmidt, R.J., and Sidebottom, O.M, 1993). All of these books are based upon the fundamental concepts of strain and stress. For example, the stress vector (also called the traction vector) is defined as

$$\boldsymbol{t} \equiv \lim_{\Delta A \to 0} \left(\frac{\Delta \boldsymbol{F}}{\Delta A}\right), \qquad (1.1)$$

where $\Delta \boldsymbol{F}$ is the force vector acting upon the area, ΔA, as shown in Fig. 1.1.[h] Eq. 1.1 only defines the traction vector with respect to a point on a particular oriented surface, and furthermore it only defines the traction vector if the outward-pointing unit normal $\hat{\boldsymbol{n}}$ is uniquely defined. Thus, Eq. 1.1 only defines the traction vector if a unique limit, as ΔA goes to zero, actually exists. For geometric domains with corners, for example, one can show that this limit does not uniquely exist. Likewise, the traction vector is undefined for points on the crack surfaces if these surfaces are not smooth, and certainly not on the crack front.

Similarly, we define strains in terms of change in length per unit length, or more mathematically, in terms of displacement derivatives, so, for example, the uniaxial normal infinitesimal[i] strain component ε_{xx} is

$$\varepsilon_{XX} \equiv \lim_{\Delta X \to 0} \left(\frac{\Delta u}{\Delta X}\right) = \frac{\partial u}{\partial X}, \tag{1.2}$$

where u is the axial displacement and X is the uniaxial reference position.

These definitions are fine as long as the limits exist to a reasonable approximation. However, particularly for materials that are materially nonlinear, in which displacement discontinuities (also called cracks) can arise during deformation, the definitions described by Eqs. 1.1 and 1.2 are rendered inapplicable because the displacement field becomes discontinuous; hence the need for the auxiliary theory of fracture mechanics. Additionally, these limits are a problem because all real materials have microstructure, which means that the limits do not uniquely exist in any real material.

This situation has recently led to the quest for *multiscale modeling*, where researchers attempt to link models at the macroscale and the next level down, called the mesoscale. Multi-scale modeling makes the models unduly complex. While multi-scale modeling might be a great way to design and understand materials, it is not a reasonable way to design and understand large engineering structures. Multiscale modeling

---

[h] All scalars are in normal font, and all vectors and tensors are in bold font in this book.

[i] Also called the Cauchy, or small-displacement strain. Other definitions of strain exist.

is not the goal of this book; instead, this book has a different objective: rigorous, robust, modeling of solids at a single scale.

One way to rectify these problems, while not disposing with continuum mechanics, is to "regularize" continuum mechanics by providing localization limiters, both upon the initial geometry of the solid and upon its strain and stress responses. However, this is a band-aid approach, and we seek a more fundamental methodology.

As we shall see, in the case of the state-based peridynamic lattice model, the localization limiter is intrinsic to the model: the particle lattice spacing in the reference configuration provides the material length scale.

1.3 Where classical solid mechanics fails

Consider a very modest problem: a simply supported beam subject to a centrally-located point load, as shown in Fig. 1.2. A more familiar structural engineering problem does not exist. Typically, this type of problem is solved analytically (by hand) using Euler-Bernoulli beam theory of bending (plane sections remain plane and normal to the deformed axis of the beam) or Timoshenko beam theory (which additionally includes a model for shear deformation).

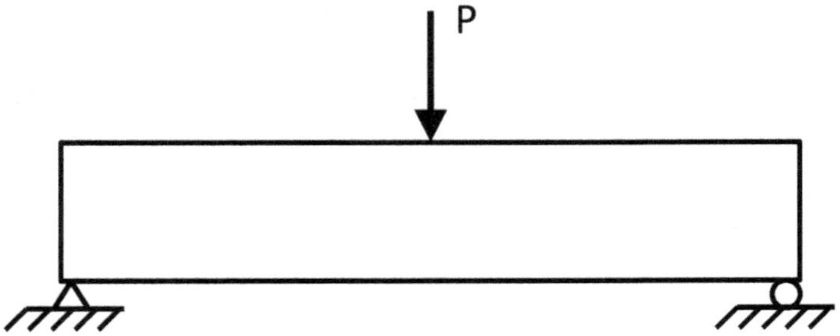

Figure 1.2. Simply-supported, point-loaded beam.

Assuming classical linear elastic stress-strain behavior, an exact closed-form analytical solution to the linear differential equations of elasticity of this boundary value problem does not exist. However, we can find a solution expressed as an infinite sum of analytical basis functions. This solution shows that singularities in the strain and stress fields exist at the supports and at the point of applied loading.

Finite element computer programs for structural analysis (such as SAP2000, and RISA) implement simplified Euler and Timoshenko beam theories, and thus ignore solution singularities. Hence, it is currently necessary for engineers to use extensive judgment in interpreting the results of both Euler and Timoshenko beam theories, and the results of finite element programs. For example, engineers must use completely different analyses to represent bearing stresses at supports and at points of applied loading.

On the other hand, when engineers employ more general continuum finite element computer programs (like ANSYS and ABAQUS) they obtain indeterminate results near re-entrant corners, again requiring engineering experience and judgement. With reference to Fig. 1.2, in an elastic analysis, as the mesh is refined, the stresses near the supports and near the location of the applied point load grow without limit. The engineer must decide when the mesh is sufficiently refined, and he or she must decide how to handle the regions where solution convergence is not achieved. A lot of hand waving is required.

An experienced computational modeler will realize that point loads and point supports are idealizations that do not actually exist in reality, and might avoid such a model as is shown in Fig. 1.2. The fact is that it is difficult to avoid spurious singularities with continuum models. It would be nice if the solid model from the start precluded the possibility of indeterminate solutions.

Researchers have attempted to "regularize" the continuum model by employing localization limiters in conjunction with a damage model. In such a case, one assumes that the stress at a point is a function of the strain averaged over some characteristic finite neighborhood of the point where the stress is to be determined. However, this approach is both awkward to implement in a finite element program, and more complex than needed, in light of the simplicity provided by the peridynamic model, which dispenses with strain and stress altogether.

With the state-based peridynamic lattice model (SPLM) presented in this book, singularities cannot arise in the solution, and mesh convergence studies are unnecessary. Hence, the solution to a problem defined in the SPLM framework is meaningful even without the necessity for (as much) extensive engineering interpretation and judgment. Of course, all engineering models, including the SPLM, are only approximations of reality, and therefore some engineering judgement will always be needed. Here we simply seek to remove several of the major obstacles to accurate computational engineering prediction.

In the end, the engineer should not be bothered with the artifices of *strain* and *stress*. Instead, the engineer should be concerned with structural deformations under conceivable design loads (and overloads). In this book, we seek, for prediction of structural deformations and velocities under prescribed mechanical loadings, a better-defined and simpler model than the continuum mechanics approach.

1.4 Introduction to the peridynamic model

The original continuum peridynamic model starts with the assumption that Newton's second law holds true on every infinitesimally small differential volume (continuum material particle) within the domain of analysis (Silling, 2000). A force density function, called the pairwise force function (or peridynamic kernel), $\boldsymbol{f}$, (with units of force per unit volume per unit volume) between each pair of infinitesimal particles is postulated to act if the particles are closer together (in the initial reference, undeformed, configuration) than some finite distance, called the material horizon, δ. If the particles in the reference configuration are farther apart than δ, the pairwise force function $\boldsymbol{f}$ is assumed to be zero. In the original (bond-based) model (Silling, 2000), the pairwise force function was assumed to be a function of the relative position (in the reference configuration) and the relative displacement between the pair of particles.

One may also assume that the pairwise force function is a function of other variables, such as temperature, plastic work, damage, and even the positions of other nearby particles within the peridynamic horizon. When the pairwise force function is a function not only of the positions of the two interacting particles, but also of the positions of other nearby particles, the peridynamic model is called "state-based" (Silling, S.A., Epton, M., Weckner, O., Xu, J., and Askari, E., 2007). In another twist, when the peridynamic particles possess rotational degrees of freedom and associated moments (in addition to translational displacement degrees of freedom and associated forces), the peridynamic model is called "micropolar" (Gerstle, W., and Sau, N., and Silling, S., 2007).

We employ a spatial integration process to determine the total internal force acting upon each infinitesimal material particle at a given time, and we use a time integration process to track the positions of the material particles as they change with time due to applied loads. As described by (Silling, 2000) and other subsequent publications, the continuum peridynamic model may be approximately implemented on a computer as an array of interacting discrete particles of finite mass. Computational implementations of continuum peridynamic models should utilize particle spacing of a size appropriate to the problem at hand, but no less than the size scale of the material fabric.

The description of the peridynamic pairwise force function is phenomenological, similar to the method by which the elastic constants of solid mechanics are determined, rather than micromechanical. Unlike a molecular dynamics model, a peridynamic model has constitutive properties that depend upon the reference (undeformed) configuration. In addition, unlike continuum mechanics, the peridynamic model does not require continuous response fields, nor, for that matter, does it require a continuous material space, as has been assumed by Silling (Silling, S.A., Epton, M., Weckner, O., Xu, J., and Askari, E., 2007).

In this book, we relax the assumption that the material particles are described by a continuous *real* Cartesian space, instead choosing to describe particles as a lattice described by an *integer* Cartesian space, as shown in Fig. 1.3.

Essentially, the peridynamic model assumes that the internal force acting upon a particular particle P_i in a solid is defined as a function of the reference locations $\boldsymbol{X}_j$ and the deformed locations $\boldsymbol{x}_j$ of the nearby particles P_j with respect to the reference location $\boldsymbol{X}_i$ and the deformed location $\boldsymbol{x}_i$ of particle P_i. There are various ways to define the inter-particle force function, also called "pairwise peridynamic force function". One specifies the material constitutive behavior by prescribing of the pairwise force function.

This is in contrast to the approach of classical elasticity and continuum mechanics, in which we express the material constitutive behavior as relations between strain and stress at a point in a continuum.

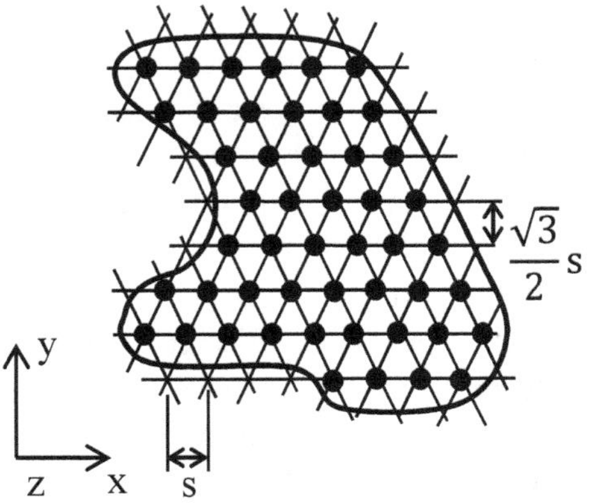

Figure 1.3. Particle lattice model for a material body.

In 2000, Silling published the first journal paper on peridynamics; the version described in that paper we now call "bond-based" peridynamics (Silling, 2000). With bond-based peridynamics, one assumes that the pairwise force between a pair of particles is a function only of the relative locations (in both the undeformed and the deformed configurations) of the two interacting particles. Silling later found the bond-based model to be too limiting, as he could not model materials with arbitrary Poisson's ratios using the bond-based theory, and in addition, materials whose behavior is sensitive to volumetric deformation modes (such as plastic materials) could not be adequately modeled.

A generalization of bond-based peridynamics to include rotational and moment degrees of freedom, called micropolar peridynamics, was introduced in 2008 (Gerstle, W., and Sau, N., and Silling, S., 2007). This generalization allows materials with arbitrary Poisson's ratio to be accurately simulated, but does not help in solving the problem of constitutive material sensitivity to volumetric deformation.

An improvement to the original bond-based theory of peridynamics, called "state-based" peridynamics was published in 2007 (Silling, S.A.,

Epton, M., Weckner, O., Xu, J., and Askari, E., 2007). A "state" is essentially a scalar, vector, or tensor field defined upon a finite neighborhood of a reference point. Silling defined physical objects called "deformation states" and "force states" that are similar to, but more general than, strain tensors and stress tensors, respectively. In state-based peridynamics, one assumes the force state is a function of the deformation state, which includes information about the relative locations (in both the undeformed and the deformed configurations) of all particles within the material horizon (as opposed to only the locations of the two interacting particles, as with bond-based peridynamics).

In both the bond-based and the state-based peridynamic models, Silling treats the reference material space as a continuum in $\mathbb{R}^3$. This makes computational implementation of the model challenging, as one must first discretize the model. In this book, we make some changes to the continuum peridynamic model, and develop the state-based peridynamic lattice model (SPLM).

The reader should study the landmark pair of papers by Silling together with this book. Silling's papers are quite mathematical and may be daunting even for graduate students. One purpose of this book is to present peridynamics in a more direct, practical, and accessible way.

1.5 What is a "solid material"?

Starting with Navier, researchers have used the words "elastic solid body" and "material" quite liberally, without ever really clearly defining them. Navier's original paper was entitled "Sur Les Lois De L'Equilibre Et Du Mouvement Des Corps Solides Elastiques" (On the Laws of Equilibrium and of Movement of Elastic Solid Bodies). In this paper, he states, "On regarde un corps solide elastique comme un assemblage de molecules materielles placees a des distances extremement petites" (One regards an elastic solid body as an assemblage of material molecules placed at extremely small distances).

Cauchy similarly spoke of "solid body", entitling one of his important papers "Recherches sur L'equilibre et le Mouvement Interieur des Corps Solides ou Fluides, Elastiques ou non Elastiques" (Research On The

Equilibrium And Interior Movement Of Solid Or Fluid, Elastic Or Non-Elastic, Bodies).

The implicit assumption of both researchers is that these solid bodies are isotropic, homogeneous and continuous at the scale of observation. Neither Navier nor Cauchy explicitly considered the fact that the assumed force function between molecules $f(r)$ should actually be integrated differently for particles that are close to a boundary, because the domain of force integration is no longer spherical for such particles. They said little explicitly about the "material horizon" being very much smaller than the least dimension of the elastic body. Indeed, for their theories to be applicable, the material horizon must be much smaller than the least dimension of the elastic body.

For many materials, like concrete and fiber-reinforced composites, the mesoscale is only slightly smaller than the least dimension of the body. For example, for commonly used four-inch (10 cm) thick concrete slabs constructed with three-quarter inch diameter (2 cm) aggregate, the macroscale is only about five times larger than the mesoscale. In such a case, one should not expect the Navier-Cauchy theory to be applicable, yet their theory is none-the-less used routinely for such problems.

If we allow the mesoscale to be only somewhat smaller than the macroscale, then the concept of "elastic solid body" is different from the solid continuum implied by the Navier-Cauchy theory. Strain and stress are inadequate measures successfully to describe such bodies.

The mesostructure (material structure) of almost all civil engineering materials is complex and random. This is true of steel, with its iron crystallites, concrete, with stone aggregate of various sizes and cement, wood, with its complex cellular structure, and masonry, which is similar to concrete.

When designing, civil engineers are typically not interested in the complexity of modeling a material in a statistical manner. This would be taxing, both computationally and conceptually. Thus, we here limit our attention to deterministic models.

The simplest deterministic model of a deformable solid that we can think of, short of modelling the mesostructure explicitly (and a deterministic description of the mesoscale structure of the material is not known), is to assume that the particles of the body, in its reference configuration, constitute a regular lattice. This leads us to the notion of "lattice-based deformable solid" as described in the next section.

1.6 Geometric modeling of solids

In current engineering practice, solids are usually described using solid models that are defined analytically in $\mathbb{R}^3$ space using constructive solid geometry (CSG) together with piecewise analytic primitive objects such as cuboids, cylinders, prisms, and nonuniform rational B-splines (NURBS), as shown at left in Fig. 1.4. In current practice, for stress analysis, one must convert such models into discretized finite element meshes for stress and deformation analysis, as shown at right in Fig. 1.4. The CAD modeling approach is usually satisfactory for mechanical design, in the sense that it precisely defines the undeformed geometry.

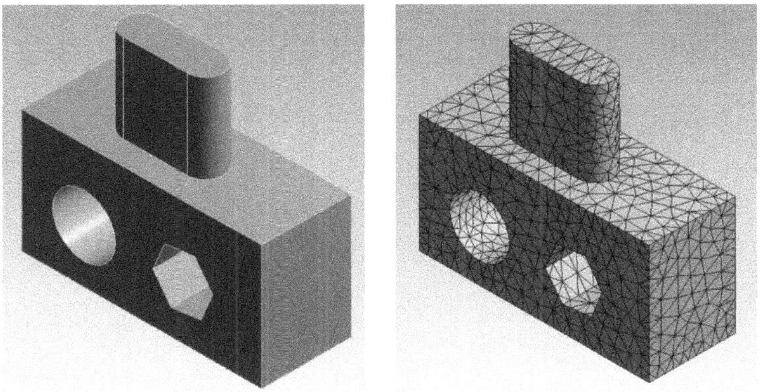

Figure 1.4. CAD CSG solid model of widget (left), and a finite element representation (right), generated using ANSYS Workbench.

However, as the body deforms, perhaps due to mechanical loading, CAD geometry is no longer a valid description of the deformed geometry. This is a principal reason that CAD geometry is not a completely general way to describe solids – particularly as they deform. The developers of CAD have simply not designed the models to support the deformed geometry – they must represent the deformed geometry using an associated finite element mesh, or some other discretization.

Also, CSG is not ideal for defining the geometry of civil engineering structures composed of real materials (like concrete), because it is possible with CSG to describe geometric features with length scales smaller than is meaningful for the material being defined. For example, if a structure is composed of concrete having aggregate of 1 inch (2.54 cm) size, it makes no sense to define a feature such as a notch whose characteristic length is ¼ inch (0.63 cm). Similarly, the crispness of the CSG allows re-entrant corners (as shown in Fig. 1.4) to be defined with zero radius of curvature, resulting in infinite stresses and strains, thus rendering continuum mechanics solutions invalid, (because any theory which admits infinities is not a quantitative theory, and is thus incomplete). Some CSG modeling applications allow users to define curved fillets, as a means of eliminating sharp re-entrant corners, but this puts a large burden on the modeler. It would be best to have a geometric model that allows only acceptable features from the get go.

For these reasons, CAD geometry may not always be the best method for our needs.[j]

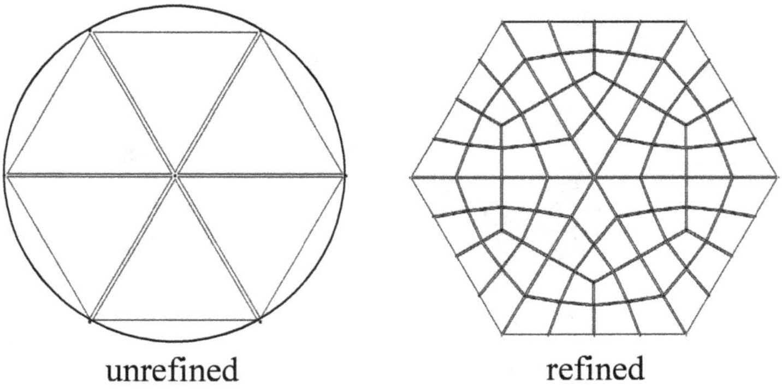

Figure 1.5. Should the finite element mesh at left, representing a circular disk, be refined as shown on the right?

---

[j] Recently, "isogeometric analysis" (Hughes, T.J.R., Cottrell, J.A., and Bazilevs, Y., 2005) has been developed. The idea is to represent the analysis domain directly using CAD geometry, rather than a finite element mesh. They make the arguable assumption that the CAD model is the "exact" geometric model.

Another way that engineers describe the geometry of a solid structure is by using a finite element mesh, using, say, triangles connected by nodes, as shown in Fig. 1.5. The topology of the triangles (using interpolation functions) together with the nodal locations defines the geometry. However, finite element problems require mesh convergence studies, and if the only geometric information one has comes from the finite element mesh, then it is unclear how the mesh is to be refined at the non-straight boundaries of the mesh (as is illustrated by Fig. 1.5).

Voxels (derived from the words "volume pixels") have been used to describe amorphous solids such as are found in the biological and medical fields, as shown in Fig. 1.6. This is a reasonable approach for describing solid domains, as long as modelers clearly define the minimum voxel size. Often, with voxel models, modelers further define the geometry by adding surface triangles to the domain, in an effort to define a "smooth" boundary of the domain.

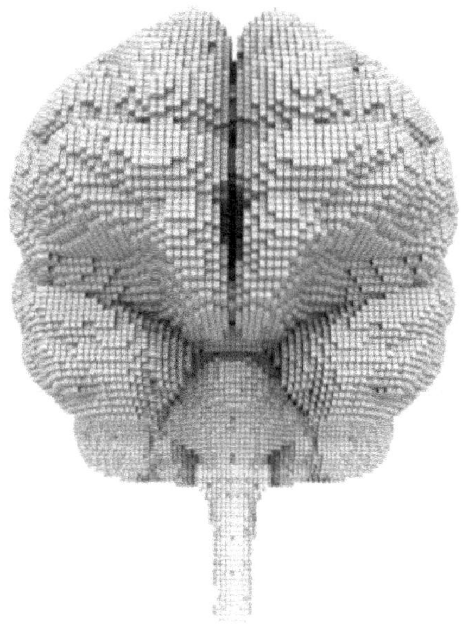

Figure 1.6. Voxel-based geometric model of the brain.
(From http://cdn2.everythingzoomer.com/wp-content/uploads/2013/10/predicting-alzheimers-brain-gettyimages.jpg).

Modelers usually convert the voxel model to a finite element or finite difference model for stress and deformation analysis. For representing the deformed geometry, the voxel model, like the CSG model, is unsuitable.

Material scientists and microbiologists have used molecular dynamics to represent materials as a collection of interacting atoms (or molecules). However, because, say, one cubic micron of copper contains approximately one hundred billion atoms, each of which has a vibrational frequency of 10^{13} hertz, this approach to solid modeling of large engineering structures is clearly beyond the ability of current computers.

We therefore propose another approach to geometric modeling of solids. We recommend using a regular array of uniformly spaced close-packed particles to describe the solid domain in its reference configuration. We call this array of particles a *lattice body*. The minimum spacing, L, will vary depending upon the material type. For concrete, for example, L must be greater than the largest aggregate size of say 1 cm. For steel, L must be larger than the grain size of perhaps 100 microns. For computational reasons, to limit the number of particles required, the particle spacing L may be much larger than the material grain size, but then the user must also be aware that this choice of particle spacing precludes features smaller than size L from both the initial model and the simulation results.

We believe that using a lattice of particles to define the reference configuration is more useful than a random distribution or ad-hoc array because a lattice is conceptually simpler, more mathematically tractable, and more computationally efficient.

By choosing to use a lattice of particles to model the solid structure, one describes the material domain boundaries only approximately. This may seem to be a limitation, but in fact, it is a feature, because boundaries of a real material may not be any more precise than the mesoscale of the material allows.

1.7 Digital aspects of geometric modeling

Because constructive solid geometry (CSG) models provide no inherent geometric minimum size scale limiter, CSG may fail when evaluated on a digital computer. This is because digital computers only use approximations of real numbers, employing instead floating-point representations of real numbers. For example, evaluating the locus of points defining the intersection between a spherical surface and a cylindrical surface, although simple to represent analytically, is a complicated, approximate, task for a digital computer.

If instead of analytical geometric models defined in $\mathbb{R}^3$, particle lattice models defined in $\mathbb{Z}^n$ and mapped through an affine (parallel line preserving) transformation to physical $\mathbb{R}^3$ reference space, are employed, one is assured that no features of the geometry will be too small to be represented using floating point numbers. The Boolean operations of constructive solid geometry are robust in $\mathbb{Z}^n$, whereas in $\mathbb{R}^3$, the operations of CSG can fail due to lack of numerical precision of floating point numbers. Evaluating conventional CSG models for graphical display is an approximate, algorithmically complicated, and computationally demanding task. On the other hand, evaluating $\mathbb{Z}^n$ lattice models is robust, and is algorithmically straightforward.

Of course, there is a price to pay in representing geometry using a particle lattice. Features emerge only by including or excluding material points from the domain. This may also be viewed as an advantage, however, as real materials cannot reliably reproduce features at scales smaller than their constituent grain size in any case. Thus, the smallest feature that can be represented is a single lattice point, which, for three-dimensional models, represents a volume of approximately L^3.

1.8 Scope of book

The purpose of this book is to present a new model for solid mechanics, called the state-based peridynamic lattice model (SPLM). In order to appreciate the SPLM, it is necessary first to understand both the history and the content of solid mechanics as it stands today. Thus, this book presents, without going into too much detail, many aspects of solid mechanics that are well developed, not to say entrenched, in engineering practice today. We critically evaluate the classical theories of solid mechanics, particularly in light of the capabilities of modern computers.

To understand why one might want to depart from a well-established classical theory, it is first important that we clearly understand the foundations of the theory. Many readers may have an incomplete understanding of the basics of continuum mechanics. For example, the first two sentences on page one of (Madenci, E. and Oterkus, E., 2014) state, "One of the underlying assumptions in the classical theory is its locality. The classical continuum theory assumes that a material point only interacts with its immediate neighbors; hence, it is a local theory." This statement may give the impression that a material point in a classical solid *has* immediate neighbors. To correct this possible misconception, Chapter 2 presents an account of the origins of the theory of elasticity, including annotated English translations of the original papers (in French) by Navier and Cauchy. These papers are so fundamental that it is well worth the effort to read them, if only to appreciate the boldness and beauty of their work, as well as to gain some appreciation for the historical development of the theory of solid mechanics. It is important to understand the simplifying assumptions that went into these theories, and to understand that they are simply engineering models; not true theories.

Chapter 3 presents the main ideas of continuum mechanics, which emerged in the 1950s as the dominant framework for both solid mechanics and fluid mechanics. While continuum mechanics is a major achievement, we find that the fundamental assumption of continuous fields is too restrictive for many situations, where fields may be essentially discontinuous.

Because of the limitations of both the classical theory of elasticity and of continuum mechanics, the discipline of fracture mechanics arose over the course of the twentieth century. Chapter 4 presents the main ideas of both linear elastic and nonlinear fracture mechanics. We show that the ideas of fracture mechanics, while instructive, are far from being directly applicable in routine engineering design and analysis. Indeed, by reframing the discipline of solid mechanics using the SPLM, we have the possibility to render the computational disciplines of continuum mechanics, fracture mechanics, and even finite element analysis superfluous (but only if computer power continues to increase sufficiently).

We present Silling's continuum bond-based peridynamic theory in Chapter 5. While this theory is appealing, it was found by Silling himself to be inadequate to model materials with Poisson's ratio other than one-quarter, and is not extensible to nonlinear constitutive modeling of, for example, isochoric plasticity.

Chapter 6 discusses the particle lattice model for solids, and makes the case that this approach is more effective than analytical geometric modeling (CSG) for many types of problems.

Chapter 7 presents several bond-based linear elastic peridynamic lattice models, but concludes that these bond-based models are insufficiently general for many applications.

Silling developed the continuum state-based peridynamic framework as a means to generalize the original continuum bond-based peridynamic theory. Chapter 8 follows the pioneering work of the continuum state-based peridynamic framework, but with important differences. In Chapter 8, the state-based peridynamic model as applied to particle lattices, rather than to a continuous material description. This renders the model finite-dimensional, rather than infinite-dimensional, and thus directly appropriate for computational implementation.

In Chapter 9, we develop the elastic state-based peridynamic lattice model (SPLM). We begin by showing how, in the context of the SPLM, classical stress relates to the force state and similarly, how classical strain relates to the deformation state. We show that the SPLM states are more general than the classical concepts of strain and stress. We derive the microelastic SPLM peridynamic parameters directly from the classical parameters (Young's modulus and Poisson's ratio) of linear isotropic elasticity. The linear elastic SPLM is comparable to the classical theory of elasticity, and indeed, from the perspective of energy, converges to the classical theory of elasticity as the lattice spacing goes to zero compared to the characteristic size of the geometric domain.

Chapter 10 presents an SPLM plasticity model, and Chapter 11 presents an SPLM damage model. We must consider geometric nonlinearity when displacements of the lattice particles are large compared to the lattice spacing in the reference configuration. Chapters 9 through 11 incorporate large deformations into the SPLM.

The subject of Chapter 12 is particle dynamics. In particular, we present a time integration algorithm, and derive the critical time step for this algorithm. In addition, we discuss various damping models.

To be useful, SPLM models often must contain thousands to millions of lattice particles. We discuss the computational simulation of such large collections of particles on parallel computers in Chapter 13.

The SPLM is particularly useful for modeling composite structures like reinforce concrete. Chapter 14 discusses the modeling of reinforced concrete structures using the SPLM.

The intent of this book is to be useful for a graduate student. Some may say that the SPLM is "just a special case of the finite element method", or "just a special case of peridynamics", or "just another particle method", or "just another lattice model". However, the author believes that peridynamics, and particularly, the SPLM, bring something new and important to the table, complementing, and perhaps in some cases even replacing, the classical models of elasticity and continuum mechanics, appropriate for the current computer age.

1.9 Exercises

1.1 Show why mass density, $\rho \equiv \lim_{\Delta V \to 0} \left(\frac{\Delta m}{\Delta V}\right)$, is undefined on the boundary of a homogeneous material body.

1.2 Describe and sketch three physical situations in which the traction vector, defined in Eq. 1.1, is undefined at a point.

1.3 Describe and sketch three physical situations in which the strain component at a point ε_{xx} defined in Eq. 1.2, is undefined.

1.4 The centrally-loaded beam shown in Fig. 1.2 has a 24" span from support to support, is 6" deep, and is 3" thick. If the maximum allowable shear stress is 15,000 PSI, what load, P, is permitted using

a. Timoshenko beam theory, ignoring stress concentrations.
b. Plane stress finite element analysis. Assume all loads and reactions are uniformly distributed tractions over a length of one inch along the beam length, and a width of 3" transverse to the beam axis.
c. Concentrated loads and reactions.

1.5 Assume that the 6" by 12" rectangular by ¼" thick glass plate shown in Fig. 1.7 is in a condition of plane stress. The plate has a 2" diameter hole at its center. The glass has a tensile strength of 10,000 PSI. If we apply equal and opposite uniformly distributed loads of P to opposite ends of the glass plate as shown, what is the load at which the glass plate will break? (Use a finite element program or a solution from the literature to solve this problem.)

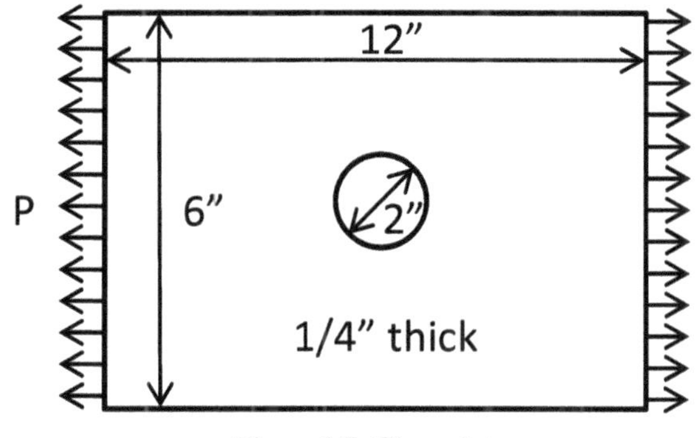

Figure 1.7. Glass plate.

1.6 Immediately after the tensile strength of the glass is reached in Problem 1.5, two cracks propagate from the top and bottom of the hole in a direction transverse to the direction of applied load P. When the two cracks are each 0.5" long, what is the maximum stress in the plate?

1.7 The plate in Fig. 1.7 is changed to mild A36 steel, with a uniaxial yield strength of $F_y = 36\,KSI$ and an ultimate strength of $F_u = 58\,KSI$. Rather than cracking, it fails by yielding plastically. What is the applied load, P_1, at which yield begins, and what is the ultimate strength, P_2, of the plate?

1.8 Using a single-precision floating-point number, what is the closest approximation to π that can be achieved? What is the percentage error?

1.9 A solid body is constructed, using CSG, as the Boolean sum of two primitive spherical bodies of radius 1 m whose centers are separated by 2 m. Describe and sketch the topology of the point of contact. Is the body continuous at the point of contact?

1.10 Use a CAD modeler such as AutoCAD or SolidWorks to create a geometric model of a solid body whose surface is described by the equation $3X^2 + 2Y^2 + Z^2 = 25$. What issues do you encounter?

1.11 Repeat Problem 1.10, this time using MatLab. Create a picture of the solid body.

Chapter 2

Beginnings of the Theory of Elasticity

Chapter objectives are to:

- review the history of the theory of elasticity;
- appreciate the assumptions made in the theory of elasticity;
- present Navier's theory of elasticity;
- understand why Navier's theory was insufficiently general;
- present Cauchy's correction to Navier's theory;
- comprehend the origins of stress and strain.

Before we attempt to revise the theory of deformable solids, it is well to understand the history of its development. Following Newton's 1687 masterpiece entitled "Philosophiæ Naturalis Principia Mathematica", (Newton, 1999), in which the concepts of force, mass, and acceleration were established, and the calculus simultaneously invented, natural philosophers, such as Euler and the Bernoulli brothers, successfully modeled elastic axial members such as beams and columns – without the concepts of strain and stress.

Navier was the first to attempt to develop a mathematical model for a three-dimensional elastic solid body. Navier was not at all interested in developing a fully general theory for the behavior of deformable solids; instead, he only wanted to develop a first-order linear elastic theory.

Navier's theory was immediately criticized and overturned by his colleague, Cauchy. It is Cauchy's notions of strain and stress that have come to define what we mean by "deformable solid" – and we now find this approach insufficiently general in light of new theories of damage and fracture and in light of the power of modern computers. We are now interested in predicting continuous deformation, as well as discontinuous fracture deformation within a single theory, using computers.

Let us review the history first.

114 DIALOGO SECONDO

sin quì dichiarate, non sarà difficile l'intender la ragione, onde auuenga, che vn Prisma, ò Cilindro solido di vetro, acciaio, legno, ò altra materia frangibile, che sospeso per longo sosterrà grauissimo peso, che gli sia attaccato, mà in trauerso (come poco fà diceuamo) da minor peso assai potrà tal volta essere spezzato, secondo che la sua lunghezza eccederà la sua grossezza. Imperò che figuriamoci il Prisma solido A B, C D fitto in vn muro dalla parte A B, e nell'altra estremità s'intenda la forza del Peso E. (intendendo sempre il muro esser eretto all'Orizonte, & il Prisma, ò Cilindro fitto nel muro ad angoli retti) è manifesto che douendosi spezzare si romperà nel luogo B, doue il taglio del muro serue per sostegno, e la B C per la parte della Leua, doue si pone la forza, e la grossezza del solido B A e l'altra parte della Leua, nella quale è posta la resistenza, che consiste nello staccamento, che s'hà da fare della parte del solido B D, che è fuor del muro, da quella che è dentro; e per le cose dichiarate il momento della forza posta in C al momento della resistenza che stà nella

Figure 2.1. Galileo's cantilever beam. Source: Galilei, Galileo, Discorsi e dimostrationi matematiche, Intorno à due nuove scienze, attenenti alla mecanica, & i movimenti locali. Bologna. Per gli HH del Dozza, 1655, 86.

2.1 Brief history of solid mechanics

Mathematical modeling of solids has a long history, beginning approximately 3000 BCE with the Babylonians and Egyptians. These early engineers were the builders of the ziggurats and pyramids who knew how to calculate distances, areas, and volumes. Around 500 BCE the Greek geometers made great mathematical advances with the Platonic solids, the Pythagorean Theorem, and Euclid's Elements, and, after a long lull, Leonardo Da Vinci and Galileo, the fathers of modern science, revived fluid and solid modeling during the Renaissance. While Galileo's simple model for a cantilever beam (Fig. 2.1) did not withstand the test of time, we still regard as being as true today as it was in 1687, when he published the Principia (Fig. 2.2), Newton's idea that the properties of fluids arise from forces acting between particles possessing mass.[a]

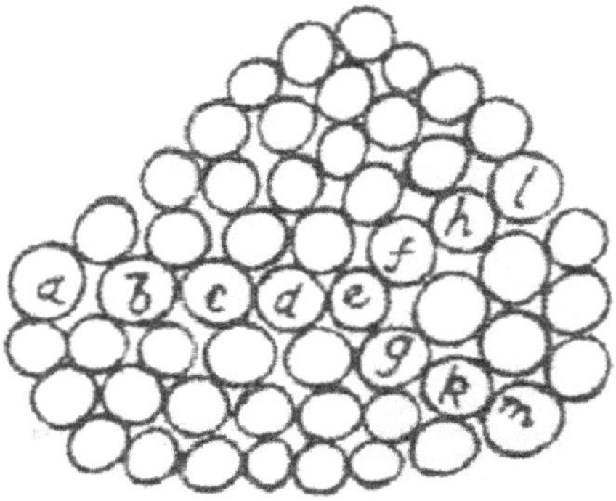

Figure 2.2. Newton's conception of a fluid as interacting particles, showing how force and displacement waves propagate in a fluid. Source: Book 2, Section 8, Proposition 41, Theorem 32, of The Principia (1687) (Newton, 1999).

Although Newton apparently did not explicitly address the deformation of solids, we can assume that, as with fluids, he understood solids to be a collection of interacting particles, as indicated by the following quote, translated from Latin by Andrew Motte, from the preface to the first edition of Isaac Newton's "Mathematical Principles of Natural Philosophy", 1687 (Newton, 1999):

> "Then from these forces, by other propositions which are also mathematical, I deduce the motions of the planets, the comets, the moon, and the sea. I wish we could derive the rest of the phenomena of nature by the same kind of reasoning from mechanical principles, for I am induced by many reasons to suspect that they may all depend upon certain forces by which the particles of bodies, by some causes hitherto unknown, are either mutually impelled towards one another, and cohere in regular figures, or are repelled and recede from one another. These forces being unknown, philosophers have hitherto attempted the search

[a] Actually, Newton's idea that a fluid consists of particles is subject to revision with the 20th century understanding that atoms possess a quantum mechanical nature that might be better thought of as a wave function describing an electron density probability distribution. Nonetheless, even atoms can be thought of as particles for the purposes of the deformation of solids and fluids at scales large compared to the atomic scale.

of Nature in vain; but I hope the principles here laid down will afford some light either to this or some truer method of philosophy."

The modern theory of elasticity progressed over the course of almost four hundred years, starting with Leonardo da Vinci (1452–1519), and followed by natural philosophers such as Galileo Galilei (1564–1642), Edme Mariotte (1620–1684), Robert Hooke (1635–1703), Isaac Newton (1642–1727), Jacob Bernoulli (1654–1705), John Bernoulli (1667–1748), Leonard Euler (1707–1783), and Joseph-Louis Lagrange (1736–1813). An authoritative history of the theory of elasticity is given by Timoshenko (Timoshenko, 1953), and Maugin provides a very readable story of the development of continuum mechanics to the present day (Maugin, 2013).

Claude-Louis Navier (1785–1836) and Augustin-Louis Cauchy (1789–1857) together developed the modern theory of elasticity, together with the concepts of strain and stress. The corrected mathematical formulation, called the Navier-Cauchy theory of elasticity, rested heavily upon assumptions of continuity and analyticity (calculus), and resulted in a set of spatio-temporal partial differential equations that they could solve in closed form for a limited number of simple problems. Others subsequently augmented their theory of elasticity, with the introduction of nonlinear elasticity, time-dependent material behavior, plasticity, large geometric deformations, and damage mechanics. In the mid-1900's, the theory of elasticity was generalized and re-branded as the theory of continuum mechanics, which now embraces both solid mechanics and fluid mechanics. Truesdell and others in the 1950s and 1960s (Truesdell, 1960) advanced the theory of continuum mechanics and placed it on a firmer mathematical foundation.

The advent of the digital computer greatly aided the solution of specific continuum mechanics problems. Computers allowed the partial differential equations of continuum mechanics to be approximately solved using the finite difference, finite element, and boundary element methods. All of these methods essentially convert the partial differential equations of continuum mechanics into approximating discrete algebraic equations, which a digital computer can evaluate in a variety of ways.

However, the linear theory of elasticity is an incomplete theory, because infinite values of the responses (strain and stress) are possible, for example at a re-entrant corner or at the tip of a crack. Any physical theory that involves infinities is not a quantitative theory, because infinity is not a quantity (number). This is not to say that infinite solutions are not meaningful: these solutions simply indicate that the theory does not produce quantitative information. For example, at the tip of a sharp crack in a classical elastic solid, infinite strain and stress are predicted (producing a singularity), indicating that the classical elastic theory is not applicable in the neighborhood of the crack tip. With infinite stress predicted at the crack tip, a stress-based failure criterion would predict that the crack must propagate even under miniscule loads, which is of course nonsense, as we know that cracked bodies can actually carry significant loads.

Thus, in the early twentieth century, because of the fundamental assumptions and limitations of theory of elasticity and continuum mechanics it became necessary to develop an entirely new branch of mechanics called fracture mechanics, which handles, quantitatively, the singularity at the tip of a crack. However, the theory of fracture mechanics is an auxiliary theory, sitting outside the theory of continuum mechanics.

Combining fracture mechanics and continuum mechanics of solids (or more specifically, the theory of elasticity), while possible, has proven in many cases to be impractical.

Ingraffea has been a leader in the field of discrete computational fracture mechanics. See, for example, (T.N. Bittencourt, P.A. Wawrzynek, A.R. Ingraffea, and J.L. Sousa, 1996). In their approach, they employed the finite element and boundary element methods, and in a sequential manner, they altered the geometry of the problem to allow a propagating discrete crack, or even several discrete cracks, to evolve. They handled the problem of determining when and where a crack will nucleate, and along what trajectory it will propagate, by the theory of fracture mechanics. Research and development issues arose because it is no simple matter to update the geometry, the finite element mesh, and the boundary conditions in a coherent manner, and discrete fracture mechanics makes several (perhaps overly) simplistic assumptions about the nature of a "crack". The traditional linear elastic fracture mechanics concept of a crack being a "surface of separation, with no tractions between the two new crack surfaces" is very far from the truth in many cases. Nonetheless, fracture mechanics has been a useful tool in addressing many practical engineering problems.

Bazant (Bazant, Z. P. and Jirasek. M., 2002), and many others, have taken another approach to computational modeling of cracks: the smeared crack approach, pioneered by Rashid (Rashid, 1968). In one approach, called nonlocal continuum damage mechanics, smeared crack models represent cracks by evolving spatially distributed damage. Nonlocal damage mechanics models assume that the stress at a point depends upon the spatially averaged strain in a finite neighborhood of the point. Another approach is the "blunt crack band approach", in which the material constitutive model is adjusted in accordance with the size of the finite element.

The other main approaches to solid modeling are the discrete element method, (Cusatis, 2001), (J. E. Bolander and G. S. Hong, 2002), (Radjaï, Farhang and Dubois, Frédéric, 2011) and the beam and frame lattice methods (Schlangen, E. and Van Mier, J.G.M., 1992).

While the discrete crack, the continuum damage, the discrete element, and the lattice approaches have each attained moderate success, it seems that so far they are all somewhat ad hoc and more complicated than necessary. While often employed by researchers and academics, neither the discrete crack approach nor the continuum damage mechanics approach has proven sufficiently robust for structural designers and forensic engineers to use them routinely in design and analysis.

In the early 1900's mathematicians and engineers discovered that the solutions of the classical equations of elasticity could be approximated using equivalent truss and frame structures (Weighardt, 1906); (Reidel, 1927); (Hrennikoff, 1941). An early finite element method was developed by the great mathematician, Richard Courant (Felippa, 1994). The purpose of the method was to solve approximately the equations of elasticity for the torsional behavior of a shaft using triangular finite elements. The finite element method for digital computers was later independently developed by Clough and his coworkers (M.J. Turner, R.W. Clough, H.C. Martin, and L.J. Topp, 1956).

Nonlocal elasticity, including investigations of lattice and continuum models, was written about in the 1960's with the work of the physicists

Krumhansl (Krumhansl, 1968), Kroner (Kroner, 1981), and Kunin (Kunin, 1983).

Eringen developed and wrote about nonlocal continuum mechanics theories starting in the 1960s. He apparently was not interested in developing a theory that was free of the spatial partial derivatives involved in defining strain (Eringen, 2002). He developed a lattice model that has similarities to the peridynamic lattice model described in this book. He also considered micropolar models.

As mentioned previously, in 2000, Silling (Silling, 2000) introduced the theory of continuum peridynamics. Peridynamics, as described by Silling, relinquishes the assumption of continuity of the displacement field, but still assumes that the material space is continuous. The paper describing the theory is notable for its mathematical rigor. However, the theory did not entirely dispense with the concept of material continuity, and Silling presented the theory as a "continuum model".

Schlangen and Van Mier (Schlangen, E. and Van Mier, J.G.M., 1992) performed truss-lattice modeling of concrete at the meso-mechanical level, with the purpose of understanding the mechanisms of fracture in cementitious materials at the mesoscale. They did not appear to be interested in using the truss lattice model as an alternative theory to continuum mechanics at the structural level (macroscale). In the same vein, Cusatis (Cusatis, 2001) used the random particle model to study concrete at the meso level, including explicit modeling of cement paste and aggregate. Bolander developed a rigid-body-spring network model for concrete members. His model constructs a random lattice of rigid bodies using Voronoi tessellations of the domain (J. E. Bolander and G. S. Hong, 2002). Nonlinear springs connect the rigid bodies together.

In this book, we present a model for solid mechanics based upon the peridynamic model, but, in recognition of the strengths and weaknesses of digital computers, our model assumes a priori that the material reference space is a discontinuous, regular, close-packed lattice of particles, such as the two-dimensional hexagonal lattice shown in Fig. 1.3. We call the model the state-based peridynamic lattice model (SPLM). Because of the SPLM's simplicity, and its clearly stated assumptions, one can define the model mathematically and unambiguously. One can also clearly define its limitations. Thus, we hope, engineers will have sufficient understanding of the model to use it confidently for routine engineering analysis and design problems. Of course, because the lattice often involves a large number of material particles, powerful computers are required to solve realistic problems, but even single-processor computers are already sufficiently powerful to solve an interesting class of useful SPLM problems, and this class will broaden as computers continue to become even more powerful.

The SPLM does not seek to solve continuum mechanics problems, except in an approximate manner. Instead, we design the SPLM to solve problems of solid mechanics directly using a rigorously defined formulation. The SPLM is useful because it allows us to model reasonably large structures – including elasticity, large deformations, plasticity, damage, and fracture – using computers, more simply and more directly than does the continuum theory of solid mechanics, and all of the afore-mentioned theories.

To provide background, we next present annotated translations of the pioneering papers by Navier and Cauchy, who developed the linear theory of elasticity, involving the concepts of strain and stress, which over time have come to aid, but simultaneously to constrain, our understanding of "material" and "deformable solid".

2.2 Navier's treatise

Perhaps because Navier wrote his seminal paper (Navier, 1827) on the theory of elasticity in French, which, to our knowledge, has not been previously translated into English, there appear to be misunderstandings regarding his basic assumptions. Therefore, we provide in this section an annotated English translation of Navier's paper, which although only published in 1827, Navier originally presented to the Royal Academy of Sciences of France in 1820.

Navier's theory contains only one elastic constant, ε, which Cauchy soon showed to be insufficiently general to represent isotropic linear elastic materials (although Navier never accepted Cauchy's corrected theory). We present Cauchy's theory in Section 2.4.

Navier's treatise is presented because it offers fundamental ideas relating to the peridynamic model. This classical paper elucidates the fundamental assumptions of the linear theory of elasticity, and validates the peridynamic model. Melissa Rose Arce Miller translated the paper from French. Annotations by the current author are included as footnotes.

TREATISE
ON THE LAWS OF THE EQUILIBRIUM AND MOVEMENT OF SOLID ELASTIC BODIES
BY MR. NAVIER

Read to the Royal Academy of Sciences, May 14, 1821

1. If one considers an elastic body, supposing that forces are applied to the points of this body and make themselves mutually in equilibrium, one can question the figure change that the body will have undergone as a result of the action of these forces. One can also question, supposing that, after the figure change, the body is left to itself, the laws of the movements of oscillation that will take place by virtue of the forces that constitute the elasticity. [b]

The solution to these questions is made up of two parts: 1° the search for differential equations[c] that express the laws of the equilibrium or

---

[b] This translation is intentionally very close to the original French prose, perhaps at the expense of fluent English idiom. In this way, it is hoped that little nuance in meaning will have been lost in translation.

[c] Note that Navier, right from the start, is explicitly seeking to develop differential equations. Thus, he assumes a certain degree of analyticity tacitly and perhaps unwittingly, a priori. This would be in keeping with Navier's time, in which calculus and real analysis was relatively new and highly valued. In the current book, we take a different turn, with the state-based peridynamic lattice model (SPLM), avoiding analyticity and assumptions of spatial continuity completely. An assumption of continuous deformation unnaturally constrains the solution space, preventing cracks from emerging. Furthermore, in consideration of modern computers, which use discrete numerical computations, one gains little if any computational advantage by using the analytical methods of calculus.

movement; 2° the integration of these equations.[d] The search for differential equations is the object of this treatise. It must be founded on an exact notion of the nature of the forces by virtue of which a body is elastic, and of the interior composition of this body.

One looks at a solid elastic body as an assembly of material molecules placed at extremely small distances.[e] These molecules exert two opposing actions on each other, namely, a force belonging to attraction, and a force of repulsion due to the principle of heat.[f] Between a molecule M, and any M' of the neighboring molecules, there exists an action P, which is the difference of these two forces[g]. In the natural state of the body, all actions P are non-existent, or destroy each other reciprocally, because the molecule M is at rest[h]. When the figure of the body has been changed, the action P has taken a different value Π, and there is equilibrium between all the forces Π and the forces applied to the body, by which the figure change has been produced[i]. One can always imagine the forces Π each divided into two parts π and π', supposing the first part π such that, if it subsisted alone, there would be equilibrium between all forces π, in the same manner that there was equilibrium between all forces P in the natural state of the body. The forces π destroying each other mutually therefore, it will be necessary that the equilibrium subsist between the remaining forces π' and the forces

---

[d] Therefore, Navier first differentiates, and then integrates. This seems more roundabout than necessary!

[e] Navier ignores the fact here, that most materials have meso structures (aggregate particles, crystallites, grains, and other structures) that are much larger than molecules and atoms.

[f] In Navier's time, the concept of electronic bonds between atoms was unknown.

[g] Navier considers the mutual force between two particles as the difference between two separate forces – an attractive force and a thermal repulsive force. Although this idea is not quite correct, it is similar to Silling's idea of "peridynamic pairwise force". Note that Navier starts out with discrete molecules (particles), M and M'.

[h] A molecule can be at rest even though self-equilibrating bond forces, P, may be nonzero.

[i] Here, Navier tacitly assumes static, rather than dynamic conditions.

applied to the body.[j] That set down, we take here as a principle that these latter forces π', developed by the figure change of the body between any two material molecules M, M', and which alone must balance out the forces applied to this body, are respectively proportional to the quantity of which the figure change (supposed very small) has made the distance MM' of two molecules vary.[k] The force π' is an attraction if the distance MM' has increased; it is a repulsion if this distance has diminished. Moreover, we look at the molecular actions in question as only subsisting between very close molecules, and as having values which decrease very rapidly, following an unknown law, for molecules more and more distant from each other.[l]

General equations of the equilibrium and movement of solid elastic bodies.

---

[j] Apparently, forces π are what we would call "self-equilibrating bond forces prior to deformation" and π' are bond forces due to deformation caused by externally-applied loads.

[k] Note that Navier assumes that the force between two molecules M and M' is a linear function only of the (very small) change in distance between these two molecules. This assumption of linearity could certainly have been generalized. For example, Navier could have assumed that the force between M and M' is a function also of the change in distance between M and another molecule, M''. This generalization would have been sufficient, even without concepts of stress and strain later introduced by Cauchy, to develop the more general two-parameter (E, ν) classical linear elasticity model. Additionally, Navier could have assumed a more general *nonlinear* force-versus-stretch function, and then specialized this to a linear function if necessary. Assumptions creep into a theory unnoticed in this way. Unlike Navier, with peridynamic theory Silling is careful to make no overly-restrictive, un-noticed, assumptions about the nature of the force-deformation relationship in his state-based theory.

[l] Thus, Navier assumes that the material horizon is infinitesimal. Peridynamics makes no such assumption. Navier does not discuss the fact that molecules sufficiently near the boundary of the domain would not see the same neighborhood of particles as interior molecules, and perhaps the limiting process that he employs makes this distinction unnecessary.

2. One represents by *a, b, c* the three rectangular coordinates of any point M in the interior of an elastic body.[m] Supposing that, by the effect of the forces applied to this body, its figure has changed, one designates by $a + x$, $b + y$, $c + z$ the coordinates of point *m* to which point M is transported; such that *x, y, z* are respectively the quantities by which point M moves parallel to the axes of *a, b, c*.[n] The quantities *x, y, z* are generally functions of *a, b, c* and of the forces applied to the body. It is a question of knowing the nature of these functions, and, in the case where the points of the body oscillate around their natural situations, of finding the values of *x, y, z* as functions of *a, b, c* and of time *t*.

Let us consider a point M′ neighbor of point M, and of which the coordinates are $a + \alpha$, $b + \beta$, $c + \gamma$. One will know the position *m'* that point M′ will take after the figure change of the body, by substituting $a + \alpha$, $b + \beta$, $c + \gamma$ in the place of *a, b, c* in the expressions of *x, y, z* as functions of *a, b, c*. Consequently, if one represents by *x', y', z'* the quantities by which point M′ is displaced in the direction of each axis, one will have[o]

$$\begin{aligned} x' = x &+ \frac{dx}{da}\alpha + \frac{1}{2}\frac{d^2x}{da^2}\alpha^2 + etc., \\ &+ \frac{dx}{db}\beta + \frac{d^2x}{dadb}\alpha\beta \\ &+ \frac{dx}{dc}\gamma + \frac{1}{2}\frac{d^2d}{db^2}\beta^2 \\ &+ \frac{d^2x}{dadc}\alpha\gamma \\ &+ \frac{d^2x}{dbdc}\beta\gamma \\ &+ \frac{1}{2}\frac{d^2x}{dc^2}\gamma^2 \end{aligned}$$

$$\begin{aligned} y' = y &+ \frac{dy}{da}\alpha + \frac{1}{2}\frac{d^2y}{da^2}\alpha^2 + etc., \\ &+ \frac{dy}{db}\beta + \frac{d^2y}{dadb}\alpha\beta \\ &+ \frac{dy}{dc}\gamma + \frac{1}{2}\frac{d^2y}{db^2}\beta^2 \\ &+ \frac{d^2y}{dadc}\alpha\gamma \\ &+ \frac{d^2y}{dbdc}\beta\gamma \\ &+ \frac{1}{2}\frac{d^2y}{dc^2}\gamma^2 \end{aligned}$$

$$\begin{aligned} z' = z &+ \frac{dz}{da}\alpha + \frac{1}{2}\frac{d^2z}{da^2}\alpha^2 + etc. \\ &+ \frac{dz}{db}\beta + \frac{d^2z}{dadb}\alpha\beta \\ &+ \frac{dz}{dc}\gamma + \frac{1}{2}\frac{d^2z}{db^2}\beta^2 \\ &+ \frac{d^2z}{dadc}\alpha\gamma \\ &+ \frac{d^2z}{dbdc}\beta\gamma \\ &+ \frac{1}{2}\frac{d^2z}{dc^2}\gamma^2 \end{aligned}$$

---

[m] Navier uses a, b, and c to define the undeformed, or reference coordinates of the particle. In modern times, we would use reference coordinates X, Y and Z.

[n] Reading the equations in this paper is rather difficult, as in modern times we normally use u, v, and w to define displacement components, and we use x, y, and z to define spatial (deformed) positions of the particle. The modern reader must deal with these naming convention clashes throughout the paper.

[o] In this next equation, Navier assumes that the displacement field and all of its derivatives are differentiable. Thus, he assumes that the displacement field is smooth, which precludes discrete cracks.

The distance of the two points M, M' before the figure change of the body is

$$\sqrt{\alpha^2 + \beta^2 + \gamma^2}\,.^{\text{p}}$$

After this change it has become

$$\sqrt{(\alpha + x' - x)^2 + (\beta + y' - y)^2 + (\gamma + z' - z)^2};$$

or, in expanding and disregarding the higher powers of the quantities *x'- x, y'- y, z'- z,* considered as very small,

$$\sqrt{\alpha^2 + \beta^2 + \gamma^2} + \frac{\alpha(x'-x)+\beta(y'-y)+\gamma(z'-z)}{\sqrt{\alpha^2+\beta^2+\gamma^2}}.^{\text{q}}$$

Thus naming *p, q,* ψ the angles between the line MM' and the axes of *a* and *b* and the axis normal to plane *ab,* one will have[r]

$$(x' - x)cos.p + (y' - y)cos.q + (z' - z)sin.\psi$$

for the expression of the increase mm' – MM' in question. The force with which point M is now attracted by point M' is proportional, according to the aforementioned principle, to this increase.[s] This force depends, moreover, on the distance[t] between the two points, and becomes very small, as soon as this distance takes a perceptible value. Thus, in naming the distance MM' ρ, and representing by f_ρ an unknown function that decreases very rapidly when ρ increases, the force in question will have to be expressed by

---

[p] Navier rarely uses equal signs, which make the paper a bit difficult for a modern reader.

[q] This is where Navier makes the assumption of small deformations.

[r] $cos.p = \cos(p)$

[s] This is an application of Hooke's law of linear elasticity.

[t] Navier means the reference distance (undeformed distance between M and M') distance here.

$$f_\rho \cdot [(x'-x)cos.p + (y'-y)cos.q + (z'-z)sin.\psi] .$$ [u]

Point M is attracted with similar forces by all the points M' that are situated around it.[v] If one writes that this point is in equilibrium by virtue of these forces, and by virtue of other forces X, Y, Z, which would be applied to this point in the direction of each axis, one will express the condition on which the figure affected by the elastic solid depends.

3. The active force following the line MM', of which one has just found the expression, being broken down in the direction of the axes of *a, b, c,* gives the three components

$$f_\rho \cdot [(x'-x)cos.^2 p + (y'-y)cos.p\ cos.q + (z'-z)cos.p\ sin.\psi] ,$$
$$f_\rho \cdot [(x'-x)cos.p\ cos.q + (y'-y)cos.^2 q + (z'-z)cos.q\ sin.\psi] ,$$
$$f_\rho \cdot [(x'-x)cos.p\ sin.\psi + (y'-y)cos.q\ sin.\psi + (z'-z)sin.^2 \psi] .$$

Let's name φ the angle that the projection of the line MM' or ρ on the plane of *ab* makes with the axis of *a*. We will have cos.*p* = cos.ψ cos.φ, cos.*q* = cos.ψsin.φ, and the preceding expressions will become[w]

$$f_\rho \cdot [(x'-x)cos.^2 \psi cos.^2 \varphi + (y'-y)cos.^2 \psi\ sin.\varphi\ cos.\varphi + (z'-z)sin.\psi\ cos.\psi cos.\varphi] ,$$
$$f_\rho \cdot [(x'-x)cos.^2 \psi\ sin.\varphi\ cos.\varphi + (y'-y)cos.^2 \psi cos.^2 \varphi + (z'-z)\ sin.\psi cos.\psi sin.\varphi] ,$$
$$f_\rho \cdot [(x'-x)sin.\psi\ cos.\psi cos.\varphi + (y'-y)sin.\psi cos.\psi sin.\varphi + (z'-z)sin.^2 \psi] .$$

The equilibrium of point M will be expressed by equaling respectively to – X, –Y, – Z, the sum of all the components parallel to the axis of a, the axis of b, the axis of c.

Let us consider in the first place the components parallel to the axis of *a*. As one has α = ρcos.ψcos.φ, *β* = ρcos.ψsin.φ, γ = ρsin.ψ, the expression of these components, by putting for *x' – x, y' – y, z' – z* their values, will become

---

[u] Thus, the pair-wise force is proportional to the change in length of the bond, as well as to the unknown function $f_\rho = f(\rho)$, which has units of force per unit length.

[v] As long as it is sufficiently far from a boundary!

[w] Here, Navier is using spherical angles defined in Fig. 2.4.

$$
f\rho.\left\{
\begin{array}{l}
\rho\left\{
\begin{array}{lll}
\frac{dx}{da}cos.^3\,\psi\,cos.^3\,\varphi & +\frac{dy}{da}cos.^3\,\psi\,sin.\,\varphi\,cos.^2\,\varphi & +\frac{dz}{da}sin.\,\psi\,cos.^3\,\psi\,cos.^2\,\varphi \\
\frac{dx}{db}cos.^3\,\psi\,sin.\,\varphi\,cos.^2\,\varphi & +\frac{dy}{db}cos.^3\,\psi\,sin.^2\,\varphi\,cos.\,\varphi & +\frac{dz}{db}sin.\,\psi\,cos.^2\,\psi\,sin.\,\varphi\,cos.\,\varphi \\
\frac{dx}{dc}sin.\,\psi\,cos.^2\,\psi\,cos.^2\,\varphi & +\frac{dy}{dc}sin.\,\psi\,cos.^2\,\psi\,sin.\,\varphi\,cos.\,\varphi & +\frac{dz}{dc}sin.^2\,\psi\,cos.\,\psi\,cos.\,\varphi
\end{array}
\right\}+ \\
\rho^2\left\{
\begin{array}{lll}
\frac{1}{2}\frac{d^2x}{da^2}cos.^4\,\psi\,cos.^4\,\varphi & +\frac{1}{2}\frac{d^2y}{da^2}cos.^4\,\psi\,sin.\,\varphi\,cos.^2\,\varphi & +\frac{1}{2}\frac{d^2z}{da^2}sin.\,\psi\,cos.^3\,\psi\,cos.^2\,\varphi \\
\frac{d^2x}{dadb}cos.^4\,\psi\,sin.\,\varphi\,cos.^3\,\varphi & +\frac{d^2y}{dadb}cos.^4\,\psi\,sin.^2\,\varphi\,cos.^2\,\varphi & +\frac{d^2z}{dadb}sin.\,\psi\,cos.^3\,\psi\,sin.\,\varphi\,cos.^2\,\varphi \\
\frac{1}{2}\frac{d^2x}{db^2}cos.^4\,\psi\,sin.^2\,\varphi\,cos.^2\,\varphi & +\frac{1}{2}\frac{d^2y}{db^2}cos.^4\,\psi\,sin.^3\,\varphi\,cos.\,\varphi & +\frac{1}{2}\frac{d^2z}{db^2}sin.\,\psi\,cos.^3\,\psi\,sin.^2\,\varphi\,cos.\,\varphi \\
\frac{d^2x}{dadc}sin.\,\psi\,cos.^3\,\psi\,cos.^3\,\varphi & +\frac{d^2y}{dadc}sin.\,\psi\,cos.^3\,\psi\,sin.\,\varphi\,cos.^2\,\varphi & +\frac{d^2z}{dadc}sin.^2\,\psi\,cos.^2\,\psi\,cos.^2\,\varphi \\
\frac{d^2x}{dbdc}sin.\,\psi\,cos.^3\,\psi\,sin.\,\varphi\,cos.^2\,\varphi & +\frac{d^2y}{dbdc}sin.\,\psi cos.^3\,\psi\,sin.^2\,\varphi\,cos.\,\varphi & +\frac{d^2z}{dbdc}sin.^2\,\psi\,cos.^2\,\psi\,sin.\,\varphi\,cos.\,\varphi \\
\frac{1}{2}\frac{d^2x}{dc^2}sin.^2\,\psi\,cos.^2\,\psi\,cos.^2\,\varphi & +\frac{1}{2}\frac{d^2y}{dc^2}sin.^3\,\psi\,cos.^2\,\psi\,sin.\,\varphi\,cos.\,\varphi & +\frac{1}{2}\frac{d^2z}{dc^2}sin.^3\,\psi\,cos.\,\psi\,cos.\,\varphi
\end{array}
\right\}+ \\
etc.
\end{array}
\right.
$$

To take the sum of all these components, it is necessary, conforming to the principles of integral calculus, to remark that the element of volume of the space that surrounds point M, using the coordinates ρ, φ and ψ, is expressed by $d\rho d\psi d\varphi.\rho^2 cos.\psi$.[x] One will multiply therefore the preceding quantity by this element, and one will integrate with respect to the three variables ρ, ψ and φ: namely, with respect to ρ from ρ = 0 to ρ = ∞; with respect to ψ from $\psi = -\frac{1}{2}\pi$ to $= \frac{1}{2}\pi$, π representing the ratio of circumference to diameter; and finally with respect to φ from φ = 0 to φ = 2π. One will encompass thus all the forces parallel to the axis of *a*, by which point M is attracted. But one easily sees that in carrying out this operation, the terms of the preceding quantity that contain an odd power of $sin.\psi$, and those that contain an odd power of $sin.\varphi$ or of $cos.\varphi$, will give zero for a result. Thus, the first term multiplied by ρ

[x] This is the differential volume in spherical coordinates, shown in Fig. 2.4.

will disappear entirely. If the term multiplied by ρ^3 was written, one would see that it would disappear also, and the same applies to all the terms that contain odd powers of ρ. The terms that contain even powers of this quantity remain alone in the result of the integration; but, as one will see later, one must limit oneself to considering the first among them. By taking only this term therefore, and not writing the quantities that become non-existent as a result of the integration, one will have

$$\int_0^\infty d\rho \int_{-\frac{1}{2}\pi}^{\frac{1}{2}\pi} d\psi \int_0^{2\pi} d\varphi \cdot \rho^4 f_\rho \left\{ \begin{array}{c} \frac{1}{2}\frac{d^2x}{da^2} cos.^5\,\psi\, cos.^4\,\varphi + \frac{d^2y}{dadb} cos.^5\,\psi\, sin.^2\,\varphi\, cos.^2\,\varphi + \frac{d^2z}{dadc} sin.^2\,\psi\, cos.^3\,\psi\, cos.^2\,\varphi \\ \frac{1}{2}\frac{d^2x}{db^2} cos.^5\,\psi\, sin.^2\,\varphi\, cos.^2\,\varphi \\ \frac{1}{2}\frac{d^2x}{dc^2} sin.^2\,\psi\, cos.^3\,\psi\, cos.^2\,\varphi \end{array} \right\}$$

for the expression of the sum of the interior forces that attract point M parallel to the axis of *a*.

To carry out the integration, one will remark that $\int_0^{2\pi} d\varphi \cos^4\varphi = \frac{3\pi}{4}$, $\int_0^{2\pi} d\varphi \sin^2\varphi \cos^2\varphi = \frac{\pi}{4}$, $\int_0^{2\pi} d\varphi \cos^2\varphi = \pi.$ Thus, by first integrating with respect to φ, we get

$$\int_0^\infty d\rho \int_{-\frac{1}{2}\pi}^{\frac{1}{2}\pi} d\psi \cdot \rho^4 f_\rho \left\{ \begin{array}{c} \frac{1}{2}\frac{3\pi}{4}\frac{d^2x}{da^2} cos.^5\,\psi + \frac{\pi}{4}\frac{d^2y}{dadb} cos.^5\,\psi + \pi\frac{d^2z}{dadc} sin.^2\,\psi\, cos.^3\,\psi \\ \frac{1}{2}\frac{\pi}{4}\frac{d^2x}{db^2} cos.^5\,\psi \\ \frac{1}{2}\pi\frac{d^2x}{dc^2} sin.^2\,\psi\, cos.^3\,\psi \end{array} \right\}$$

One remarks then that $\int_{-\frac{1}{2}\pi}^{\frac{1}{2}\pi} d\psi \sin^2\psi \cos^3\psi = \frac{4}{15}$, $\int_{-\frac{1}{2}\pi}^{\frac{1}{2}\pi} d\psi \cos^5\psi = \frac{4\cdot 4}{15}$; such that the integration with respect to ψ will give

$$\int_0^\infty d\rho \cdot \rho^4 f_\rho \cdot \frac{\pi}{4}\frac{4\cdot 4}{15}\left(\frac{3}{2}\frac{d^2x}{da^2} + \frac{1}{2}\frac{d^2x}{db^2} + \frac{1}{2}\frac{d^2x}{dc^2} + \frac{d^2y}{dadb} + \frac{d^2z}{dadc}\right).$$

If therefore, to shorten [it], one defines

$$\int_0^\infty d\rho \cdot \rho^4 f_\rho \cdot \frac{\pi}{4} \cdot \frac{4 \cdot 4}{15} \cdot \frac{1}{2} = \varepsilon\,,$$

ε being an unknown constant that depends on the intensity of the elastic force of the body, one will have, to express the equilibrium of the forces that attract point M parallel to the axis of *a*, the equation

$$-X = \varepsilon\left(3\frac{d^2x}{da^2} + \frac{d^2x}{db^2} + \frac{d^2x}{dc^2} + 2\frac{d^2y}{dadb} + 2\frac{d^2z}{dadc}\right).$$

One has said above that the terms containing powers of ρ higher than two had to be disregarded. This proposition may be rendered very noticeable by specifying the function f_ρ, and giving it the characteristic of a function that decreases very rapidly when ρ increases. If, for example, one takes for this function $e^{-k\rho}$, *e* being the number of which the Napierian logarithm is the unit, and *k* a constant coefficient, one will have

$$\int_0^\infty d\rho \cdot \rho^4 e^{-k\rho} = \frac{1.2.3.4}{k^5},\ \int_0^\infty d\rho \cdot \rho^6 e^{-k\rho} = \frac{1.2.3.4.5.6}{k^7}, etc.$$

But in order that the quantity $e^{-k\rho}$ should decrease with a very great rapidity when ρ increases, it is necessary to suppose that the coefficient *k* is a very large number. The successive terms of the series are therefore affected after the integration with respect to ρ by coefficients that decrease with an extreme rapidity, and must all be disregarded in relation to the first. This remark makes evident the spirit of the type of analysis that one uses here. It shows how, the forces that produce the elasticity being only exerted at extremely small distances, as well as indicate all the known phenomena, it will be permitted to disregard the terms of the higher orders; and how the laws of equilibrium are then expressed with accuracy by means of equations that contain only terms of the second order.

In carrying out an absolutely similar calculation, which one believes unnecessary to explain in detail, on the expressions of the forces that attract point M parallel to the axes of b and c, one obtains results analogous to the preceding; such that one has definitively for the expression of conditions of equilibrium of the point in question, the three following equations, where the forces X, Y, Z are considered as positive when they tend to increase the displacements,

$$-X = \varepsilon\left(3\frac{d^2x}{da^2} + \frac{d^2x}{db^2} + \frac{d^2x}{dc^2} + 2\frac{d^2y}{dadb} + 2\frac{d^2z}{dadc}\right),$$

$$-Y = \varepsilon\left(\frac{d^2x}{da^2} + 3\frac{d^2y}{db^2} + \frac{d^2y}{dc^2} + 2\frac{d^2x}{dadb} + 2\frac{d^2z}{dbdc}\right),$$

$$-Z = \varepsilon\left(\frac{d^2z}{da^2} + \frac{d^2z}{db^2} + 3\frac{d^2z}{dc^2} + 2\frac{d^2x}{dadc} + 2\frac{d^2y}{dbdc}\right).$$

4. These equations respond to that which one ordinarily names the *indefinite equations*.[y] They indicate a characteristic common to all the points of the body, which must present the analytical expressions that will give the values of x, y, z, as functions of a, b, c, X, Y, Z. But one knows, by the questions of the same type that have been already solved, that there exists furthermore conditions particular to the points situated at the limits of the body, conditions that the preceding analysis does not permit to determine easily. The following analysis, which gives the same indefinite equations that have just been obtained, has furthermore the advantage of leading at the same time to the particular conditions in question.[z]

---

[y] These are the three Navier equations of elasticity in terms of the three unknown displacement components x, y, and z. Note particularly that there is no concept of stress or of strain. However, in hindsight, we can understand the second-order displacement derivatives as first order derivatives of what we now call strain components. The single parameter ε represents the elastic stiffness of the material, similar to Young's modulus. In such a model, the effective Poisson's ratio would be ¼.

[z] Navier is now concerned with developing the boundary conditions, equations that must be satisfied on the boundary of the body, in addition to the partial differential equations that must be satisfied at all points on the interior of the body. Thus, in the remainder of

Let us come back to the consideration of the neighboring points M, M', such as it is used in n° 2. Since the actions that are exerted between these points only have perceptible values at extremely small distances, one must only take for point M' points extremely nearby to point M. The quantities α, β, γ, which are the coordinates of point M' taken into consideration starting from point M, are therefore extremely small, and one can disregard their powers and products. So one has simply[aa]

$$x' - x = \frac{dx}{da}\alpha + \frac{dx}{db}\beta + \frac{dx}{dc}\gamma\,,$$
$$y' - y = \frac{dy}{da}\alpha + \frac{dy}{db}\beta + \frac{dy}{dc}\gamma\,,$$
$$z' - z = \frac{dz}{da}\alpha + \frac{dz}{db}\beta + \frac{dz}{dc}\gamma\,.$$

But the projections on each of the axes of the distance *mm'*, to which points M, M' are placed after the figure change of the body, are respectively $\alpha + x' - x$, $\beta + y' - y$, $\gamma + z' - z$. The expression of this distance is therefore

$$\sqrt{(\alpha + x' - x)^2 + (\beta + y' - y)^2 + (\gamma + z' - z)^2}\,;$$

or, by substituting the preceding values,

$$\sqrt{\left(\alpha + \frac{dx}{da}\alpha + \frac{dx}{db}\beta + \frac{dx}{dc}\gamma\right)^2 + \left(\beta + \frac{dy}{da}\alpha + \frac{dy}{db}\beta + \frac{dy}{dc}\gamma\right)^2 + \left(\gamma + \frac{dz}{da}\alpha + \frac{dz}{db}\beta + \frac{dz}{dc}\gamma\right)^2}\,.$$

The figure change of the body being supposed very small, the quantities *x, y, z* and their partial differential coefficients are very small, and one can disregard their powers and products. The preceding expression will therefore take the form

---

the paper, he re-derives the equations of elasticity using what we would now call a virtual displacement approach.

[aa] This is a Taylor series expansion, neglecting higher-order terms. The equations are not correct at points located on the boundary of the domain.

$$\sqrt{\alpha^2+\beta^2+\gamma^2+2\left[\frac{dx}{da}\alpha^2+\left(\frac{dx}{db}+\frac{dy}{da}\right)\alpha\beta+\left(\frac{dx}{dc}+\frac{dz}{da}\right)\alpha\gamma+\frac{dy}{db}\beta^2+\left(\frac{dy}{dc}+\frac{dz}{db}\right)\beta\gamma+\frac{dz}{dc}\gamma^2\right]},$$

or by expanding in a series, and always disregarding the higher powers of the differential coefficients,

$$\sqrt{\alpha^2+\beta^2+\gamma^2}+\frac{1}{\sqrt{\alpha^2+\beta^2+\gamma^2}}\left[\frac{dx}{da}\alpha^2+\left(\frac{dx}{db}+\frac{dy}{da}\right)\alpha\beta\right.$$
$$\left.+\left(\frac{dx}{dc}+\frac{dz}{da}\right)\alpha\gamma+\frac{dy}{db}\beta^2+\left(\frac{dy}{dc}+\frac{dz}{db}\right)\beta\gamma+\frac{dz}{dc}\gamma^2\right].$$

The first term is the original value of the distance MM' of the two points that one is considering, which has been represented above by ρ. The second term represents therefore the variation that this distance has undergone as a result of the figure change of the body, and to which the force that acts from M' on M is proportional. If one replaces α, β, γ with the values $\alpha = \rho\, cos.\psi\, cos.\varphi$, $\beta = \rho\, cos.\psi\, sin.\varphi$, $\gamma = \rho\, sin.\psi$, this variation will become

$$\rho\left[\frac{dx}{da}cos.^2\psi\, cos.^2\varphi+\left(\frac{dx}{db}+\frac{dy}{da}\right)cos.^2\psi\, sin.\varphi\, cos.\varphi\right.$$
$$+\left(\frac{dx}{dc}+\frac{dz}{da}\right)cos.\psi\, sin.\psi\, cos.^2\varphi+\frac{dy}{db}cos.^2\psi\, sin.^2\varphi$$
$$\left.+\left(\frac{dy}{dc}+\frac{dz}{db}\right)sin.\psi\, cos.\psi\, sin.\varphi+\frac{dz}{dc}sin.^2\psi\right].$$

To abbreviate, let us represent this quantity by *f*.[bb] The force with which point M' attracts M will therefore be proportional to *f*. The moment of this force, this expression being taken in the same sense as in *la Mécanique analytique,* will be evidently proportional to $f\ \delta\ f$, or

[bb] *f* here is not a force; rather, it is a variation in length of the bond due to displacements *x*, *y*, and *z*. Not to be confused with f_ρ.

to $\frac{1}{2}\delta f^2$.[cc] Consequently, if one multiplies $\frac{1}{2}\delta f^2$ by $d\rho\, d\psi\, d\varphi . \rho^2 cos\, \psi \cdot f_\rho$; if one moves the symbol δ in front of the symbols of integration related to ρ, ψ and φ, which is permitted; and if one integrates between the same limits that one has done in n° 3: one will have a quantity proportional to the sum of the moments[dd] of all the interior forces by which point M is attracted. This quantity is therefore

$$\frac{1}{2}\delta \int_0^\infty d\rho \int_{-\frac{1}{2}\pi}^{\frac{1}{2}\pi} d\psi \int_0^{2\pi} d\varphi \cdot \rho^4 cos.\psi$$

$$\cdot f_\rho \left[\frac{dx}{da} cos.^2 \psi\, cos.^2 \varphi + \left(\frac{dx}{db}+\frac{dy}{da}\right) cos.^2 \psi\, sin.\varphi\, cos.\varphi + \left(\frac{dx}{dc}+\frac{dz}{da}\right) cos.\psi\, sin.\psi\, cos.\varphi + \frac{dy}{db} cos.^2 \psi\, sin.^2 \varphi + \left(\frac{dy}{dc}+\frac{dz}{db}\right) sin.\psi\, cos.\psi\, sin.\varphi + \frac{dz}{dc} sin.^2 \psi\right]^2 .$$

By raising to the appropriate square, and not writing the terms containing odd powers of sin.φ and cos.φ, which will give zero as a result of integration, this expression becomes

$$\frac{1}{2}\delta \int_0^\infty d\rho \int_{-\frac{1}{2}\pi}^{\frac{1}{2}\pi} d\psi \int_0^{2\pi} d\varphi \cdot \rho^4 cos.\psi \cdot$$

$$f_\rho \left\{ \begin{array}{c} \frac{dx^2}{da^2} cos.^4 \psi\, cos.^4 \varphi + \left(\frac{dx}{db}+\frac{dy}{db}\right) + 2\frac{dx}{da}\frac{dy}{db}\Big\} cos.^4 \psi\, sin.^2 \varphi\, cos.^2 \varphi \\ \left(\frac{dx}{dc}+\frac{dz}{da}\right)^2 + 2\frac{dx}{da}\frac{dz}{dc}\Big\} sin.^2 \psi\, cos.^2 \psi\, cos.^2 \varphi + \frac{dy^2}{db^2} cos.^4 \psi\, sin.^4 \varphi \\ \left(\frac{dy}{dc}+\frac{dz}{db}\right)^2 + 2\frac{dy}{db}\frac{dz}{dc}\Big\} sin.^2 \psi\, cos.^2 \psi\, sin.^2 \varphi + \frac{dz^2}{dc^2} sin.^4 \psi \end{array} \right\}$$

[cc] Here, Navier is using a virtual displacement approach. The "moment if the force" is in todays terms, the variation in the force.

[dd] Here, "moments" means "variations".

By integrating first with respect to φ, one will have

$$\frac{1}{2}\delta\int_0^{\infty} d\rho \int_{-\frac{1}{2}\pi}^{\frac{1}{2}\pi} d\psi\cdot\rho^4$$

$$\cdot f_\rho\frac{\pi}{4}\left\{\begin{array}{c} 3\frac{dx^2}{da^2}cos.^5\,\psi+\left(\frac{dx}{db}+\frac{dy}{da}\right)^2+2\frac{dx}{da}\frac{dy}{db}\Big\}cos.^5\,\psi \\ 4\left(\frac{dx}{dc}+\frac{dz}{da}\right)^2+4\cdot 2\frac{dx}{da}\frac{dz}{dc}\Big\}sin.^2\,\psi\; cos.^3\,\psi+3\frac{dy^2}{db^2}cos.^5\,\psi \\ 4\left(\frac{dy}{dc}+\frac{dz}{db}\right)^2+4\cdot 2\frac{dy}{db}\frac{dz}{dc}\Big\}sin.^2\,\psi\; cos.^3\,\psi+3\frac{dz^2}{dc^2}sin.^4\,\psi cos.\,\psi \end{array}\right\}$$

By integrating next with respect to ψ, then representing by ε the coefficient that will remain after the integration with respect to ρ, it will come definitively for the expression of the sum of moments in question

$$\frac{1}{2}\varepsilon\cdot\delta\left[\begin{array}{c} 3\left(\frac{dx^2}{da^2}\right)+\left(\frac{dx}{db}+\frac{dy}{da}\right)^2+2\frac{dx}{da}\frac{dy}{db}+\left(\frac{dx}{dc}+\frac{dz}{da}\right)^2+2\frac{dx}{da}\frac{dz}{dc} \\ +3\frac{dy^2}{db^2}+\left(\frac{dy}{dc}+\frac{dz}{db}\right)^2+2\frac{dy}{db}\frac{dz}{dc}+3\frac{dz^2}{dc^2} \end{array}\right].$$

5. The value of this sum is the same for all the points included in the element of the body of which the volume is *da db dc*. One will have therefore the total sum of similar moments by multiplying the preceding expressing by *da db dc,* and then integrating in the full expanse of the body. According to that, let's designate as [we did] above by X, Y, Z the forces applied to the points included in the same element, these quantities being weights added on to the unit of volume. Let us represent furthermore by X', Y', Z' the forces applied parallel to the axes of *a, b, c*, to the point of the surface of the body of which the coordinates are *a', b', c',* these quantities designating weights added to the unit of surface, and by *ds* the element of the surface situated at the same point: one will have X'*ds,* Y'*ds,* Z'*ds* for the values of the forces applied to this element

in the direction of each axis.[ee] The equation that will express the equilibrium of the interior forces originating from the elasticity and of the forces applied to the body will be therefore

$$
\begin{gathered}
0 = \\
\varepsilon \iiint dadbdc \left\{ \begin{array}{l}
3\left(\frac{dx}{da}\frac{\delta dx}{da}\right) + \left(\frac{dx}{db}\frac{\delta dx}{db}\right) + \left(\frac{dx}{db}\frac{\delta dy}{da}\right) + \left(\frac{dy}{da}\frac{\delta dx}{db}\right) + \left(\frac{dy}{da}\frac{\delta dy}{da}\right) + \left(\frac{dx}{da}\frac{\delta dy}{db}\right) + \left(\frac{dy}{db}\frac{\delta dx}{da}\right) \\
\left(\frac{dx}{da}\frac{\delta dx}{dc}\right) + \left(\frac{dx}{dc}\frac{\delta dz}{da}\right) + \left(\frac{dz}{da}\frac{\delta dx}{dc}\right) + \left(\frac{dz}{da}\frac{\delta dz}{da}\right) + \left(\frac{dx}{da}\frac{\delta dz}{dc}\right) + \left(\frac{dz}{dc}\frac{\delta dx}{da}\right) + 3\left(\frac{dy}{db}\frac{\delta dy}{db}\right) \\
\left(\frac{dy}{dc}\frac{\delta dy}{dc}\right) + \left(\frac{dy}{dc}\frac{\delta dz}{db}\right) + \left(\frac{dz}{db}\frac{\delta dy}{dc}\right) + \left(\frac{dz}{db}\frac{\delta dz}{db}\right) + \left(\frac{dy}{db}\frac{\delta dz}{dc}\right) + \left(\frac{dz}{dc}\frac{\delta dy}{db}\right) + 3\left(\frac{dz}{dc}\frac{\delta dz}{dc}\right)
\end{array} \right\} \\
- \iiint dadbdc(X\delta x + Y\delta y + Z\delta z) - \int ds(X'\delta x' + Y'\delta y' + Z'\delta z'),
\end{gathered}
$$

where one has expanded the appropriate squares, and carried out the differentiation marked by δ, in the term that contains the moments of the interior forces.[ff] The forces X', Y', Z' are considered as positive when they aim to increase the displacements.

It is now necessary to shift the *d* in front of the δ in the term of which one has just spoken and to carry out the integrations by parts that have as a purpose to make the differentials of the variations disappear.[gg] This operation will change this term into

$$
-\varepsilon \iiint dadbdc \left\{ \begin{array}{l}
\left(3\frac{d^2x}{da^2} + \frac{d^2x}{db^2} + \frac{d^2x}{dc^2} + 2\frac{d^2y}{dadb} + 2\frac{d^2z}{dadc}\right)\delta x \\
\left(\frac{d^2y}{da^2} - 3\frac{d^2y}{db^2} + \frac{d^2y}{dc^2} + 2\frac{d^2x}{dadb} + 2\frac{d^2z}{dbdc}\right)\delta y \\
\left(\frac{d^2z}{da^2} + \frac{d^2z}{db^2} + 3\frac{d^2z}{dc^2} + 2\frac{d^2x}{dadc} + 2\frac{d^2y}{dbdc}\right)\delta z
\end{array} \right\}
$$

[ee] Here, Navier is defining surface traction components (force components per unit surface area).

[ff] This equation can be recognized as an equation of virtual displacements.

[gg] Integration by parts is a theorem that relates the integral of a product of functions to the integral of their derivative and indefinite integral.

$$+\varepsilon\left[\iint db'dc'\left(3\frac{dx'}{da'}+\frac{dy'}{db'}+\frac{dz'}{dc'}\right)+\iint da'dc'\left(\frac{dx'}{db'}+\frac{dy'}{da'}\right)+\iint da'db'\left(\frac{dx'}{dc'}+\frac{dz'}{da'}\right)\right]\delta dx'$$
$$+\varepsilon\left[\iint db'dc'\left(\frac{dx'}{db'}+\frac{dy'}{da'}\right)+\iint da'dc'\left(\frac{dx'}{da'}+3\frac{dy'}{db'}+\frac{dz'}{dc'}\right)+\iint da'db'\left(\frac{dy'}{dc'}+\frac{dz'}{db'}\right)\right]\delta dy'$$
$$+\varepsilon\left[\iint db'dc'\left(\frac{dx'}{dc'}+\frac{dz'}{da'}\right)+\iint da'dc'\left(\frac{dy'}{dc'}+\frac{dz'}{db'}\right)+\iint da'db'\left(\frac{dx'}{da'}+\frac{dy'}{db'}+3\frac{dz'}{dc'}\right)\right]\delta dz'.$$

To abbreviate, one only writes once the terms belonging to the limits of the integrals, and that relate to the points of the limits of the body. But it will be necessary to pay attention that when considering these terms as belonging to the points of the first limit, it is necessary to assign them the – sign, and that when considering them as belonging to the points of the opposite limit, it is necessary to take them positively.

The equation of equilibrium above will therefore give first, in equaling the coefficients of the variations separately to zero, the three indefinite equations

$$-X=\varepsilon\left(3\frac{d^2x}{da^2}+\frac{d^2x}{db^2}+\frac{d^2x}{dc^2}+2\frac{d^2y}{dadb}+2\frac{d^2z}{dadc}\right),$$
$$-Y=\varepsilon\left(\frac{d^2y}{da^2}+3\frac{d^2y}{db^2}+\frac{d^2y}{dc^2}+2\frac{d^2x}{dadb}+2\frac{d^2z}{dbdc}\right),$$
$$-Z=\varepsilon\left(\frac{d^2z}{da^2}+\frac{d^2z}{db^2}+3\frac{d^2z}{dc^2}+2\frac{d^2x}{dadc}+2\frac{d^2y}{dbdc}\right),$$

which are the same equations that have been obtained in another manner in n° 3. Furthermore, let us name *l, m, n* the angles that the plane tangent to the surface in the point of which the coordinates are *a', b', c'* form with the planes of *bc, ac,* and *ab* : one will be able to replace, in the terms that refer to the points of the surface, *db' dc'* by *ds* cos *l, da' dc'* by *ds* cos *m,* and *da' db'* by *ds* cos *n* (see *la Mécanique analytique,* v. I, p. 205).

These terms will supply therefore the determined equations

$$X'=\varepsilon\left[\cos.l\left(3\frac{dx'}{da'}+\frac{dy'}{db'}+\frac{dz'}{dc'}\right)+\cos.m\left(\frac{dx'}{db'}+\frac{dy'}{da'}\right)+\cos.n\left(\frac{dx'}{dc'}+\frac{dz'}{da'}\right)\right],$$
$$Y'=\varepsilon\left[\cos.l\left(\frac{dx'}{db'}+\frac{dy'}{da'}\right)+\cos.m\left(\frac{dx'}{da'}+3\frac{dy'}{db'}+\frac{dz'}{dc'}\right)+\cos.n\left(\frac{dy'}{dc'}+\frac{dz'}{db'}\right)\right],$$
$$Z'=\varepsilon\left[\cos.l\left(\frac{dx'}{dc}+\frac{dz'}{da'}\right)+\cos.m\left(\frac{dy'}{dc'}+\frac{dz'}{db'}\right)+\cos.n\left(\frac{dx'}{da'}+\frac{dy'}{db'}+3\frac{dz'}{dc'}\right)\right],$$

which give the value of the forces that must be applied to the points of the surface of the body, or of the tractions exerted on the obstacles that would hold fixed these points.[hh] The second members must be assigned the – sign for the points related to the first limit of this body, and taken positively for the points related to the opposite limit.[ii]

The preceding equations contain, under the differential form, all that is possible to state in a general manner on the conditions of the equilibrium of a solid elastic body. In order to make new steps in the search for this equilibrium, it is indispensable to specify the figure of the body. One will show in another treatise that the preceding equations may be integrated in a great number of cases, for which one attains a complete knowledge of the state of equilibrium in question.

6. One passes easily from general equations of equilibrium to those which express the laws of vibrations. The forces X, Y, Z are supposed positive when they increase the quantities x, y, z. If the points of the body perform movements of oscillation around their original positions, the accelerating forces, to which these movements will be due, will aim to diminish these same quantities. Naming Π the weight of the unit of volume of the body, g the speed that gravity imparts to heavy bodies in the unit of time, the accelerating forces, to which the movements of point M in the direction of each axis will be due, will be respectively $\frac{\Pi}{g}\frac{d^2x}{dt^2}$, $\frac{\Pi}{g}\frac{d^2y}{dt^2}$, $\frac{\Pi}{g}\frac{d^2z}{dt^2}$, dt being the element of time. One must therefore add respectively these three quantities to the first members of the preceding equations; and if one supposes that no force is applied to the interior points of the body, these equations will be simply

[hh] Navier uses the term "efforts", which has been translated as "tractions". "Stresses" might be another translation.

[ii] The equations, both on the domain interior and on the boundary appear to be correct. The details of this derivation are obscure, as several important steps have been omitted.

$$\frac{\Pi}{g}\frac{d^2x}{dt^2} = \varepsilon\left(3\frac{d^2x}{da^2} + \frac{d^2x}{db^2} + \frac{d^2x}{dc^2} + 2\frac{d^2y}{dadb} + 2\frac{d^2z}{dadc}\right),$$
$$\frac{\Pi}{g}\frac{d^2y}{dt^2} = \varepsilon\left(\frac{d^2y}{da^2} + 3\frac{d^2y}{db^2} + \frac{d^2y}{dc^2} + 2\frac{d^2x}{dadb} + 2\frac{d^2z}{dbdc}\right),$$
$$\frac{\Pi}{g}\frac{d^2z}{dt^2} = \varepsilon\left(\frac{d^2z}{da^2} + \frac{d^2z}{db^2} + 3\frac{d^2z}{dc^2} + 2\frac{d^2x}{dadc} + 2\frac{d^2y}{dbdc}\right).$$

7. In order to form an exact notion of the nature of the constant ε which enters into the preceding equations, one may consider the simple case of a solid of an indefinite expanse included between the plane of *bc*, and another plane parallel to this last [plane]. One will admit that all the points of the face that coincides with the plane of *bc* are fixed, and that all the points of the opposite face are pulled by equal forces perpendicular to this face.[jj] In such a system, it is evident that the points of the body do not move in the direction of *b* and *c* : therefore $y = 0$, $z = 0$, $\frac{dx}{db} = 0$, $\frac{dx}{dc} = 0$. Furthermore, [with] no force being applied to the interior points of the body, X = 0, Y = 0, Z = 0. Therefore, the last two indefinite equations disappear, and the first is reduced to

$$0 = \varepsilon \cdot 3\frac{d^2x}{da^2},$$

of which the integral is

$$x = \mathrm{P}a,$$

P being an arbitrary constant, since $x = 0$ when $a = 0$.

As for the defined equations, one has $l = 0$, $m = \frac{1}{2}\pi$, $n = \frac{1}{2}\pi$ for the two faces of the solid; and regarding the preceding remarks, these equations for these faces become

$$\mathrm{X}' = \varepsilon \cdot 3\frac{dx'}{da'},\ \mathrm{Y}' = 0,\ \mathrm{Z}' = 0.$$

[jj] This is a case of uniaxial normal strain in the *a* direction, with all other strain components zero.

The first gives the value of the force applied to the points of the faces of the solid for a superficial unit. As one has moreover $P = \frac{dx'}{da'}$, the constant P is defined, and $= \frac{X'}{3\varepsilon}$. The equation that gives the displacement of the points of the solid is therefore

$$x = \frac{X'}{3\varepsilon} \cdot a \, .$$

One draws from this equation

$$\varepsilon = \frac{X'}{3} \cdot \frac{a}{x} \, ;$$

from where on concludes that the constant ε is equal to the weight which, in a solid such as this one that one has considered, has been distributed on the superficial unit, multiplied by a third of the ratio of the length of the solid to the lengthening that this weight has produced.

As in making $x = a$, one has $\varepsilon = \frac{X'}{3}$, one may define the constant ε in a simpler manner, by saying that it is equal to a third of the weight that, distributed on the superficial unit, has lengthened the solid in a manner to double its length. It is understood moreover that this last definition is purely abstract, the preceding notions are only in agreement with the natural effects as long as x is very small in relation to a.

Thus concludes Navier's ground-breaking paper. One can only marvel at the high quality of both the mathematics and the type-setting of the equations, all without the help of computers.

2.3 Discussion of Navier's paper

Silling's bond-based peridynamics is strikingly similar to the one-elastic-parameter, ε, theory of Navier presented in the previous section.

Navier's one-parameter "molecular" theory was quickly shown by Cauchy (Cauchy, De la pression ou tension dans un corps solide, 1827) to be insufficiently general, because two parameters, not one, are required to fully characterize an isotropic elastic deformable solid. Navier was defensive of his work, and apparently never acknowledged that his theory was incomplete.

In a parallel situation, Silling (Silling, 2000) quickly realized that bond-based peridynamics was insufficiently general, and self-corrected his own theory by creating state-based peridynamics (Silling, S.A., Epton, M., Weckner, O., Xu, J., and Askari, E., 2007).

In this section, we review Navier's theory and discuss how it could have been modified to better satisfy the demands of modern engineering analysis (including a capability to model elasticity, damage, plasticity, and fracture).

Navier's original paper is rather difficult to follow due to unfamiliar notation. Several books present abbreviated summaries of Navier's theory, for example (Timoshenko, 1953) and (Grattan-Guinness, 1990). In this section, we summarize the Navier theory, and suggest modifications to the theory, using notation more familiar to modern readers.

We focus our attention on two distinct points in an elastic body, $\boldsymbol{P}$ at reference (undeformed) location (X, Y, Z) and $\boldsymbol{P}'$ at reference location (X', Y', Z'), as shown in Fig. 2.3. We consider these points as position vectors with respect to the origin of the (X, Y, Z) Cartesian coordinate system. The relative position vector between the two points in the undeformed configuration is given by $\boldsymbol{R} \equiv \boldsymbol{P}' - \boldsymbol{P}$. These two points move to locations $\boldsymbol{p}$ and $\boldsymbol{p}'$, respectively, as the body deforms. (We indicate all of these points, being position vectors with respect to the origin of the coordinate system, by bold font.)

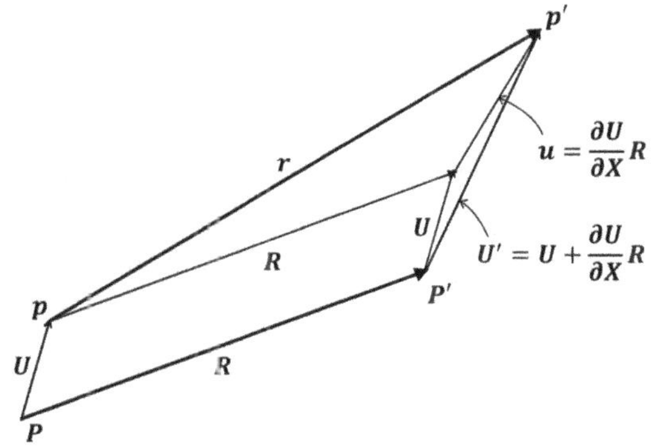

Figure 2.3. Updated terminology for derivation of Navier theory of elasticity.

Under deformation, point $\boldsymbol{P}$ is displaced by displacement vector $\boldsymbol{U}$ to position $\boldsymbol{p}$, and point $\boldsymbol{P}'$ is displaced by $\boldsymbol{U}'$ to position $\boldsymbol{p}'$. Assuming continuous deformations, we can represent the displacement field using a Taylor series expansion,

$$U'_i = U_i + \sum_{j=1:3} \frac{\partial U_i}{\partial X_j} R_j + O(R^2) \quad \text{or} \quad \boldsymbol{U}' \cong \boldsymbol{U} + \frac{\partial \boldsymbol{U}}{\partial \boldsymbol{X}} \boldsymbol{R} . \tag{2.1}$$

We define the linearized relative displacement vector between $\boldsymbol{p}'$and $\boldsymbol{p}$ as

$$\boldsymbol{u} = \boldsymbol{U}' - \boldsymbol{U} = \frac{\partial \boldsymbol{U}}{\partial \boldsymbol{X}} \boldsymbol{R} \ , \tag{2.2}$$

or, using indicial notation $u_i = \frac{\partial U_i}{\partial X_j} R_j = U_{i,j} R_j$.

Note that $\frac{\partial \boldsymbol{U}}{\partial \boldsymbol{X}}$, called the displacement gradient, is a tensor, represented as a 3x3 matrix of partial derivatives.

This linearized relative displacement between the two points is valid only for displacement vectors $\boldsymbol{u}$ whose magnitudes are small relative to the magnitude of $\boldsymbol{R}$. Also, the magnitude of $\boldsymbol{R}$ must be small relative to the ratio of the norm of the first derivative to the norm of the second derivative of the displacement field:

$$\|\boldsymbol{R}\| \ll \left(\left\|\frac{\partial U}{\partial X}\right\|\right) \Big/ \left(\left\|\frac{\partial^2 U}{\partial X^2}\right\|\right) ; \tag{2.3}$$

in other words, the higher-order terms of the Taylor series expansion must be small compared to the retained terms. In addition, the displacement field, $\boldsymbol{U}$, must be differentiable. Thus, $\boldsymbol{U}$ must be continuous. Finally, of course, $\|\boldsymbol{R}\|$ cannot be so large that it allows points $\boldsymbol{P}$ or $\boldsymbol{P}'$ that lie outside of the body. So, there are several important fundamental assumptions limiting the applicability of this theory. This is the reason for the alternative theory of solid mechanics that we describe in this book, the state-based peridynamic lattice-base model (SPLM). In the SPLM, to avoid having to make an assumption of a continuous displacement field, we use finite differences between lattice particles (rather than partial derivatives) to determine the forces between particles.

Navier assumes that the force between the particles located at deformed configurations $\boldsymbol{p}$ and $\boldsymbol{p}'$ is proportional to the change in length of the bond $\boldsymbol{R}$ due to deformation between the two points. The change in length of the bond is *not* equal to the magnitude of $\boldsymbol{u}$. Instead, the change in length of the bond, $\boldsymbol{R}$, is given, approximately, for small deformations, by the *projection* of relative displacement vector $\boldsymbol{u}$ in the direction of $\boldsymbol{R}$:

$$\Delta R = \boldsymbol{u} \cdot \hat{\boldsymbol{e}}_{\boldsymbol{R}} = \boldsymbol{u} \cdot \left(\frac{\boldsymbol{R}}{\|\boldsymbol{R}\|}\right) . \tag{2.4}$$

Note, in particular, that Navier's theory assumes that the displacement magnitude $\|\boldsymbol{u}\|$ is very small compared with the bond length $\|\boldsymbol{R}\|$ which itself is assumed to be infinitesimally small compared to other dimensions of the body. Again, with the SPLM, we can afford to relax these restrictions because with the computational power of modern computers, we can obtain nonlinear numerical solutions relatively easily.

Back to our explanation of Navier's model, in rectangular Cartesian coordinates, the change in length ΔR of bond $\boldsymbol{R}$ is given (approximately) by

$$\Delta R = {(u_1 R_1 + u_2 R_2 + u_3 R_3)}\Big/{R} = {(u_i R_i)}\Big/{\sqrt{R_j R_j}} , \tag{2.5}$$

where the last term in Eq. 2.5 is expressed using indicial notation. Using Eqs. 2.2 and 2.5,

$$\Delta R = {(U_{i,k}R_k)R_i}\Big/{\sqrt{R_jR_j}}\ , \tag{2.6}$$

which when expanded gives

$$\Delta R = \begin{pmatrix} U_{1,1}R_1^2 + (U_{1,2} + U_{2,1})R_1R_2 + (U_{1,3} + U_{3,1})R_1R_3 \\ +U_{2,2}R_2^2 + (U_{2,3} + U_{3,2})R_2R_3 + U_{3,3}R_3^2 \end{pmatrix} \Big/ R\ . \tag{2.7}$$

Navier then assumed that the force-per-unit-volume acting upon $\boldsymbol{p}$, F, associated with volume, dV, which is associated with particle $\boldsymbol{p}$, due to the relative displacement of $\boldsymbol{p}'$ is proportional to the change in length, ΔR, of the bond between the two particles:

$$F = G(R)\Delta R dV'\ , \tag{2.8}$$

where $G(R)$ is a rapidly decaying function only of $R = \|\boldsymbol{R}\|$, and dV' is the differential volume associated with particle $\boldsymbol{p}'$. Thus, the "spring constant" can vary with distance between the particles, and it drops quite rapidly with increasing R. Note that $G(R)$ must have units of force per (length to the seventh power).

In contrast, with the SPLM, we relax the assumption that the force F and the change in bond length ΔR are linearly related.

Navier next converted Eq. 2.7 to the spherical coordinate system shown in Fig. 2.4, using the coordinate transformations shown in the figure. Note that the partial derivatives $U_{1,2}$, etc., are assumed constant with respect to $\boldsymbol{R}$ and, to be integrated spatially, do not need to be transformed into spherical coordinates. After transformation to spherical coordinates, Eq. 2.7 becomes:

$$\Delta R = R \begin{pmatrix} U_{1,1}(cos\psi cos\varphi)^2 - (U_{1,2} + U_{2,1})(cos^2\psi cos\varphi sin\varphi) \\ +(U_{1,3} + U_{3,1})cos\psi cos\varphi sin\psi + U_{2,2}(cos\psi sin\varphi)^2 \\ +(U_{2,3} + U_{3,2})cos\psi sin\varphi sin\psi + U_{3,3}(sin\psi)^2 \end{pmatrix}. \tag{2.9}$$

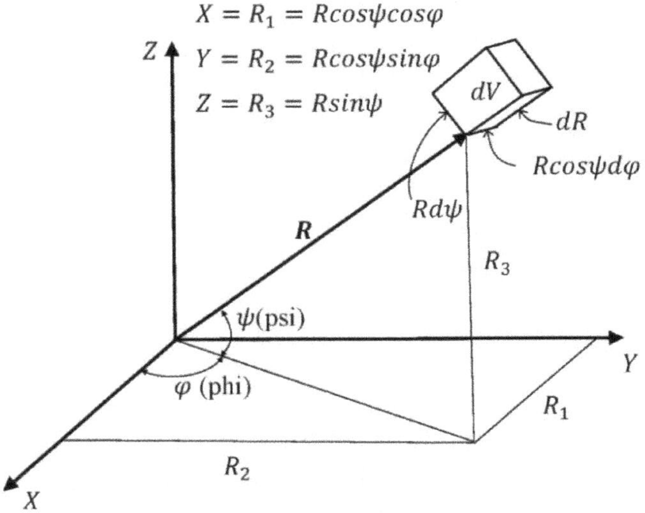

Figure 2.4. Cartesian and spherical coordinate systems.

The internal virtual work (per unit volume) done by each bond in going through a virtual elongation $\delta\Delta R$ is

$$\delta W_{bond} = \delta\Delta R \cdot F = \delta\Delta R \cdot G(R)\Delta R dV' = \frac{1}{2}\delta(\Delta R^2)G(R)dV', \tag{2.10}$$

where we have used the variational calculus identity

$$\Delta R \cdot \delta\Delta R = \frac{1}{2}\delta(\Delta R^2)\,. \tag{2.11}$$

Integrating the virtual work per unit volume done by all bonds attached to particle P over the entire spatial domain,

$$\delta W_{all\ bonds} = \int_{\varphi=0}^{2\pi}\int_{\psi=\frac{-\pi}{2}}^{\frac{\pi}{2}}\int_{R=0}^{\infty}\frac{1}{2}\delta(\Delta R^2)G(R)R^2\cos(\psi)\,dR d\psi d\varphi. \tag{2.12}$$

After integrating Eq. 2.12 in the φ and ψ coordinate directions, following Navier, we obtain

$$\delta W_{all\ bonds} = \frac{1}{2}\varepsilon\delta\Big[3U_{1,1}{}^2 + \left(U_{1,2} + U_{2,1}\right)^2 + 2U_{1,1}U_{2,2} + \left(U_{1,3} + U_{3,1}\right)^2 + 2U_{1,1}U_{3,3} + 3U_{2,2}{}^2 + \left(U_{2,3} + U_{3,2}\right)^2 + 2U_{2,2}U_{3,3} + 3U_{3,3}{}^2\Big] \quad (2.13)$$

where ε represents the integral

$$\varepsilon \equiv \left(\frac{2\pi}{15}\right)\int_0^\infty R^4 G(R)\,dR\,. \quad (2.14)$$

Navier next assumes body forces $\boldsymbol{B} = (B_X, B_Y, B_Z)$ act per unit volume, Ω, and surface tractions $\boldsymbol{T} = (T_X, T_Y, T_Z)$ act per unit boundary surface area, Γ. If the body is in equilibrium, then the total virtual work due to virtual displacement field $\delta\boldsymbol{U}$ must be zero:

$$\iiint_\Omega \frac{1}{2}\delta W_{all\ bonds} d\Omega - \iiint_\Omega B_i\delta U_i d\Omega - \iint_\Gamma T_i\delta U_i d\Gamma = 0. \quad (2.15)$$

(The factor of ½ in the first term of Eq. 2.15 needs to be included. This factor is present because each bond is shared by two particles, and should not be double-counted.) Upon taking the variation of Eq. 2.13 and expanding Eq. 2.15, following Navier, we obtain

$$\begin{aligned}\varepsilon\iiint_\Omega \{&3U_{1,1}\delta U_{1,1} + U_{1,2}\delta U_{1,2} + U_{1,2}\delta U_{2,1} + U_{2,1}\delta U_{1,2} \\ &+ U_{2,1}\delta U_{2,1} + U_{1,1}\delta U_{2,2} + U_{2,2}\delta U_{1,1} \\ &+ U_{1,1}\delta U_{1,3} + U_{1,3}\delta U_{3,1} + U_{3,1}\delta U_{1,3} \\ &+ U_{3,1}\delta U_{3,1} + U_{1,1}\delta U_{3,3} + U_{3,3}\delta U_{1,1} \\ &+ 3U_{2,2}\delta U_{2,2} + U_{2,3}\delta U_{2,3} + U_{2,3}\delta U_{3,2} \\ &+ U_{3,2}\delta U_{2,3} + U_{3,2}\delta U_{3,2} + U_{2,2}\delta U_{3,3} \\ &+ U_{3,3}\delta U_{2,2} + 3U_{3,3}\delta U_{3,3}\}d\Omega \\ - \iiint_\Omega &B_i\delta U_i d\Omega - \iint_\Gamma T_i\delta U_i d\Gamma = 0.\end{aligned} \quad (2.16)$$

(In taking the variation of a function, the rules are quite like differentiation using the chain rule; thus $\delta\left({U_{1,1}}^{2}\right) = 2U_{1,1}\delta U_{1,1}$, for example.)

To integrate Eq. 2.16, it is necessary to integrate by parts a function of several variables. As a reminder, integration by parts states that if Ω is an open bounded set of $\mathbb{R}^n$ with a piecewise smooth boundary Γ and if u and v are two continuously differentiable functions on the closure of Ω, then

$$\int_{\Omega} \frac{\partial u}{\partial x_i} v\, d\,\Omega = \int_{\Gamma} uvn_i d\,\Gamma - \int_{\Omega} u \frac{\partial v}{\partial x_i} d\,\Omega\,, \tag{2.17}$$

where n_i is the ith component of the outward unit surface normal to Γ, $\boldsymbol{n}$.

Upon differentiating Eq. 2.16 using Eq. 2.17, Navier obtains

$$\begin{aligned}
0 = -\varepsilon \iiint_{\Omega} &\left\{ \begin{matrix} (3U_{,XX}+ U_{,YY}+ U_{,ZZ}+ 2V_{,XY}+ 2W_{,XZ})\delta U \\ +(V_{,XX}+ 3V_{,YY}+ V_{,ZZ}+ 2U_{,XY}+ 2W_{,YZ})\delta V \\ +(W_{,XX}+ W_{,YY}+ 3W_{,ZZ}+ 2U_{,XZ}+ 2V_{,YZ})\delta W \end{matrix} \right\} d\Omega \\
&- \iiint_{\Omega} \left\{ \begin{matrix} B_X \delta U \\ B_Y \delta V \\ B_Z \delta W \end{matrix} \right\} d\Omega \\
+\varepsilon &\left[\iint_{\Gamma} (3U_{,X}+ V_{,Y}+ W_{,Z})n_X + (U_{,Y}+ V_{,X})n_Y + (U_{,Z}+ W_{,X})n_Z \right] d\Gamma \delta U \\
+\varepsilon &\left[\iint_{\Gamma} (U_{,Y}+ V_{,X})n_X + (U_{,X}+ 3V_{,Y}+ W_{,Z})n_Y + (V_{,Z}+ W_{,Y})n_Z \right] d\Gamma \delta V \\
+\varepsilon &\left[\iint_{\Gamma} (U_{,Z}+ W_{,X})n_X + (V_{,Z}+ W_{,Y})n_Y + (U_{,X}+ V_{,Y}+ 3W_{,Z})n_Z \right] d\Gamma \delta W \\
&- \iint_{\Gamma} \left\{ \begin{matrix} T_X \delta U \\ T_Y \delta V \\ T_Z \delta W \end{matrix} \right\} d\Gamma
\end{aligned} \tag{2.18}$$

Because δU, δV, and δW are *arbitrary* virtual displacement fields, Eq. 2.18 can only be true for all virtual displacements if each of the terms multiplying them in the volume integrals are independently zero; thus:

$$\begin{aligned}-B_X &= \varepsilon(3U_{,XX} + U_{,YY} + U_{,ZZ} + 2V_{,XY} + 2W_{,XZ}) , \\ -B_Y &= \varepsilon(V_{,XX} + 3V_{,YY} + V_{,ZZ} + 2U_{,XY} + 2W_{,YZ}) , \\ -B_Z &= \varepsilon(W_{,XX} + W_{,YY} + 3W_{,ZZ} + 2U_{,XZ} + 2V_{,YZ}) .\end{aligned} \tag{2.19}$$

An equivalent expression to Eq. 2.19 is, using indicial notation

$$-B_i = \varepsilon\left(U_{i,jj} + 2U_{j,ij}\right) . \tag{2.20}$$

These are three second-order linear partial differential equations that must be satisfied, if equilibrium exists, at each point within the domain, Ω, of the problem. Navier also realized that the equations of dynamics could be obtained by including the inertial forces, according to D'Alembert's (born 1717, died 1783) principle:

$$\begin{aligned}\rho U_{tt} &= \varepsilon(3U_{,XX} + U_{,YY} + U_{,ZZ} + 2V_{,XY} + 2W_{,XZ}) + B_X , \\ \rho V_{tt} &= \varepsilon(V_{,XX} + 3V_{,YY} + V_{,ZZ} + 2U_{,XY} + 2W_{,YZ}) + B_Y , \\ \rho W_{tt} &= \varepsilon(W_{,XX} + W_{,YY} + 3W_{,ZZ} + 2U_{,XZ} + 2V_{,YZ}) + B_Z ,\end{aligned} \tag{2.21}$$

where ρ is the mass density of the material.

The terms in the surface integrals in Eq. 2.18 multiplying each of the independent virtual displacement components must also be independently zero:

$$\begin{aligned}T_X &= \varepsilon[(3U_{,X} + V_{,Y} + W_{,Z})n_X + (U_{,Y} + V_{,X})n_Y + (U_{,Z} + W_{,X})n_Z] , \\ T_Y &= \varepsilon[(U_{,Y} + V_{,X})n_X + (U_{,X} + 3V_{,Y} + W_{,Z})n_Y + (V_{,Z} + W_{,Y})n_Z] , \\ T_Z &= \varepsilon[(U_{,Z} + W_{,X})n_X + (V_{,Z} - W_{,Y})n_Y + (U_{,X} + V_{,Y} + 3W_{,Z})n_Z] .\end{aligned} \tag{2.22}$$

An expression equivalent to Eq. 2.22 is, using indicial notation

$$T_i = \varepsilon\left[\left(U_{i,j} + U_{j,i}\right)n_j + U_{j,j}n_i\right] . \tag{2.23}$$

Thus concludes our derivation of Navier's equation of elasticity, which includes only one elastic parameter ε and thus cannot simulate materials with Poisson's ratios other than one-quarter.

In Chapter 8, we will generalize Navier's theory so that it provides the correct Navier-Cauchy equation of linear elasticity. However, for now, we continue on to Cauchy's response to Navier's treatise.

2.4 Cauchy's treatise

Cauchy's rebuttal paper (Cauchy, 1823), presented here in English for the first time, corrected a shortcoming of Navier's elasticity model. More importantly, Cauchy's paper introduced the concepts of strain and stress. The paper translated next is really a high-level review paper describing a treatise including the results of several more detailed mathematical papers on elasticity. More mathematical detail is found in (Cauchy, De la pression ou tension dans un corps solide, 1827) and (Cauchy, Sur la condensation et la dilitation des corps solides, 1827).

The paper has been translated from French by Melissa Rose Arce Miller. Annotations by the current author are included as footnotes.

RESEARCH
ON THE EQUILIBRIUM AND INTERIOR MOVEMENT
OF SOLID OR FLUID,
ELASTIC OR NON-ELASTIC, BODIES.
By Augustin Cauchy

Bulletin of the Philomathic Society, p. 9-13; 1823.

This research has been undertaken on the occasion of a Treatise published by Mr. Navier on August 14, 1820[kk]. In order to establish the equation of equilibrium of the elastic plane, the author[ll] had considered two types of forces produced, some by dilation or contraction, the others

---

[kk] (Navier, 1827) Although Cauchy refers to Navier's "Treatise" of 1820, Navier's paper was actually only published seven years later, in 1827.

[ll] Here, it is unclear if Cauchy is referring to Navier or to himself. By "flexion", Cauchy means tension, apparently.

by the flexion of this same plane.[mm] Moreover, in his calculations, he had assumed both types perpendicular to the lines or to the faces against which they are exerted.[nn] It seemed to me that these two types of forces could be reduced to a single, which must be constantly called tension or pressure, and which was of the same nature as the hydrostatic pressure exerted by a fluid at rest against the surface of a solid body. Only the new pressure remained neither always perpendicular to the faces that were subjected to it, nor the same in every direction at a given point[oo]. In developing this idea, I soon arrived at the following conclusions.

If in an elastic or non-elastic solid body, one comes to make rigid and invariable a small element of the volume enclosed by any faces, this little element will experience on its different faces, and at each point of each of them, a determined pressure or tension. This pressure or tension will be similar to the pressure that a fluid exerts against an element of the envelope of a solid body, with this sole difference, that the pressure exerted by a fluid at rest against the surface of a solid body, is directed perpendicularly to this surface from the outside to the inside, and independent at each point of the slope of the surface in relation to the coordinate planes, while the pressure or tension exerted at a given point of a solid body against a very small element of surface passing by this point, may be directed perpendicularly or obliquely to this surface, sometimes from the outside to the inside, if there is condensation, sometimes from the inside to the outside, if there is dilation, and may depend on the slope of the surface in relation to the planes in question.[pp] Moreover, the pressure or tension exerted against any plane may be deduced very easily, in size as well as in direction, from the pressures or

---

[mm] The French word "plan" has here been translated as "plane". It is unclear what Cauchy intended by the word.

[nn] This entire introduction is rather unclear. Apparently, Cauchy was careless and vague in his writing style.

[oo] This appears to be the first clearly described "traction vector" or a "stress vector" acting upon an imagined plane within a solid body.

[pp] This sentence (somewhat awkwardly) describes the idea of a differential volume element subjected to general stress components.

tensions exerted against three given rectangular planes[qq]. I was at this point, when Mr. Fresnel, coming to speak to me about the work on light in which he was engaged, and of which he had thus far only presented a part to the Institute, informed me that, on his side, he had obtained a theorem analogous to mine on the laws, according to which elasticity varies in the various directions that emanate from a unique point. Nevertheless, the theorem in question was far from sufficing to me for the purpose that I was intending, from that time, to form general equations of the equilibrium and interior movement of a body; and it is only recently that I managed to establish new assumptions suitable to guide me to this result, and which I am going to make known.

From the theorem stated above, it follows that the pressure or tension at each point is equivalent to the unit divided by the radius vector of an ellipsoid.[rr] To the three axes of this ellipsoid correspond three pressures or tensions that we will name *principal*, and one can prove[ss] that each of them is perpendicular to the plane against which it is exerted. Among these principal pressures or tensions are the *maximum* pressure or tension and the *minimum* pressure or tension.[tt] The other pressures or tensions are distributed symmetrically around the three axes. Moreover, the pressure or tension normal to each plane, that is to say, the component, perpendicular to a plane, of the pressure or tension exerted against this plane is reciprocally proportional to the square of the radius vector of a second ellipsoid. Sometimes this second ellipsoid is replaced by two hyperboloids, one with one layer, the other with two layers, which have the same center, the same axes, and are touched at infinity by the same

---

[qq] This sentence describes the "stress transformation freebody" found in any standard textbook on mechanics of solids.

[rr] Cauchy here is referring to what we today call the "stress tensor".

[ss] The remark that we make here is in agreement with the recent research of Mr. Fresnel (*See* the *Bulletin* of May 1822)[Cauchy's original footnote].

[tt] These would be the principal stresses in today's jargon. Note that Cauchy uses the words "pressure or tension" instead of today's jargon: "normal stress".

conical surface of the second degree, of which the edges indicate the directions for which the normal pressure or tension reduces to zero[uu].

This set down, if one considers a solid body variable in shape and subjected to any accelerating forces, in order to establish the equations of equilibrium of this solid body, it will suffice to write that there is equilibrium between the driving forces that attract an infinitesimally small element in the direction of the coordinate axes, and the orthogonal components of the exterior pressures or tensions that act against the faces of this element.[vv] One will obtain thus three equations of equilibrium that include, as a particular case, those of the equilibrium of fluids. But, in the general case, these equations contain six unknown functions of the coordinates *x, y, z.*[ww] It remains to determine the values of these six unknowns; but the solution of this last problem varies according to the nature of the body and its more or less perfect elasticity. Let us explain now how one manages to solve it for elastic bodies.

When an elastic body is in equilibrium by virtue of any accelerating forces, one must suppose each molecule displaced from the position it occupied when the body was in its natural state. By virtue of displacements of this type, there is around each point different condensations or dilatations in the different directions.[xx] But it is clear that each dilatation produces a tension and each condensation a pressure. Moreover, I show that the diverse condensations or dilations around a point, diminished or increased by the unit, become equal in absolute value, to the radius vectors of an ellipsoid.[yy] I call *principal condensations* or *dilatations* those which take place following the axes of this ellipsoid, around which all the others are symmetrically distributed.

[uu] Cauchy is here summarizing, in words, results from a more mathematical handling of the problem of representing the stress tensor.

[vv] Here, Cauchy is describing the static equilibrium of a differential volume element.

[ww] Presumably, these six unknown functions are the six stress components, each of which varies with position.

[xx] What Cauchy calls "condensations" and "dilations" are what we call "strains". Note the tacit assumption that these deformations arise only from continuous deformations.

[yy] Cauchy recognizes here that "condensations" and "dilatations" are of a tensor nature (although tensors had not yet been invented in 1823).

This set down, it is clear that in an elastic solid, the tensions or pressures depend only on the condensations or dilatations; the principal tensions or pressures will be directed in the same directions as the principal condensations or dilatations.[zz] Moreover, it is natural to suppose, at least when the displacements of the molecules are very small, that the principal tensions or pressures are respectively proportional to the principal condensations or dilatations.[aaa] In accepting this assumption, one arrives immediately at the equations of equilibrium of an elastic body. In the case of very small displacements[bbb], the component, perpendicular to a plane, of the pressure or tension exerted against this plane, always retains the same relationship with the condensation or dilatation that has taken place in the direction of this component, and the formulas of equilibrium come down to four equations with partial derivatives of which one determines separately the condensation or dilatation of the volume, while each of the others serves to determine the displacement parallel to one of the coordinate axes.

The equations of the equilibrium of an elastic body being formed, it is easy to deduce from them the equations of the movement by ordinary methods. These latter [equations] are again four in number, and each of them is a linear equation with partial differentials with a final variable term.[ccc] They integrate by the methods set out in our previous Treatise. One of these equations contains only the unknown that represents the condensation or dilation of the volume.[ddd] In the particular case where the accelerating force becomes constant and retains everywhere the same direction, this equation reduces to that which determines the propagation of sound in air, with the only difference, that the constant that it contains, instead of depending on the height of the atmosphere, assumed homogenous, depends on the linear dilatation or condensation of a body

---

[zz] This statement appears to be based upon a consideration of symmetry.

[aaa] This assumption of linearity is a generalization of Hooke's law.

[bbb] Note the assumption of "very small displacements" necessary to produce a linear differential equation.

[ccc] It is unclear why Cauchy came up with four differential equations of equilibrium, rather than three.

[ddd] In modern terms, volumetric strain.

under a given pressure. One must conclude from there that the speed of sound in an elastic solid is constant, as in air, but varies from one body to the other according to the substance of which it is composed. This constancy is especially remarkable, as the displacements of the molecules considered successively in elastic solids and fluids follow different laws.

My Treatise concludes with the formation of the equations of the interior movement of solid bodies entirely devoid of elasticity. In order to achieve this, it suffices to suppose that in these bodies the pressures or tensions around a point in motion no longer depend on the complete condensations or dilatations that correspond to the absolute displacements taken into consideration starting from the initial positions of the molecules, but only, at the end of some time, very small condensations or dilations, which correspond to the respective displacements of the different points during a very short moment. One finds then that the condensation of volume is determined by an equation similar to that of heat, which creates a remarkable analogy between the propagation of the caloric and the propagation of the vibrations of a body entirely devoid of elasticity.[eee]

In another Treatise, I will give the application of the formulas that I have obtained for the theory of elastic plates and blades.[fff]

Thus concludes Cauchy's paper.

2.5 Summary

The papers of Navier and Cauchy have come to define the concepts of "solid body" and "material". Most engineers think of the deformation of solid bodies in terms of "strain" and "stress". However, the concepts of strain and stress are insufficient to describe discontinuous material

---

[eee] Cauchy is explaining "caloric" as a mechanical phenomenon. Caloric is an early, and wrong, concept for heat.

[fff] "lamés", translated as "blades" probably refers to what we today call "shells".

behavior like fracture. Additionally, strain and stress fields are awkward to express on digital computers.

In the history of solid mechanics, we see that the pioneers started with very specific and limited assumptions (small deformations, Hooke's law, infinitesimal material horizon). These assumptions have gradually been relaxed over the years, resulting in the more general discipline of continuum mechanics, Silling's theory of peridynamics, and finally culminating with the current SPLM, which dispenses with spatial continuity completely.

We will define peridynamic states in Chapter 8. The peridynamic concept of *tensor state* – used to define "force state" and "deformation state" – is much more general than the associated concept of *tensor*, used to define "stress" and "strain".

As we shall see, the concept of "tensor state" is more general than necessary for convenient computational implementation, and thus we have developed the SPLM in this book.

In the next chapter, because we must first clearly understand existing theory before we move on to a new one, we provide a brief overview of the discipline of *continuum mechanics*, which engineers and mathematicians developed subsequent to the pioneering papers of Navier and Cauchy.

2.6 Exercises

2.1 Navier states that "Between a molecule M, and any M' of the neighboring molecules, there exists an action P, which is the difference of these two forces. In the natural state of the body, all actions P are non-existent, or destroy each other reciprocally, because the molecule M is at rest." Describe and sketch how the pair-wise bond forces, P, between particles need not necessarily be zero in a body at rest, even without external loading.

2.2 Navier's paper has no sketches to help with understanding. Draw a sketch, showing the material point M and its neighbor M', both before and after deformation. Annotate the sketch, using the geometric and kinematic symbols used in Navier's paper.

2.3 Navier develops an expression for the change in distance between the two points M and M' due to deformation as:

$$\Delta s = (x' - x)cos.p + (y' - y)cos.q + (z' - z)sin.\psi$$

Using modern notation, draw a sketch and derive an equivalent expression for the change in length of the bond Δs.

2.4 Navier's theory has one elastic constant, ε. Show that this constant is equal to $\frac{2E}{5}$, where E is Young's modulus.

2.5 Derive Eq. 2.5: $\Delta R = {(u_1R_1 + u_2R_2 + u_3R_3)}/{R} = {(u_iR_i)}/{\sqrt{R_jR_j}}$.

2.6 Derive Eq. 2.7 from Eq 2.6

2.7 Derive Eq. 2.9

2.8 Derive Eqs. 2.13 and 2.14 from Eq. 2.12.

2.9 Derive Eq. 2.16.

2.10 Derive Eq. 2.18.

2.11 Show that Eqs. 2.22 and 2.23 are identical.

2.12 Show that Navier's model predicts that Poisson's ratio is equal to one-quarter.

Chapter 3

Continuum Mechanics

Chapter objectives are to:

- understand continuum mechanics and it limitations;
- review the laws of physics governing material bodies;
- appreciate the mathematics of vectors and tensors;
- develop continuum kinematics of deformation;
- examine continuum kinetics (stress);
- develop an understanding of continuum constitutive relations.

The reason for including this chapter is to help in understanding the similarities and differences between continuum mechanics, the state based peridynamic model, and the state based peridynamic lattice model (SPLM).

Textbooks devoted to the topic provide comprehensive developments of continuum mechanics; see for example (Tadmor, E. B.; Miller, R. E.; and Elliot, R. S., 2012). Continuum mechanics is an imposing intellectual edifice that has certainly dominated engineering thinking, as is evident from the following quote from Tadmor et al.:

> "Continuum mechanics is in many ways the 'grand unified theory' of engineering science. As long as the fundamental continuum assumptions are valid and relativistic effects are negligible, the governing equations of continuum mechanics provide the most general description of the behavior of materials (solid and fluid) under any arbitrary loading."

Most books on continuum mechanics start out with a few paragraphs providing caveats about the limitations of continuum mechanics. For example, Tadmor et al. begin the chapter on the kinematics of deformation with a discussion as follows:

> "Continuum mechanics deals with the change of shape (deformation) of bodies subjected to external mechanical and thermal loads ... Kinematics does not deal with predicting the deformation resulting from a given loading, but rather with the machinery for describing all possible deformations a body can undergo." (Underlined text is mine.)

Why would one limit kinematics only to continuous deformations? One sees evidence of discontinuous deformation, in the form of cracks, in all types of civil engineering structures including those made of steel, concrete, masonry, and wood. For example, Fig. 3.1 shows a photograph of a cracked concrete sidewalk. We certainly cannot consider the deformation of this sidewalk to be continuous.

Figure 3.1. Concrete sidewalk with cracks. (Photograph by Walter Gerstle.)

Additionally, Tadmor et al. state:

> "In situations where [continuum mechanics] breaks down, it is necessary to resort to multiscale methods that combine lower-level microscopic models with continuum models."

Nonetheless, one can avoid multiscale methods and microscopic models, with all their complexity, if one avoids making the kinematic assumption of continuity in the first place. There is no reason to assume that materials deform continuously, apart from the fact that in many regimes of behavior, continuity is indeed an appropriate assumption, and that continuum models are often more economically described using the analytical methods of calculus than discontinuous models.

Our approach in this book is to depart from an initial assumption of spatially continuous kinematic behavior. By choosing to represent the material body as a discrete lattice in the reference configuration, we develop a comparatively simple material model. In so doing, with the help of digital computers, we are able to simulate both continuous and discontinuous kinematic behavior.

Of course, in choosing to model the material as a particle lattice, we are still making certain assumptions. We are "lumping" into a single lattice particle mass that might actually have quite complex physical characteristics at a smaller scale. The lattice model has no chance of describing kinematics that might happen at a length scale smaller than the lattice spacing. However, we can also construe this discretization technique as an advantage, in the sense that we intrinsically hide the unwanted complexity of smaller scales. Thus, the use of a lattice provides a means for defining the smallest scale that the model represents. In contrast, continuum mechanics, which provides no intrinsic length scale, fails to place an intrinsic lower limit on the model length scale.

The governing partial differential equations of continuum mechanics (the Navier Cauchy equations for linear elastic solids, for example) are difficult to solve exactly. Except for a handful of simple problems, closed-form analytical solutions do not exist. Thus, in today's world, we usually discretize and solve continuum mechanics problems approximately using digital computers. Usually, we use the finite element method, boundary element method, or the finite difference method to solve continuum mechanics problems.

Refer to Fig. 3.2 for a comparison of the continuum and the peridynamic lattice procedures. The left column of Fig. 3.2 shows the current computational engineering approach, which is quite indirect, as depicted by Rube Goldberg machine at lower right. Navier started with discrete particles (atoms); then created a continuum particle using the concept of density. Cauchy then reconsidered the behavior of these infinitesimal particles to create a stress-strain archetype; resulting in the governing Navier-Cauchy differential equations, which are usually considered to be the starting point for the mechanics of deformable solid bodies. We then convert the Navier-Cauchy equations back into a weak form (integral equations) for solution using finite elements. The finite element model results in numerical matrix equations, which we must still solve using perhaps an implicit nonlinear algorithm like the Newton-Raphson method, or an explicit method like the Newmark-beta method. Adding to the Rube Goldberg-like complexity, to represent cracks, "localization limiters" or discrete crack propagation methods are required. Continuum mechanics certainly employs beautiful mathematics, and is essential in many situations, but its connection to the reality of material behavior is limited.

Classical Continuum Mechanics	**State-Based Peridynamic Lattice**
Physical Solid Body	Physical Solid Body
Discrete Particle	Discrete Particle Lattice
Continuum Particle	Peridynamic Constitutive Model
Integral Equations	Apply Newton's Laws
Governing Differential Equations	Numerical Solution
Convert to Weak Form (Integral Equations)	Self-Operating Napkin 
Finite Element Discretization	
Algebraic Matrix Equations	
Solve Nonlinear Equations	
Handle Crack Propagation (?)	
Numerical Solution	

Figure 3.2. Comparison of classical continuum mechanics, and state-based peridynamic lattice model (SPLM). In comparison to the SPLM, the current computational approach resembles the Rube Goldberg machine pictured at lower right. Artwork Copyright © and TM Rube Goldberg Inc. All Rights Reserved. RUBE GOLDEBERG® is a registered trademark of Rube Goldberg Inc. All materials used with permission. Rubegoldberg.com.

However, there is a shortcut: the state-based peridynamic lattice method, as shown in the right column of Fig. 3.2. By employing SPLM, we avoid vast swaths of theoretical machinery, including tensor analysis, continuum mechanics, fracture mechanics and the finite element method.

It is fair to say that SPLM is a more fundamental application of Newton's laws to material bodies than continuum mechanics.

With this rather ominous introduction, we now present a sketch, using broad brush strokes, of the concepts of continuum mechanics, which is, after all, a beautiful and powerful theory (or rather, model) that should be appreciated in its own right, and furthermore is helpful in developing the peridynamic theory (or better, model). As the following sections are necessarily brief, the reader is encouraged to study the covered topics in further detail as time allows and interest dictates.

3.1 Newton, vectors, calculus, and continuity

Newton, in his 1687 treatise *Philosophiae Naturalis Principia Mathematica* developed the laws of motion assuming an "inertial reference frame" – the stage of space and time, considered absolute, upon which the laws take place. Translated from Latin into English, Newton's laws of motion are:

1. Every body remains in a state, resting or moving uniformly in a straight line, except insofar as forces compel it to change its state.
2. The [rate of] change of momentum is proportional to the motive force impressed, and is made in the direction of the straight line in which the force is impressed.
3. To every action there is always opposed an equal reaction.

To implement these laws mathematically, Newton developed and employed the mathematics of vectors and real calculus, which is essentially a continuum version of discrete mathematics. The roots of continuum mechanics go directly back to Newton's calculus, which in turn relies heavily upon the notions of Descartes, who invented the Cartesian coordinate system of real numbers, and N_R-dimensional real Cartesian space, denoted as $\mathbb{R}^{N_R}$.

Newton's second law (assuming a point-like free body, of constant mass, m) is usually stated mathematically as

$$\boldsymbol{F} = \frac{d}{dt}\left(m\frac{d\boldsymbol{v}}{dt}\right) = m\mathbf{a}\,, \tag{3.1}$$

where t is time, $\boldsymbol{F}$ is the force vector, $\boldsymbol{v}$ is the velocity vector, and $\mathbf{a}$ is the acceleration vector of a particle of mass m.[a] The acceleration $\mathbf{a}$ is the second derivative with respect to time t of the spatial position vector $\boldsymbol{x}$:

$$\mathbf{a} = \frac{\partial^2 \boldsymbol{x}}{\partial t^2} = \ddot{\boldsymbol{x}}\,, \tag{3.2}$$

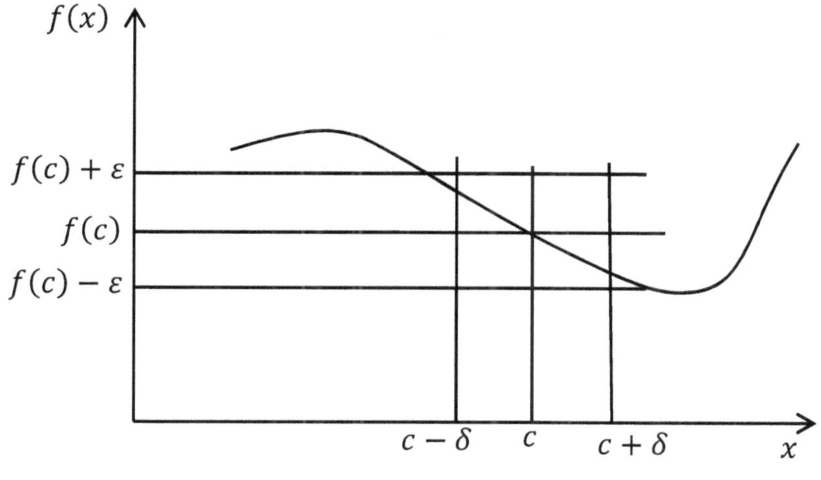

Figure 3.3. Concept of a "continuous function".

Thus, Newton already thought about differentiation of vectors, although not using current terminology. Calculus relies heavily upon the concept of continuous functions, illustrated graphically in Fig. 3.3, and defined next:

Definition 3.1 Continuous function. Given a function $f: D \to \mathbb{R}$, defined on a domain, $D \in \mathbb{R}$, and a real number $c \in D$, f is said to be *continuous* at c if for any number $\varepsilon > 0$, however small, there

---

[a] Vectors and tensors are indicated in bold font.

exists some number $\delta > 0$ such that for all $x \in D$ with $c - \delta < x < c + \delta$, the value of $f(x)$ satisfies

$$f(c) - \varepsilon < f(x) < f(c) + \varepsilon.$$

The concept of continuity extends to multidimensional spaces and domains, and is one of the fundamental aspects of the mathematical discipline of topology. Roughly speaking, a continuous function is "connected", but not necessarily "smooth" at every point in its domain.

As it happens, the physical positions and velocities of particles having constant mass are, according to Newton's laws, continuous functions of time (assuming the particle does not break apart and that the particle masses and applied forces are finite). According to Newton's second law, if a force is suddenly applied, the acceleration will be a discontinuous function of time. However, velocities and positions, being the first and second integrals of acceleration, will be continuous.

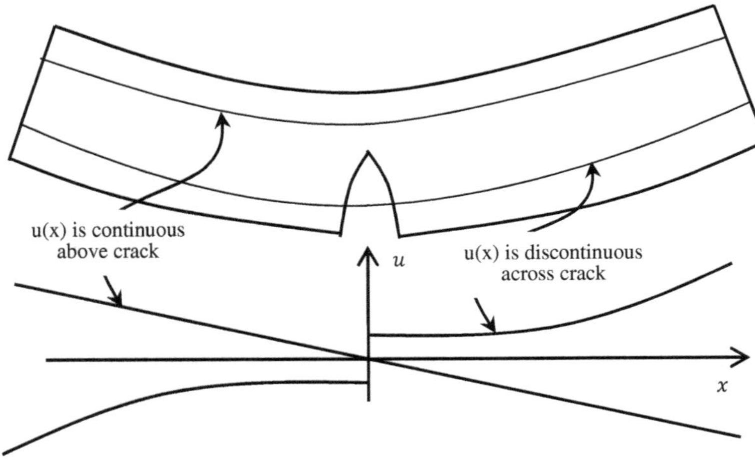

Figure 3.4. The x-component, u, of displacement is discontinuous across the crack.

However, positions and velocities may be discontinuous functions of *spatial* location, as, for example, when a crack forms. For this reason, continuum mechanics assumes more continuity than is often physically manifested in material bodies. For example, Fig. 3.4 shows that the x-component of displacement is discontinuous in a deformed beam with a crack.

3.2 Vectors and tensors

In this section, we very briefly review the mathematics of vectors, and more generally, that of tensors. Roughly speaking, a vector is a mathematical object, having a magnitude and a direction, both of which are independent of coordinate system, that resides in an N_R-dimensional real space $\mathbb{R}^{N_R}$. However, not all objects having magnitude and direction are vectors. Thus, again roughly speaking, a vector is an entity that knows nothing about coordinate systems. Coordinate systems are man-made constructs used to help in describing vectors and tensors.

Newton thought about forces, positions, velocities, and accelerations before mathematicians had defined "vector" as a mathematical object. Researchers only later developed the mathematics of vectors and tensors, over the course of the nineteenth and early twentieth centuries, with contributions from Carl Friedrich Gauss, William Rowan Hamilton, Woldemar Voigt, Gregorio Ricci-Curbastro, Tullio Levi-Civita, and Albert Einstein.

More precisely, we define a vector as an object within a *vector space*. We next give a modern mathematical definition for "vector space":

Definition 3.2 A *vector space* over a field F is a set, V, together with the operations of vector addition and scalar multiplication, that satisfy the eight axioms listed below. We call the elements of vector space V *vectors*. We call the elements of field F *scalars*. Let $\boldsymbol{u}$, $\boldsymbol{v}$ and $\boldsymbol{w}$ be arbitrary vectors in V, and a and b scalars in F. The eight axioms are:

1) Associativity of addition: $\boldsymbol{u} + (\boldsymbol{v} + \boldsymbol{w}) = (\boldsymbol{u} + \boldsymbol{v}) + \boldsymbol{w}$
2) Commutativity of addition: $\boldsymbol{u} + \boldsymbol{v} = \boldsymbol{v} + \boldsymbol{u}$

3) Identity element of addition[b]: $\exists \mathbf{0} \in V \ : \ \boldsymbol{v} + \mathbf{0} = \boldsymbol{v} \ \ \forall \boldsymbol{v} \in V$
4) Inverse element of addition: $\forall \boldsymbol{v} \in V, \exists (-\boldsymbol{v}) \in V : \ \ \boldsymbol{v} + (-\boldsymbol{v}) = \mathbf{0}$
5) Associativity of multiplication: $a(b\boldsymbol{v}) = (ab)\boldsymbol{v}$
6) Scalar identity element of multiplication: $1\boldsymbol{v} = \boldsymbol{v}$
7) First distributivity scalar/vector property: $a(\boldsymbol{u} + \boldsymbol{v}) = a\boldsymbol{u} + a\boldsymbol{v}$
8) Second distributivity scalar/vector property: $(a + b)\boldsymbol{v} = a\boldsymbol{v} + b\boldsymbol{v}$.

Many spaces satisfy the definition of being vector spaces. For our purposes, the most important vectors are those defined in a Euclidean space, $\mathbb{R}^{N_R}$. A Euclidean space, in addition to being a vector space, must have an *inner product*, which we may use to describe the angle between two vectors, and a *norm*, which gives the length, or magnitude, of a vector. The norm of a vector $\boldsymbol{v}$, $\|\boldsymbol{v}\|$, is a scalar which must satisfy the property that it is positive if $\boldsymbol{v} \neq \mathbf{0}$, and $\|\boldsymbol{v}\| = 0$ if $\boldsymbol{v} = \mathbf{0}$.

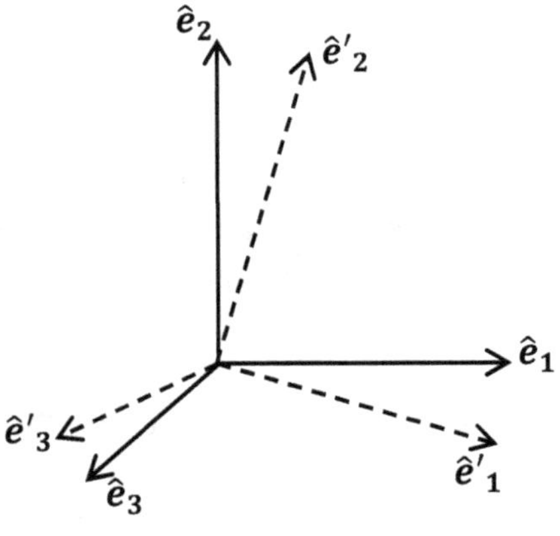

Figure 3.5. Two different Cartesian coordinate systems, represented by unit basis vectors $\hat{\boldsymbol{e}}_i$ and $\hat{\boldsymbol{e}}'_i$, can be used to represent all vectors in a Euclidean space.

We often represent Euclidean vectors using *orthonormal coordinate systems*. A coordinate system is a set of basis vectors $\hat{\boldsymbol{e}}_i$ that spans the space $\mathbb{R}^{N_R}$, as depicted in Fig. 3.5.[c] Hence, according to the

[b] This mathematical statement, translated into English, states "There exists a null vector, $\mathbf{0}$, element of the vector space, V, such that the vector $\boldsymbol{v}$ plus the null vector $\mathbf{0}$ is equal to $\boldsymbol{v}$ for all vectors elements $\boldsymbol{v}$ of the vector space V".

[c] The caret symbol ^ over the vector $\boldsymbol{e}$ means "unit vector".

mathematical discipline of linear algebra, any vector within $\mathbb{R}^{N_R}$ can be represented as a linear combination of basis vectors. Using indicial notation, one can represent a vector $\boldsymbol{a}$ in component form with respect to a Euclidean orthonormal coordinate system $\hat{\boldsymbol{e}}_i$ as:

$$\boldsymbol{a} = \textstyle\sum_{1}^{N_R} a_i \hat{\boldsymbol{e}}_i = a_i \hat{\boldsymbol{e}}_i = \lfloor a_i \rfloor \{\hat{\boldsymbol{e}}_i\} \ , \tag{3.3}$$

where the second-to-right-most expression uses indicial notation, in which terms with repeated indices imply summation of product terms unless otherwise stated, and the right-most expression uses matrix multiplication to express the same sum of product terms.

Definition 3.3 A *Euclidean vector space* is a vector space, defined over the field of real numbers, together with a norm and a dot product. A *Euclidean vector* is an element of a Euclidean vector space.

Following the Pythagorean Theorem, we define the norm (magnitude, or length) of a Euclidean vector as the square root of the sum of the squares of its components:

$$\|\boldsymbol{a}\| = \sqrt{\textstyle\sum_{1}^{N_R} (a_i{}^2)} = \sqrt{a_i a_i} \ . \tag{3.4}$$

The inner product is the dot product, where θ is the angle between the two vectors:

$$\boldsymbol{a} \cdot \boldsymbol{b} = \|\boldsymbol{a}\| \|\boldsymbol{b}\| cos(\theta) \ . \tag{3.5}$$

Thus, the dot product, $\boldsymbol{a} \cdot \hat{\boldsymbol{e}}$, where $\hat{\boldsymbol{e}}$ is a unit vector, gives the projection of vector $\boldsymbol{a}$ in the direction of vector $\hat{\boldsymbol{e}}$, and therefore the dot product is sometimes called a "projection operator". For a Cartesian orthonormal coordinate system, we can show that

$$\boldsymbol{a} \cdot \boldsymbol{b} = a_i b_i \ . \tag{3.6}$$

Usually, we use orthonormal Cartesian coordinate systems to represent vectors in terms of their components, and we employ orthonormal Cartesian coordinate systems exclusively in this book.

An orthonormal coordinate system has the property that the dot product between any two basis vectors satisfies the equation $\hat{\boldsymbol{e}}_i \cdot \hat{\boldsymbol{e}}_j = \delta_{ij}$, where δ_{ij} is the Kronecker delta.[d]

We can span a vector space using an infinite number of different sets of orthonormal basis vectors. Accordingly, we define the orthonormal transformation matrix $\boldsymbol{Q}$ as

$$\hat{\boldsymbol{e}}'_i = \sum_1^{N_R} Q_{ji}\hat{\boldsymbol{e}}_j = Q_{ji}\hat{\boldsymbol{e}}_j \,, \tag{3.7}$$

where $Q_{ji} = \hat{\boldsymbol{e}}'_i \cdot \hat{\boldsymbol{e}}_j$ is the projection of basis vector $\hat{\boldsymbol{e}}'_i$ onto the basis vector $\hat{\boldsymbol{e}}_j$. More compactly, we write $\hat{\boldsymbol{e}}' = \boldsymbol{Q}^T\hat{\boldsymbol{e}}$. It can be shown that if both $\hat{\boldsymbol{e}}_i$ and $\hat{\boldsymbol{e}}'_i$ are orthonormal bases, then the inverse of $\boldsymbol{Q}$ is equal to its transpose: $\boldsymbol{Q}^{-1} = \boldsymbol{Q}^T$. Also, it can be shown that for orthonormal transformations, $det([\boldsymbol{Q}]) = \pm 1$. If the determinant of $\boldsymbol{Q}$ is +1, then $\boldsymbol{Q}$ is called a *proper* orthonormal transformation.[e]

The components of a vector transform as the basis coordinate system changes as follows:

$$\boldsymbol{a} = a_i\hat{\boldsymbol{e}}_i = a'_j\hat{\boldsymbol{e}}'_j \,, \tag{3.8}$$

but from Eq. 3.7 (with a change of dummy indexes), $\hat{\boldsymbol{e}}'_j = Q_{kj}\hat{\boldsymbol{e}}_k$ and therefore

$$\boldsymbol{a} = a_i\hat{\boldsymbol{e}}_i = a'_j Q_{kj}\hat{\boldsymbol{e}}_k = a'_j Q_{ij}\hat{\boldsymbol{e}}_i \,, \tag{3.9}$$

where the repeated dummy index k has been replaced by i, so

$$\left(a_i - a'_j Q_{ij}\right)\hat{\boldsymbol{e}}_i = \boldsymbol{0} \,. \tag{3.10}$$

---

[d] The Kronecker delta is defined so that $\delta_{ij} = 1$ if $i = j$; otherwise, $\delta_{ij} = 0$.

[e] For three-dimensional Euclidian spaces, a *proper* transformation preserves the right-hand rule relationship between components.

Because the orthonormal basis vectors $\hat{\boldsymbol{e}}_i$ are linearly independent, the scalar coefficients of Eq. 3.10 must all be null, so

$$a_i = Q_{ij} a'_j \,, \tag{3.11}$$

or, in matrix form,

$$\{a\} = [Q]\{a'\} \,. \tag{3.12}$$

Using the fact that $[Q]^{-1} = [Q]^T$ for orthonormal coordinate systems, we see that

$$\{a'\} = [Q]^T\{a\} \,. \tag{3.13}$$

One way to *define* a Euclidean vector is to state that if its components transform according to Eq. 3.13, under proper orthonormal coordinate system rotation $[Q]$, then it is a Euclidean vector.

Continuum mechanics is mathematically expressed using tensors. Our purpose here is not to provide a detailed exposition on tensors, which would take too many pages; many continuum mechanics textbooks serve this purpose. Nonetheless, it is useful to review, in broad outline, tensors.

Definition 3.4 An *nth-order tensor* is a real-valued *m*-linear function of vectors.

For background context, an m-linear function $f: V \times \cdots (n\ times) \cdots \times V \to \mathbb{R}$ satisfies:

$$f[\boldsymbol{a}_1, \cdots, \lambda \boldsymbol{a}_i + \mu \boldsymbol{a}'_i, \cdots, \boldsymbol{a}_m] = \lambda f[\boldsymbol{a}_1, \cdots, \boldsymbol{a}_i, \cdots, \boldsymbol{a}_m] + \mu f[\boldsymbol{a}_1, \cdots, \boldsymbol{a}'_i, \cdots, \boldsymbol{a}_m] \quad \forall \boldsymbol{a}_i, \boldsymbol{a}'_i \in V \text{ and } \forall \lambda, \mu \in \mathbb{R}$$

In other words, *m*th-order tensor $\boldsymbol{T}$ is an *m*-linear function that takes *m* vectors in $\mathbb{R}^{N_R}$ and maps them to a real number:[f]

$$\boldsymbol{T}: \mathbb{R}^{N_R} \times \cdots \times \mathbb{R}^{N_R} \to \mathbb{R}. \tag{3.14}$$

---

[f] Actually, a more general definition, involving covectors, exists. The symbol × is the Cartesian product.

Thus, a second-order tensor $\boldsymbol{T}$ is a bilinear function of two vector arguments $\boldsymbol{a}$ and $\boldsymbol{b}$ that produces a real scalar, c, expressed in various ways as $c = \boldsymbol{T}(\boldsymbol{a}, \boldsymbol{b}) = \boldsymbol{a} \cdot \boldsymbol{T} \cdot \boldsymbol{b}$, in matrix form as $c = \lfloor a \rfloor [T]\{b\}$, and in indicial notation as $c = a_i T_{ij} b_j = a_i b_j T_{ij}$.

We can express a second-order tensor in terms of its basis tensor components, just as a vector can be expressed in terms of basis vectors. Thus, given two vectors $\boldsymbol{a} = a_i \hat{\boldsymbol{e}}_i$ and $\boldsymbol{b} = b_j \hat{\boldsymbol{e}}_j$, the tensor function can be evaluated in terms of tensor and vector components as[g]

$$\begin{aligned} \boldsymbol{T}(\boldsymbol{a}, \boldsymbol{b}) &= (a_n \hat{\boldsymbol{e}}_n) \cdot \left(T_{ij} \hat{\boldsymbol{e}}_i \otimes \hat{\boldsymbol{e}}_j\right) \cdot (b_k \hat{\boldsymbol{e}}_k) \\ &= T_{ij} [(a_n \hat{\boldsymbol{e}}_n) \cdot \hat{\boldsymbol{e}}_i] \otimes \left[\hat{\boldsymbol{e}}_j \cdot (b_k \hat{\boldsymbol{e}}_k)\right] \\ &= T_{ij} [a_n \hat{\boldsymbol{e}}_n \cdot \hat{\boldsymbol{e}}_i] \otimes \left[b_k \hat{\boldsymbol{e}}_j \cdot \hat{\boldsymbol{e}}_k\right] \\ &= T_{ij} [a_i] \otimes \left[b_j\right] = a_i b_j T_{ij}. \end{aligned} \tag{3.15}$$

Thus, as required by Definition 3.4, the output $a_i b_j T_{ij}$ of the second-order tensor $\boldsymbol{T}$ acting on two vectors, $\boldsymbol{a}$ and $\boldsymbol{b}$, is a scalar, independent of coordinate system, $\hat{\boldsymbol{e}}_i$. If we let $\boldsymbol{a}$ be the ith unit coordinate basis vector, $\hat{\boldsymbol{e}}_i$, and $\boldsymbol{b}$ be the jth unit coordinate basis vector $\hat{\boldsymbol{e}}_j$, then we see that these two vectors map, under the tensor transformation, to the scalar T_{ij}. Thus, each component T_{ij} of a second order tensor is associated with a dyadic pair of unit vectors $\hat{\boldsymbol{e}}_i \otimes \hat{\boldsymbol{e}}_j$.

A "tensor transformation" also provides a linear map from one tensor to another. For example, the component σ_{ij} of the second order stress tensor, $\boldsymbol{\sigma}$, returns the (scalar) component of the traction vector, $\boldsymbol{t}$, (force per unit area) in direction $\hat{\boldsymbol{e}}_i$ acting on a face whose outward unit normal, $\hat{\boldsymbol{n}}$, is $\hat{\boldsymbol{e}}_j$:

$$\boldsymbol{t} = \boldsymbol{\sigma}(\hat{\boldsymbol{n}}) = [\boldsymbol{\sigma}] \cdot [\hat{\boldsymbol{n}}] \hat{\boldsymbol{e}}_j = \sigma_{ij} n_j \hat{\boldsymbol{e}}_i \,. \tag{3.16}$$

Tensor analysis is very involved, including the concept of principal values and directions (eigenvalues and eigenvectors), and the calculus of tensor fields, in which tensors are assumed to vary continuously with

---

[g] $\otimes$ represents the dyadic, tensor, or outer, product, of two vectors. Its output is a pair of vectors.

spatial position. We do not cover these topics here; rather we refer the reader to (Tadmor, E. B.; Miller, R. E.; and Elliot, R. S., 2012).

Our primary intent here is to discuss continuum mechanics and peridynamics, not tensor mathematics. Furthermore, we will see, in Chapter 8, that continuum mechanics, and tensors, are insufficiently general for the purpose of peridynamics, and we will generalize the concept of "tensor" to the concept of "tensor state".

The following section summarizes the kinematics of continuum mechanics. For those that have already taken a continuum mechanics course, the section summarizes ideas with which the reader is already familiar. For those who are not familiar with continuum mechanics, the section serves mainly as an introduction to continuum kinematics. We have omitted many details, and the reader should not be too troubled if some things are difficult to understand. In the author's experience, to appreciate fully these topics, both tensor analysis and continuum mechanics require multiple passes at differing levels of detail, and from differing perspectives, over the duration of a career.

3.3 Kinematics

What we commonly call *kinematics* is a major part of the continuum mechanics model. Kinematics is the study of geometry and motion. Let us start out by describing the basis of the continuum mechanics model.

Truesdell (Truesdell, The Elements of Continuum Mechanics, 1965), states that the primitive elements of mechanics are: 1. bodies; 2. motions; and 3. forces.

Truesdell defines a continuum body as follows:

> "A *body*, B, is a manifold[h] of particles, denoted by X. These particles are primitive elements of mechanics in the sense that numbers are primitive elements in analysis. Bodies are sets of particles. In continuum mechanics

[h] A manifold is a topological space that resembles Euclidean space in the infinitesimal neighborhood of each point. Thus, a manifold does not include its bounding, or limit, points.

the body manifold is assumed to be smooth, that is, a diffeomorph[i] of a domain in Euclidean space. Thus, by assumption, the particles X can be set into a one-to-one correspondence with triples of real numbers, X_1, X_2, X_3, where the X_α run over a finite set of closed intervals.[j] Such triples are sometimes called 'intrinsic coordinates' of the particles, but I shall not need to use them explicitly. The mapping from the manifold to the domain is assumed differentiable as many times as required, usually two or three times, without further mention. The level of rigor here is that common in classical differential geometry, where the context makes it clear how many derivatives are assumed to exist."

Interestingly, Truesdell considers the *body* to be the primitive object; not the *continuum particle*. The body, as defined, does not to include its boundary, because a continuum particle on the boundary is not a diffeomorph of Euclidean space. Truesdell says little about the nature of the boundary of the body. Note also, the underlined sentence limits the "holes" in the body to a finite number. Also, there is a contradiction in the above definition: the particles run over closed intervals[k], but this contradicts the definition of "continuum particle", which is undefined at the end of a closed interval. Thus, Truesdell's definition of "continuum body" is self-contradictory. (If, in the underlined sentence, we change the word "closed" to "open", the contradiction goes away.)

Truesdell continues:

"The body B is assumed also to be a σ-finite measure space with non-negative measure $M(p)$ defined over a σ-ring of subsets p, which are called parts of B. Henceforth any subset of B to which we shall refer will be assumed to be measurable. The measure $M(p)$ is called the mass distribution in B. It is assigned, once and for all, to any body we shall consider.

Bodies are available to us only in their configurations, the regions they happen to occupy in Euclidean space at some time. These configurations are not to be confused with bodies themselves. In analytical dynamics, the

---

[i] A smooth, invertible mapping

[j] The underline is mine, not Truesdell's.

[k] A closed interval includes all of its limit points.

masses are discrete and hence stand in one-to-one correspondence with the numbers 1, 2, ... , n . Nobody ever confuses the sixth particle with the number 6, or with the place the sixth particle happens to occupy at some time. The number 6 is merely a label attached to the particle, and any other would do just as well. Similarly, in continuum mechanics a body B may occupy infinitely many different regions, none of which is to be confused with the body itself.

The two fundamental properties of a continuum B are, then:

1. B consists of a finite number of parts which can be mapped onto cubes in Euclidean space.
2. B is a measure space.

These are the simplest, most fundamental things we can say in summary of our experience with actual bodies and with the older theories designed to represent physical bodies in mathematical terms."

Truesdell further states that "analytical dynamics" (meaning rigid body dynamics) will not emerge as a special case of continuum mechanics.

It seems that the continuum body is not a simple thing to define mathematically, at least using the mathematics commonly studied by engineering students. A finite set of discrete point masses is far easier to understand. Nonetheless, we continue in this section to describe the foundations of continuum kinematics, if only to provide an understanding of some of its complexities and limitations.

As Truesdell points out, the *material description* and the *initial reference description* are in conceptually different spaces. Most books on continuum mechanics make no clear distinction, and assume that the initial reference configuration is the same as the material description of the body. Usually, books assume that the reference configuration is identical to the configuration of the body prior to loading; we call this the initial reference configuration. Thus, we measure all particle displacements with respect to this initial reference configuration. We represent the position of a continuum particle P in the initial reference configuration by its initial spatial position:

$$\boldsymbol{X} = \boldsymbol{X}(P)\ . \tag{3.17}$$

Thus, in continuum mechanics, the position of the particle in the initial reference configuration $\boldsymbol{X}$ is used as a label for the particle, distinguishing it from other particles. In continuum mechanics, the deformed configuration of particle P, with initial reference position $\boldsymbol{X}$, is called the current spatial position $\boldsymbol{x}$. The current spatial positon $\boldsymbol{x}$ is described by a continuous (both in space and in time) function $\boldsymbol{\varphi}$ called the deformation mapping function. (The assumption of spatial continuity is the root of the reason why continuum mechanics is insufficiently general to represent fracture.) At a given time, the deformation function maps the initial reference position of every particle to its deformed position at that time:

$$\boldsymbol{x} = \boldsymbol{\varphi}(\boldsymbol{X}) \,, \tag{3.18}$$

or, in component form,

$$x_i = \varphi_i(X_j), \quad i = 1,2,3; \quad j = 1,2,3 \,. \tag{3.19}$$

Note that the deformation function $\boldsymbol{\varphi}$ maps one position vector to another, but $\boldsymbol{\varphi}$ is not a linear map.[1] Thus, $\boldsymbol{\varphi}$ is not a tensor.

It is assumed that no two continuum reference particles $\boldsymbol{X}$ can occupy the same deformed location $\boldsymbol{x}$, so that no self-overlap of the body is allowed as a consequence of deformation. Furthermore, no continuum particle in the reference configuration $\boldsymbol{X}$ can map to two different continuum particles in the deformed configuration $\boldsymbol{x}$ which means that "cracks" are not allowed to emerge. The consequence of these fundamental (and arguable on physical grounds) assumptions is that the deformation mapping function $\boldsymbol{\varphi}$ is one-to-one bijective, so it is invertible:

$$\boldsymbol{X} = \boldsymbol{\varphi}^{-1}(\boldsymbol{x}) \,. \tag{3.20}$$

[1] It would be more direct to say that $\boldsymbol{x} = \boldsymbol{\varphi}(P)$; this is what we do in Chapter 8, when we define the material body to be a particle lattice.

In continuum mechanics, “tensor fields” are usually defined as functions of either the initial reference coordinates $\boldsymbol{X}$ or of deformed coordinates $\boldsymbol{x}$. For example, the temperature T is a zeroth-order tensor:

$$T = T_1(\boldsymbol{x}, t) = T_2(\boldsymbol{X}, t)\,, \tag{3.21}$$

where T_1 and T_2 are two very different functions; $T_2(\boldsymbol{X}, t) = T_1(\boldsymbol{x} = \boldsymbol{\varphi}(\boldsymbol{X}, t),\ t)$. Note that the function T_1 provides the temperature at a particular position in space $\boldsymbol{x}$, regardless of which material particle $\boldsymbol{X}$ is occupying that spatial position, while the function T_2 gives the temperature of the material particle originally located at $\boldsymbol{X}$.

The mapping $T = T_1(\boldsymbol{x}, t)$ is called the *spatial* or *Eulerian* description of the temperature field, while $T = T_2(\boldsymbol{X}, t)$ is called the *material, referential*, or *Lagrangian* description of the same field.

The Lagrangian description is preferable for modeling of deformable solids, in which the forces acting between particles are functions of *displacements* from the reference configuration. The Eulerian description is often better for modeling fluid flow, in which particles can travel very far from their reference positions over the duration of the simulation, and in which the forces between particles have nothing to do with their initial positions, but instead depend upon the current relative *velocities* between particles.

Thus, tensor fields are different when defined with respect to a spatial (Eulerian) coordinate system than when we define them with respect to a material (Lagrangian) coordinate system. However, for small-deformation problems, there is little difference between $\boldsymbol{X}$ and $\boldsymbol{x}$, and therefore no distinction between Eulerian and Lagrangian coordinate systems need be made.

We define the *reference vector*, between two initial reference points $\boldsymbol{X}$ and $\boldsymbol{X}'$ as

$$\Delta\boldsymbol{X} = \boldsymbol{X}' - \boldsymbol{X}\,, \tag{3.22}$$

which is different than the *spatial vector*

$$\Delta\boldsymbol{x} = \boldsymbol{x}' - \boldsymbol{x}\,. \tag{3.23}$$

Differentiation of tensor fields is complicated by the fact that it is important, for finite deformations, to distinguish between differentiation with respect to $\boldsymbol{X}$ and differentiation with respect to $\boldsymbol{x}$.

With these cautions, we now move on to a description of the "local" deformation of a continuum particle. The deformation mapping $\boldsymbol{\varphi}$ can be differentiated (if it is smooth enough) to give a measure of local deformation. Fig. 3.6 shows a differential element of a body in the reference and deformed configurations. Under deformation, the infinitesimal spherical neighborhood of particle $\boldsymbol{X}$ is deformed into an infinitesimal ellipsoid surrounding particle $\boldsymbol{x}$. Taking the total differential of Eq. 3.18, we call the mapping $\boldsymbol{F}$ from $d\boldsymbol{X}$ to $d\boldsymbol{x}$ the deformation gradient:

$$d\boldsymbol{x} = \frac{\partial \boldsymbol{\varphi}}{\partial \boldsymbol{X}} d\boldsymbol{X} = \boldsymbol{F} d\boldsymbol{X} \,, \tag{3.24}$$

which is an application of the chain rule of differential calculus using the Jacobian matrix. The deformation gradient is a second-order tensor, and transforms under the usual rules of tensor transformation. If the deformation gradient $\boldsymbol{F}$ does not vary spatially, we call $\boldsymbol{x} = \boldsymbol{\varphi}(\boldsymbol{X})$ a *homogeneous deformation* field.

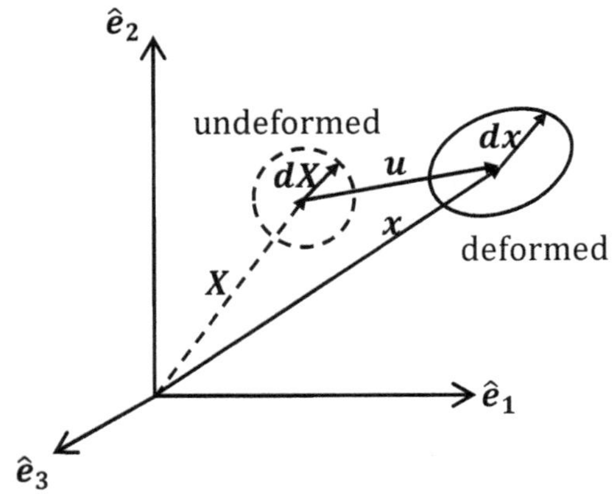

Figure 3.6. The local neighborhood of the material point $\boldsymbol{X}$ in the reference configuration maps to the neighborhood $\boldsymbol{x}$ in the deformed configuration.

The deformation gradient $\boldsymbol{F}$ captures the local spatial rate of deformation (in an infinitesimal neighborhood) of a point. $\boldsymbol{F}$ provides an affine mapping[m] of the neighborhood of a material particle from the initial reference configuration to the deformed configuration. However, it also includes information, like rigid body rotation, beyond what is necessary to characterize the shape change (stretching and shearing) at that point.

Thus, for example, we calculate the local change in volume due to deformation by

$$\frac{dV}{dV_0} = J = det\boldsymbol{F} \, , \tag{3.25}$$

where dV_0 is the original differential volume associated with material point $\boldsymbol{X}$ and dV is the deformed volume.

We use the polar decomposition theorem, following, to separate the rigid body rotational part of the deformation from the shape change part of the deformation:

Polar Decomposition Theorem: We can uniquely express any deformation gradient tensor $\boldsymbol{F}$ with positive determinant as

$$\boldsymbol{F} = \boldsymbol{R}\boldsymbol{U} = \boldsymbol{V}\boldsymbol{R} \, , \tag{3.26}$$

called the *right and left polar decompositions* of $\boldsymbol{F}$. $\boldsymbol{R}$ is the proper orthogonal transformation (rigid body rotation) and $\boldsymbol{U}$ and $\boldsymbol{V}$ are symmetric positive definite tensors called, respectively, the *right and left stretch tensors*.

In the right polar decomposition, the continuum particle is first changed in shape via the $\boldsymbol{U}$ transformation, and then rotated as a rigid body via the $\boldsymbol{R}$ transformation: $d\boldsymbol{x} = \boldsymbol{F}\, d\boldsymbol{X} = \boldsymbol{R}(\boldsymbol{U}d\boldsymbol{X})$. On the other hand, in the left polar decomposition, the continuum particle is first rotated as a rigid body via the $\boldsymbol{R}$ transformation, and then changed in shape via the $\boldsymbol{V}$ transformation: $d\boldsymbol{x} = \boldsymbol{F}\, d\boldsymbol{X} = \boldsymbol{V}(\boldsymbol{R}d\boldsymbol{X})$.

---

[m] In an *affine mapping*, sets of parallel lines remain parallel after transformation.

The polar decomposition tensors $\boldsymbol{R}$, $\boldsymbol{U}$, and $\boldsymbol{V}$ can be uniquely determined if $\boldsymbol{F}$ is known. The details are beyond the scope of this overview of continuum mechanics.

A measure of deformation (excluding the rigid-body rotation $\boldsymbol{R}$) is called a "strain". There are various definitions of strain in continuum mechanics. The right Cauchy-Green deformation tensor, $\boldsymbol{C} \equiv \boldsymbol{F}^T \boldsymbol{F}$, or in indicial form, $C_{ij} = F_{ki} F_{kj}$, is usually used as the basis for defining strain in solid mechanics.

The change in length of a material vector $\boldsymbol{dX}$ is related to the components of $\boldsymbol{C}$:

$$\begin{aligned} dS^2 &= dX_i dX_i \,, \\ ds^2 &= dx_i dx_i = \left(F_{ij} dX_j\right)\left(F_{ik} dX_k\right) \\ &= \left(F_{ij} F_{ik}\right) dX_j dX_k = C_{jk} dX_j dX_k \,. \end{aligned} \tag{3.27}$$

so

$$ds^2 - dS^2 = \left(C_{ij} - \delta_{ij}\right) dX_i dX_j \,.^{\text{n}} \tag{3.28}$$

Here, dS is the length of an infinitesimal undeformed bond, and ds is its length after deformation.

We define the Lagrangian strain tensor $\boldsymbol{E}$ as

$$\boldsymbol{E} = \frac{1}{2}(\boldsymbol{C} - \boldsymbol{I}) = \frac{1}{2}(\boldsymbol{F}^T \boldsymbol{F} - \boldsymbol{I}) \,. \tag{3.29}$$

so that

$$ds^2 - dS^2 = 2E_{ij} dX_i dX_j \,. \tag{3.30}$$

Thus we see that a measure of the change in length of a differential reference vector $\boldsymbol{dX}$ is a quadratic function of the components of $\boldsymbol{dX}$.

A similar measure of change of shape is the *Euler-Almansi strain tensor*:

---

[n] δ_{ij} is the Kronecker Delta, which is the same as the identity matrix $[I_{ij}]$, with ones on the diagonal and zeros off the diagonal.

$$e = \frac{1}{2}\left(I - B^{-1}\right) = \frac{1}{2}\left(I - F^{-T}F^{-1}\right). \tag{3.31}$$

so that

$$ds^2 - dS^2 = \left(\delta_{ij} - B_{ij}^{-1}\right)dx_i dx_j = 2e_{ij}dx_i dx_j \ . \tag{3.32}$$

Thus, the Euler-Almansi strain tensor $\boldsymbol{e}$ provides a quadratic map of a differential vector $\boldsymbol{dx}$ in the *deformed* configuration to a measure of change in length.

Both the Lagrangian strain tensor and the Euler-Almansi strain tensor are nonlinear functions of deformation $\boldsymbol{x}$. However, often it is advantageous to linearize these strain measures so that problems in solid mechanics can be incrementally solved in a piecewise linear manner. Indeed, the original Cauchy-Navier equation, developed in Chapter 2, is such a linear approximation, making judicious assumptions about small deformations and linear stress-strain behavior (Hooke's law).

In conventional implicit finite element methods, we usually solve nonlinear problems by linearizing the finite element equations and solving via a Newton-Raphson type of approach. Thus, the deformation gradient and the associated strain measures, $\boldsymbol{E}$ and $\boldsymbol{e}$ need to be linearized with respect to a given configuration $\boldsymbol{x}$. If we linearize these strain measures about the initial reference configuration, we obtain the small-strain (Cauchy) theory of elasticity.

Mathematically, the linearization of kinematics about a certain state in the deformed configuration is achieved by considering an additional small increment of deformation, call the *displacement field*, $\boldsymbol{u}(\boldsymbol{x})$, so that

$$\boldsymbol{\varphi}(\boldsymbol{X},\ t + \Delta t) = \boldsymbol{x}(\boldsymbol{X},\ t + \Delta t) = \boldsymbol{x}(t) + \boldsymbol{u}\,. \tag{3.33}$$

Let $G[\boldsymbol{\varphi}]$ be a kinematic functional of the field, $\boldsymbol{\varphi}$. (A functional is a function of a function). We can approximate the value of the functional G by a Taylor series expansion:

$$\begin{aligned} G[\boldsymbol{\varphi} + \boldsymbol{u}] &\approx G[\boldsymbol{\varphi}] + \boldsymbol{\nabla}_{\varphi} G[\boldsymbol{\varphi}] \cdot \boldsymbol{u} \\ &= G[\boldsymbol{\varphi}] + \frac{d}{d\eta} G[\boldsymbol{\varphi} + \eta \boldsymbol{u}]\Big|_{\eta=0} \,. \end{aligned} \tag{3.34}$$

where $\boldsymbol{\nabla}_{\varphi} G[\boldsymbol{\varphi}]$ is the variation of G in the direction of the vector field, $\boldsymbol{u}$.

For example, to linearize the deformation gradient, $\boldsymbol{F}$:

$$\begin{aligned}\frac{d}{d\eta}\boldsymbol{F}[\varphi_i+\eta u_i]\bigg|_{\eta=0} &= \frac{d}{d\eta}\left[\frac{\partial}{\partial X_j}(\varphi_i+\eta u_i)\right]\bigg|_{\eta=0} \\ &= \frac{\partial u_i}{\partial X_j} = \boldsymbol{\nabla}_0\boldsymbol{u}\,.\end{aligned} \tag{3.35}$$

To linearize the right Cauchy-Green deformation tensor, $\boldsymbol{C}$:

$$\begin{aligned}\frac{d}{d\eta}\boldsymbol{C}[\varphi_i+\eta u_i]\bigg|_{\eta=0} &= \frac{d}{d\eta}\left(\boldsymbol{F}^T[\varphi_i+\eta u_i]\boldsymbol{F}[\varphi_i+\eta u_i]\right)\bigg|_{\eta=0} \\ &\frac{d}{d\eta}\left[\frac{\partial}{\partial X_I}(\varphi_i+\eta u_i)\frac{\partial}{\partial X_J}(\varphi_i+\eta u_i)\right]\bigg|_{\eta=0} \\ &= \frac{\partial u_i}{\partial X_I}F_{iJ}+\frac{\partial u_i}{\partial X_J}F_{iI} = \boldsymbol{F}^T\boldsymbol{\nabla}_0\boldsymbol{u}+(\boldsymbol{F}^T\boldsymbol{\nabla}_0\boldsymbol{u})^T\,.\end{aligned} \tag{3.36}$$

and similarly, to linearize the Lagrangian strain tensor $\boldsymbol{E}$:

$$\boldsymbol{E} = \frac{1}{2}[\boldsymbol{F}^T\boldsymbol{\nabla}_0\boldsymbol{u}+(\boldsymbol{F}^T\boldsymbol{\nabla}_0\boldsymbol{u})^T]\,. \tag{3.37}$$

Note that for small deformations, $\boldsymbol{\varphi}=\boldsymbol{X}$; $\boldsymbol{F}=\boldsymbol{I}$, and the conventional Cauchy small strain tensor, $\boldsymbol{\epsilon}$, is obtained:

$$\boldsymbol{\epsilon} = \frac{1}{2}[\boldsymbol{\nabla}_0\boldsymbol{u}+(\boldsymbol{\nabla}_0\boldsymbol{u})^T];\quad \epsilon_{ij} = \frac{1}{2}\left(u_{i,j}+u_{j,i}\right). \tag{3.38}$$

While *rates of change* of kinematic variables with respect to time are also important, more so for fluids than for solids, we do not discuss kinematic rates further in this book.

Having briefly reviewed continuum kinematics, which deals only with time and physical space, (not forces), we now explore the physical laws of mechanics having to do with mass, momentum, and thermodynamics.

3.4 Physical laws of mechanics and thermodynamics

From the viewpoint of engineering mechanics, the six physical laws that any material body must follow are:

1. Conservation of mass.
2. Conservation of linear momentum.
3. Conservation of angular momentum.
4. Thermal equilibrium (0th law of thermodynamics)
5. Conservation of energy (1st law of thermodynamics).
6. Entropy never decreases (2nd law of thermodynamics)

Let's discuss each of these physical laws in turn.

Firstly, in continuum systems, material particles have only infinitesimal mass, but the mass density us finite. The mass density ρ, defined as mass per unit volume of a material particle, changes as the associated volume changes while the mass of each particle remains unchanged. Therefore,

$$dm = \rho_0 dV_0 = \rho dV \; ; \tag{3.39}$$

from which, together with $\frac{dV}{dV_0} = J = det(\boldsymbol{F})$ from Eq. 3.25 gives

$$\rho_0 = \rho \frac{dV}{dV_0} = \rho J = \rho det(\boldsymbol{F}) \; ; \tag{3.40}$$

If instead of focusing upon an individual material particle, $\boldsymbol{X}$, which moves with time, we focus upon a fixed spatial location, $\boldsymbol{x}$, in physical space, which does not move, the conservation of mass equation is much more involved:

$$\frac{D\rho}{Dt} + \rho(div\, \boldsymbol{v}) = 0 \, , \tag{3.41}$$

or equivalently,

$$\frac{\partial \rho}{\partial t} + div(\rho \boldsymbol{v}) = 0 \, . \tag{3.42}$$

The reader may be puzzled by the differences between Eqs. 3.41 and 3.42. In Eq. 3.41, $\frac{D\rho}{Dt}$ is understood to be the time rate of change of the density of the particle originally located at material point $\boldsymbol{X}$. On the other hand, in Eq. 3.42, $\frac{\partial \rho}{\partial t}$ is the partial derivative with respect to time of the density at the spatial location, $\boldsymbol{x}$. Thus, $\frac{D\rho}{Dt}$ is known as a "material derivative", while $\frac{\partial \rho}{\partial t}$ is the "local rate of change" of density at a fixed location. [o] Both Eqs. 3.41 and 3.42 represent the fact that density at a fixed spatial location can change not only to due to changes in volume, but also due to transport of particles of fixed mass into and out of a particular spatial location.

Secondly, the balance of linear momentum equation for a continuum is given by

$$\int_B \rho \ddot{\boldsymbol{x}} \, dV = \int_B \rho \boldsymbol{b} \, dV + \int_{\partial B} \boldsymbol{t} \, dA \,, \tag{3.43}$$

where B is the deformed volumetric domain of the body, ρ is the density of spatial location $\boldsymbol{x}$ currently located at differential volume dV, $\ddot{\boldsymbol{x}}$ is the acceleration of the particle currently located at spatial location $\boldsymbol{x}$, $\boldsymbol{b}$ is the applied body force per (current) unit mass ρdV, ∂B is the surface of the domain B, and $\boldsymbol{t}$ is the applied traction per unit deformed area, dA.

In practice, Eq. 3.43 is difficult to apply to solids, because the domains B and ∂B are unknown and changing with time, as indeed is the applied traction $\boldsymbol{t}$. In addition, $\boldsymbol{t}$ is not uniquely defined at sharp corners of B, as the limit depends upon how one approaches the corner. This is

[o] The differences between material and spatial representations and their material time derivatives and local rates of change are conceptually difficult. Fortunately, with the Lagrangian SPLM model, particle mass is automatically conserved without resorting to satisfying a partial differential equation. This is one of many reasons that the Lagrangian SPLM is attractive. We do not investigate Eulerian SPLM models here, but nonetheless we expect them to be simpler than Eulerian continuum models.

another manifestation of the fact that continuum mechanics requires all functions to be continuous, including ∂B and $\boldsymbol{t}$.[p]

In order to make the integration of Eq. 3.43 easier, the equation is "pulled back" to the initial reference space, so that integration will be over the fixed domain B_0 rather than the changing domain B. In addition, we will define various stress measures including the Cauchy stress and the first and second Piola-Kirchhoff stresses in the next section.

The third principle is the conservation of angular momentum, which is the statement that the change in angular momentum of a system of particles is equal to the external moment applied to the system. For a continuum, we express this mathematically as

$$\frac{D}{Dt}\int_B \boldsymbol{x} \times (\rho\dot{\boldsymbol{x}})\, dV = \int_B \boldsymbol{x} \times (\rho\boldsymbol{b})\, dV + \int_{\partial B} \boldsymbol{x} \times \boldsymbol{t}\, dA\,, \qquad (3.44)$$

where $\frac{D}{Dt}$ is the material derivative, meaning that the particle $\boldsymbol{X}$, rather than the spatial position $\boldsymbol{x}$, is held constant for the partial derivative. We can show that if all forces between particles within a system are equal, opposite, and collinear, then angular momentum of the system is conserved.

The last three laws are the three laws of thermodynamics, which deal with heat and energy, stated as follows.

Zeroth Law of thermodynamics: Given three thermodynamic systems, A, B, and C, each internally in thermodynamic equilibrium, then if A and B are in thermodynamic equilibrium, and B and C are in thermodynamic equilibrium, then it follows that A and C are in thermodynamic equilibrium.

First Law of thermodynamics: The total energy of a thermodynamic system and its surroundings is conserved.

Second Law of thermodynamics: The entropy of an isolated system can never decrease in any process.

Although thermal effects are certainly important, we do not explicitly deal with problems of thermodynamics in this book, choosing instead to

---

[p] So, it would seem that any finite element model that includes sharp corners and discontinuous tractions is not, by definition, a continuum model.

simplify our study to purely mechanical systems, in which we assume decoupling of the temperature and heat effects from the mechanical effects. Thus, with dissipative processes like damping, plasticity, and damage, it would appear that mechanical (kinetic and strain) energy is lost to the system, but this energy is actually converted into heat and surface energy, which is not explicit in our models.

3.5 Traction and stress

The *traction vector* $\boldsymbol{t}$ is defined as the force vector $\Delta\boldsymbol{F}$ acting upon a specific surface in the limit as the surface area ΔA goes to zero:

$$\boldsymbol{t} = \lim_{\Delta A \to 0} \left(\frac{\Delta F}{\Delta A}\right). \tag{3.45}$$

We must specify the surface area for the traction vector to have any clear meaning. Particularly, we must make clear whether the surface area is in the reference or in the deformed configuration. The concept of traction depends closely upon concepts of continuity, so the limit in Eq. 3.45 must exist, which in turn implies that the force vector must vary continuously over the surface area, which in turn must have a unique outward-pointing normal.

Cauchy's stress principle states that material interactions across an *internal* surface in a body can be described as a (continuous) distribution of tractions in the same way that the effect of *external* forces on physical surfaces of the body are described.

Cauchy did not differentiate between the deformed and the undeformed configurations, as both he and Navier assumed small deformations. However, the Cauchy definition of traction has come to mean force per unit area in the *deformed, or current* configuration.

Tractions acting on opposing faces of a planar surface within a material body must be equal and opposite. We show this by applying the conservation of linear momentum equation, Eq. 3.43, to a cylindrical sub-body, P, cut out of a larger material body, B, as shown in Fig. 3.7.

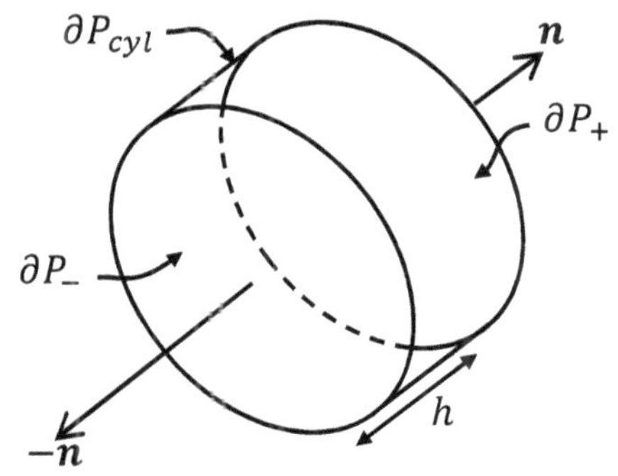

Figure 3.7. Cylindrical sub-body P.

If we let the length of the cylinder, h, go to zero, the volume of the body goes to zero, as well as the area of the curved wall of the cylinder. Thus, from Eq. 3.43,

$$\begin{aligned} \int_B \rho \ddot{\boldsymbol{x}}\, dV &= \int_B \rho \boldsymbol{b}\, dV + \int_{\partial P_{cyl}} \boldsymbol{t}\, dA + \int_{\partial P_+} \boldsymbol{t}\, dA + \int_{\partial P_-} \boldsymbol{t}\, dA, \\ \text{or} \quad 0 &= \quad 0 \quad + \quad 0 \quad + \int_{\partial P_+} \boldsymbol{t}\, dA + \int_{\partial P_-} \boldsymbol{t}\, dA. \end{aligned} \tag{3.46}$$

Letting the diameter of the cylinder go to zero, and assuming that the traction distributions on both ends of the cylinder are continuous, and by using the mean value theorem of calculus, we see that

$$\boldsymbol{t}(\boldsymbol{x}, \boldsymbol{n}) = -\boldsymbol{t}(\boldsymbol{x}, -\boldsymbol{n}), \tag{3.47}$$

so that at a given point $\boldsymbol{x}$ within a body, the tractions on opposing sides of any surface defined by unit normal vector $\boldsymbol{n}$ are equal and opposite. We can also derive Eq. 3.47 directly from Newton's third law.

Cauchy was the first to generalize the concept of "stress", which he called "pression ou tension" (Cauchy A.-L. , 1827). Cauchy considered as a free body an infinitesimal tetrahedron, conceptually removed, as a free body, from its surrounding body, as shown in Fig. 3.8.

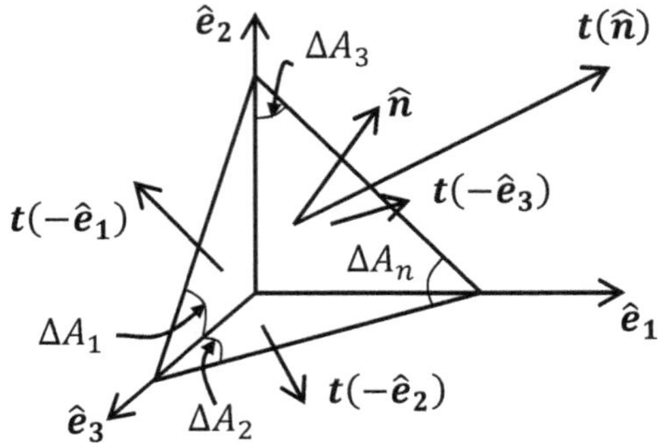

Figure 3.8. Tetrahedral sub-body, *T*.

As with all continuum fields, Cauchy's concept of stress is only valid if the stress distribution is *continuous* – otherwise the limit in Eq. 3.45 does not exit uniquely, and the stress at a point is indeterminate.

Applying balance of linear momentum (Eq. 3.43) to the tetrahedron shown in Fig. 3.8,

$$\int_T \rho \ddot{\boldsymbol{x}}\, dV = \int_T \rho \boldsymbol{b}\, dV + \int_{\Delta A_1} \boldsymbol{t}\, dA + \int_{\Delta A_2} \boldsymbol{t}\, dA + \int_{\Delta A_3} \boldsymbol{t}\, dA + \int_{\Delta A_n} \boldsymbol{t}\, dA\,, \tag{3.48}$$

and the volumetric integrals disappear as ΔA_n goes to zero, so

$$\boldsymbol{0} = \int_{\Delta A_1} \boldsymbol{t}(-\hat{\boldsymbol{e}}_1) dA + \int_{\Delta A_2} \boldsymbol{t}(-\hat{\boldsymbol{e}}_2)\, dA + \int_{\Delta A_3} \boldsymbol{t}(-\hat{\boldsymbol{e}}_3)\, dA + \int_{\Delta A_n} \boldsymbol{t}(\hat{\boldsymbol{n}}) dA\,. \tag{3.49}$$

With reference to Fig. 3.8, the tetrahedron's face areas are related geometrically as

$$\Delta A_i = \Delta A_n n_i, \quad i = 1,2,3\,. \tag{3.50}$$

Finally, as the volume of the tetrahedron shrinks to zero, the integrals in Eq. 3.49 become simple products, because the integrands approach being spatially constant over ΔA:

$$\boldsymbol{0} = \boldsymbol{t}(-\hat{\boldsymbol{e}}_1)\Delta A_1 + \boldsymbol{t}(-\hat{\boldsymbol{e}}_2)\Delta A_2 + \boldsymbol{t}(-\hat{\boldsymbol{e}}_3)\Delta A_3 + \boldsymbol{t}(\hat{\boldsymbol{n}})\Delta A_n\,. \tag{3.51}$$

Dividing through by ΔA_n, and using Eqs. 3.50 and 3.51 and rearranging,

$$\boldsymbol{t}(\hat{\boldsymbol{n}}) = \boldsymbol{t}_1 n_1 + \boldsymbol{t}_2 n_2 + \boldsymbol{t}_3 n_3 = \boldsymbol{t}_j n_j \ , \tag{3.52}$$

where we define $\boldsymbol{t}_j = \boldsymbol{t}(\hat{\boldsymbol{e}}_j)$. Dotting both sides of Eq. 3.52 with $\hat{\boldsymbol{e}}_i$ gives:

$$\hat{\boldsymbol{e}}_i \cdot \boldsymbol{t}(\hat{\boldsymbol{n}}) = t_i(\hat{\boldsymbol{n}}) = (\hat{\boldsymbol{e}}_i \cdot \boldsymbol{t}_j) n_j = \sigma_{ij} n_j \ , \tag{3.53}$$

where $\sigma_{ij} \equiv (\boldsymbol{e}_i \cdot \boldsymbol{t}_j)$ is the ith component of the traction vector $\boldsymbol{t}_j$. Thus,

$$t_i(\hat{\boldsymbol{n}}) = \sigma_{ij} n_j \quad \text{or} \quad \boldsymbol{t}(\hat{\boldsymbol{n}}) = \boldsymbol{\sigma} \cdot \hat{\boldsymbol{n}} \ . \tag{3.54}$$

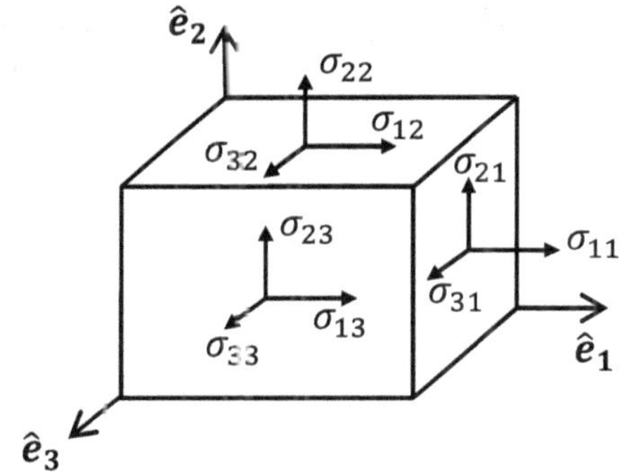

Figure 3.9 Components of the Cauchy stress tensor.

It can be proven that σ_{ij} are the components of a second-order tensor $\boldsymbol{\sigma}$ called the Cauchy stress tensor. We show the components of the Cauchy stress tensor pictorially in Fig. 3.9.

The Cauchy stress tensor $\boldsymbol{\sigma}$ gives the traction vector acting upon a unit area of material in the *deformed* configuration.

We often find the need to decompose the Cauchy stress tensor into a volumetric and a deviatoric component:

$$\sigma_{ij} = s_{ij} - p\delta_{ij} \quad \text{or} \quad \boldsymbol{\sigma} = \boldsymbol{s} - p\boldsymbol{I}, \tag{3.55}$$

where the hydrostatic part of the Cauchy stress tensor is called the pressure p:

$$p = -\frac{\sigma_{kk}}{3} \quad \text{or} \quad p = \frac{-tr(\sigma)}{3}\,,\text{[q]} \tag{3.56}$$

and the deviatoric component of the Cauchy stress tensor $\boldsymbol{s}$ is defined as

$$s_{ij} = \sigma_{ij} + p\delta_{ij} \quad \text{or} \quad \boldsymbol{s} = \boldsymbol{\sigma} + p\boldsymbol{I}\ . \tag{3.57}$$

We now develop the differential equations of motion in terms of Cauchy stress from the balance of linear momentum for an arbitrary sub-body E of a body B. Rearranging the balance of linear momentum equation, Eq. 3.43:[r]

$$\int_E \rho(\ddot{\boldsymbol{x}} - \boldsymbol{b})\, dV = \int_{\partial E} \boldsymbol{t}\, dA = \int_{\partial E} \boldsymbol{\sigma}\cdot\boldsymbol{n}\, dA = \int_E (div\,\boldsymbol{\sigma}) dV, \tag{3.58}$$

or

$$\int_E (div\,\boldsymbol{\sigma} + \rho\boldsymbol{b} - \rho\ddot{\boldsymbol{x}})\, dV = \boldsymbol{0}. \tag{3.59}$$

Since Eq. 3.59 must hold true for all sub-bodies E within the body B, the integrand must be null:

$$div\,\boldsymbol{\sigma} + \rho\boldsymbol{b} - \rho\ddot{\boldsymbol{x}} = \boldsymbol{0}\,, \tag{3.60}$$

and we arrive at the governing differential equations of dynamic equilibrium, which must be satisfied at all spatial points $\boldsymbol{x}$ within body B:

$$div\,\boldsymbol{\sigma} + \rho\boldsymbol{b} = \rho\ddot{\boldsymbol{x}}\,, \quad \text{or} \quad \sigma_{ij,j} + \rho b_i = \rho\ddot{x}_i \quad \forall\boldsymbol{x}\in B\ . \tag{3.61}$$

---

[q] The trace of a tensor, $tr(\boldsymbol{\sigma}) = \sigma_{11} + \sigma_{22} + \sigma_{33} = \sigma_{kk}$.

[r] Here, the divergence theorem, $\int_{\partial E} \boldsymbol{\sigma}\cdot\boldsymbol{n}\, dA = \int_E (div\,\boldsymbol{\sigma}) dV$, has been used. The divergence of a second order tensor, σ_{ij}, is given by $div\,\boldsymbol{\sigma} = \boldsymbol{\nabla}\cdot\boldsymbol{\sigma} = \sigma_{i1,1} + \sigma_{i2,2} + \sigma_{i3,3} = \sigma_{ij,j}$. The del operator is defined as $\boldsymbol{\nabla} = \frac{\partial}{\partial X_1} + \frac{\partial}{\partial X_2} + \frac{\partial}{\partial X_3}$.

Without proof, although the proof is not difficult, we can also show that balance of angular momentum is assured if distributed body moments are absent and if the Cauchy stress tensor is symmetric:

$$\sigma_{ij} = \sigma_{ji}\ , \quad \text{or} \quad \boldsymbol{\sigma} = \boldsymbol{\sigma}^T \tag{3.62}$$

When solving specific problems, for example using the finite element method, the governing differential equations of motion, Eq. 3.61, (called the "strong form") are normally not used – they would need to be satisfied at an infinite number of points $\boldsymbol{x}$, which is computationally impossible. Instead, an integral expression, like Eq. 3.43, for balance of linear momentum is used. We call this the "weak form" of the equations of motion. We rewrite the balance of linear momentum equation, Eq. 3.43, now in terms of Cauchy stress as:

$$\int_E \rho\ddot{\boldsymbol{x}}\,dV = \int_E \rho\boldsymbol{b}\,dV + \int_{\partial E} \boldsymbol{\sigma}\cdot\boldsymbol{n}\,dA\ . \tag{3.63}$$

Eq. 3.63 is in terms of spatial coordinates $\boldsymbol{x}$ which are elements of a changing subdomain (of finite size) E as the body deforms. Thus, the integral is very difficult to solve in this form. However, if we "pull back" Eq. 3.63 to the initial reference coordinate system $\boldsymbol{X}$ then the finite element of integration E_0 remains fixed as motion proceeds. However, we must now express all functions within the integrands as functions of material (or referential) coordinates $\boldsymbol{X}$ rather than in terms of spatial coordinates $\boldsymbol{x}$.

To enable this transformation from spatial coordinates to material coordinates, we make the following substitutions:

$$\rho dV = \rho_0 dV_0\ ; \tag{3.64}$$

$$\ddot{\boldsymbol{x}} = \ddot{\boldsymbol{\varphi}}(\boldsymbol{X})\ ;\ \text{and} \tag{3.65}$$

$$\boldsymbol{b} = \breve{\boldsymbol{b}}(\boldsymbol{X})\ ; \tag{3.66}$$

where $\breve{\boldsymbol{b}}(\boldsymbol{X})$ is defined to be the body force per unit reference (not deformed) volume.
Without proof,

$$\int_{\partial E} \boldsymbol{\sigma} \cdot \boldsymbol{n}\, dA = \int_{\partial E_0} \boldsymbol{P} \cdot \boldsymbol{N}\, dA_0 \,, \tag{3.67}$$

where $\boldsymbol{P}$ is the *first Piola-Kirchoff stress tensor*

$$\boldsymbol{P} = \boldsymbol{\tau} \boldsymbol{F}^{-\boldsymbol{T}} \,, \tag{3.68}$$

$\boldsymbol{\tau}$ is the *Kirchhoff stress tensor*

$$\boldsymbol{\tau} = J\boldsymbol{\sigma} = det(\boldsymbol{F})\boldsymbol{\sigma} \,, \tag{3.69}$$

and $\boldsymbol{N}$ is the outward-pointing normal in the referential (material, undeformed) configuration.

The first Piola-Kirchhoff stress tensor gives the force per unit *undeformed* area, while the Cauchy stress tensor gives the force per unit *deformed* area. Thus, we commonly call the Cauchy stress the *true stress*, while the first Piola-Kirchhoff corresponds to the *engineering stress*.

The first Piola-Kirchhoff stress tensor is unsymmetric, however, and thus a second stress measure, called the *second Piola-Kirchhoff stress tensor* $\boldsymbol{S}$, which is symmetric, is more commonly used in large-deformation finite element programs.

$$\boldsymbol{S} = \boldsymbol{F}^{-1} \boldsymbol{P} \,. \tag{3.70}$$

With the second Piola-Kirchhoff stress tensor, the force has also been "pulled back" to the reference configuration. The physical meaning of the second Piola-Kirchhoff stress tensor is obscure. It is sufficient to say that if we know the deformation gradient, then we can calculate the Cauchy stress as

$$\boldsymbol{\sigma} = \frac{1}{J}\boldsymbol{F}\boldsymbol{S}\boldsymbol{F}^T \,. \tag{3.71}$$

In terms of the second Piola-Kirchhoff stress tensor, the governing differential equations of motion become

$$div\,\big(\boldsymbol{F}(\boldsymbol{X})\boldsymbol{S}(\boldsymbol{X})\big) + \rho_0(\boldsymbol{X})\breve{\boldsymbol{b}}(\boldsymbol{X}) = \rho_0(\boldsymbol{X})\overline{\ddot{\boldsymbol{x}}(\boldsymbol{X})} \quad \boldsymbol{X} \in B_0 \,, \tag{3.72}$$

where $\breve{\boldsymbol{b}}$ is the body force per unit *reference* volume in the reference configuration, and $\breve{\ddot{\boldsymbol{x}}}$ is the acceleration of reference particle, $\boldsymbol{X}$.

These various stress measures are meaningful in large-deformation mechanics of continuous media, but certainly are difficult for a practicing engineer to keep track of. A simpler model for large-deformation analysis of solids would certainly be welcome. A computational lattice model, avoiding the concept of stress altogether, may provide the desired simplification.

3.6 Constitutive relations

So far in this chapter, we have discussed kinematics, assuming continuous deformation (which has no basis in physical law, but rather, is an engineering simplification of reality). Also, we have presented six physical laws that must be obeyed: conservation of mass, balance of linear momentum, balance of angular momentum, and the three laws of thermodynamics. Nothing further has been said yet about the material behavior, aside from the fact a material has mass density.

Material constitive relations provide the remaining equations necessary to solve engineering problems. With continuum mechanics, several assumptions are made about constitutive relations:

1. *Principle of determinism* – Past events determine the present state of a system. The current state of a system cannot depend upon future states of the system.
2. *Principle of local action* – Material response at a point depends only upon the conditions within an arbitrarily small neighborhood of that point.

3. *Second law of thermodynamics* – Entropy can never decrease.
4. *Principle of material frame-indifference (objectivity)* – All physical variables for which constitutive relations are required must be objective tensors. An objective tensor is an object that is the same in all frames of reference.
5. *Material symmetry* – A constitutive relation must respect any symmetry that the material possesses, such as isotropy or orthotropy.

In addition, two common assumptions about materials include:

6. Only materials without memory and without aging are considered.
7. Only materials whose internally stored energy depends solely upon the entropy and the deformation gradient are considered.

The *principle of local action* prevents the continuum model from being able correctly to model softening behavior associated with strain localization and cracks, because there is no mechanism to limit strain localization to a finite volume. A crack can grow with no energy dissipation, because a zero-volume fracture process zone dissipates finite internal strain energy density, resulting in zero fracture energy, which is unrealistic.

The most common form of constitutive relation for a solid material is the *hyperelastic* model[s], which we can show to satisfy the required seven constitutive requirements in the above list. In what we call a *simple material*, we assume that a specific internal energy[t] function exists:

$$\mathrm{u} = \bar{\mathrm{u}}(\boldsymbol{F}) \,. \tag{3.73}$$

Based upon thermodynamic considerations (Tadmor, E. B.; Miller, R. E.; and Elliot, R. S.), it can be shown that the elastic component of the Cauchy stress is given, in terms of $\boldsymbol{F}$, by

---

[s] In contrast, a *hypoelastic* material is an elastic material that has a constitutive model that is independent of finite strain measures. Hypoelastic material models are distinct from hyperelastic material models in that, with some exceptions, they are not derivable from strain energy density functions.

[t] Also called a strain energy density function.

$$\boldsymbol{\sigma} = \rho \frac{\partial \tilde{u}}{\partial \boldsymbol{F}} \boldsymbol{F}^T \text{ or, in component form,} \quad \sigma_{ij} = \rho \frac{\partial \tilde{u}}{\partial F_{ik}} F_{jk}. \tag{3.74}$$

Alternative stress-strain measures are possible. For the Lagrangian description, the first and second Piola-Kirchhoff stress measures are used. Since the rigid-body rotation component of the deformation gradient cannot contribute to the specific internal energy, the Green-Lagrange strain tensor $\boldsymbol{E}$ is usually used to represent deformation, rather than $\boldsymbol{F}$. Thus,

$$u = \tilde{u}(\boldsymbol{E}) . \tag{3.75}$$

and

$$\boldsymbol{S} = \rho_0 \frac{\partial \tilde{u}}{\partial \boldsymbol{E}} . \tag{3.76}$$

We say that the second Piola-Kirchhoff stress and the Green-Lagrange strain are *strain energy-conjugate*. Other energy-conjugate pairs include the Almansi strain and the Cauchy stress, and the deformation gradient and the first Piola-Kirchhoff stress.

The linearized stress-strain equations are the most frequently used. In linearized elasticity, we relate the increment in the second Piola-Kirchhoff stress $d\boldsymbol{S}$ to the increment in the Green-Lagrange strain $d\boldsymbol{E}$ by a fourth-order material elasticity tensor $\boldsymbol{C}$ as follows:

$$d\boldsymbol{S} = \boldsymbol{C}: d\boldsymbol{E}, \quad \text{or} \quad dS_{ij} = C_{ijkl} dE_{kl} . \tag{3.77}$$

The constants in the material elasticity tensor must satisfy certain symmetries depending upon the symmetries of the material. For a linear elastic isotropic material,

$$C_{ijkl} = \lambda \delta_{ij} \delta_{kl} + \mu\left(\delta_{ik}\delta_{jl} + \delta_{il}\delta_{jk}\right) , \tag{3.78}$$

where λ and are μ called the Lamé constants. In terms of Young's, modulus E and Poisson's ratio ν, the Lamé constants are

$$\mu = \frac{E}{2(1+\nu)} \text{ and } \lambda = \frac{\nu E}{(1+\nu)(1-2\nu)} \tag{3.79}$$

We have discussed only elastic constitutive models in this section. We defer discussion of damage and plasticity models to Chapters 10 and 11.

3.7 Solving the continuum equations

So far, we have developed a set of governing differential equations for continuum mechanics problems. In general, these differential equations are nonlinear, but for some engineering problems, we can linearize them to provide useful solutions. Solving the nonlinear continuum mechanics equations analytically in closed-form is very difficult for all but the most trivial problems.

If we can linearize the problem, given simple-enough boundary conditions, we can solve it analytically in closed-form using standard methods for solving linear partial differential equations. For example, a linear elastic uniaxial bar statically loaded at its ends provides a simple homogeneous stress solution ($\sigma = {}^{P}/_{A}$; $\epsilon = {}^{P}/_{EA}$, etc.).

In a less trivial example, a point load applied statically within an infinite linear elastic domain (the Kelvin solution) is an important analytical solution that we can use in solving other problems of static linear elasticity using integral methods.

Homogeneous deformations (in which the deformation gradient $\boldsymbol{F}$ is spatially constant) always provide solutions to continuum mechanics problems. However, for nonlinear problems, these homogeneous solutions may not be stable. For example, consider a uniaxial prismatic hyperelastic bar, of length L, and with initial elastic modulus E and cross-sectional area A, fixed at its left end and loaded at its right end with a gradually increasing applied displacement, Δ, as shown in Fig. 3.10.

Assume the uniaxial stress σ versus uniaxial strain ϵ relationship first increases and then decreases to zero,

$$\sigma = E\epsilon e^{-\left(\frac{\epsilon}{k}\right)}, \tag{3.80}$$

where k is the strain at peak stress, which is $\sigma_{max} = Eke^{-1}$. We show the stress versus strain curve in Fig. 3.10.

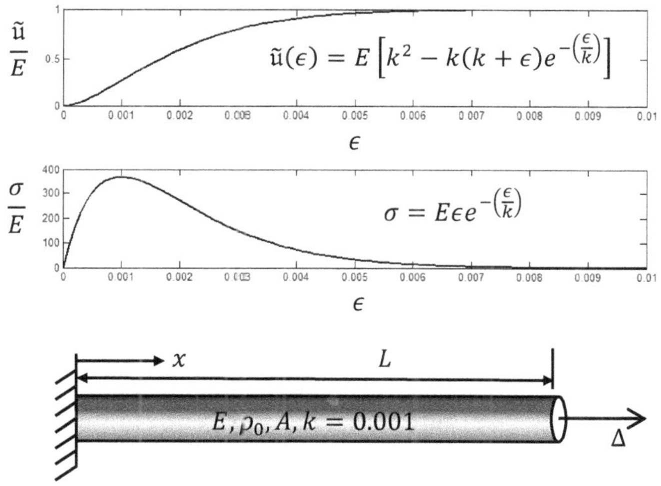

Figure 3.10. Uniaxial bar with nonlinear stress-strain relationship.

The corresponding strain energy density function ũ is found by integrating Eq. 3.80 with respect to ϵ:

$$\tilde{u}(\epsilon) = \int_{\epsilon} \sigma d\epsilon = E\left[k^2 - k(k+\epsilon)e^{-\left(\frac{\epsilon}{k}\right)}\right]. \tag{3.81}$$

As the displacement $\Delta < \text{kL}$ is gradually increased to peak load, the stress increases nonlinearly to the maximum value $\sigma_{max} = Eke^{-1} = 0.3679Ek$. However, as we increase the applied displacement Δ further, and the stress begins to decrease, the strain response becomes unstable. In part of the bar the strain decreases, decreasing the strain energy density, while in another other part of the bar the strain increases, with consequent strain energy density increase. The strain at one cross-section will rapidly increase elastically to infinity, while the rest of the bar will unload elastically. Thus, "strain localization" occurs, where infinite strain occurs over a null volume of the bar, and the rest of the bar unloads elastically. The key point is that although a discontinuity in the displacement field (a crack) has appeared, no energy has been absorbed in the process of creating the crack. To "fix" the problem of zero-energy discontinuities, engineers have invented the notion of "localization limiter". By preventing the width of the post-peak unloading zone from becoming too narrow, the strain is kept continuous and is prevented from going to infinity.

Even for linearized continuum mechanics problems, it is necessary to use numerical methods to solve nontrivial boundary value problems. The finite element method, the boundary element method, and the finite difference method are among the most common computational methods.

For nonlinear continuum mechanics problems, computational methods for solution of specific non-trivial boundary value problems are essential. Engineers generally accomplish these nonlinear solutions as iterative and incremental combinations of piecewise linearized solutions.

When the material constitutive model is such that strain-localizations are possible, then the implementation of a localization limiter is necessary. Usually, this means that we must make the finite element mesh, at least in the area of the localization, fine enough to characterize the strain distribution within the domain of localization.

Actual materials have a "peak" stress, beyond which the stress decreases to zero with increasing strain. The difficulty lies in the fact that after the peak stress has been reached, both the stress and the strain become ill defined. This is where continuum mechanics fails, and where fracture mechanics becomes necessary.

Solving the problem explicitly in the time domain does not resolve the problem of material instability. The formation of cracks with zero energy dissipation occurs even with dynamical problems.

3.8 Summary

In this chapter, we have taken a whirlwind tour of the engineering discipline of continuum mechanics. We have seen that while the mathematics of continuum mechanics is very elegant, the basic assumption of continuity is often unrealistic when it applies to spatial material behavior.

The main problem is that a well-posed continuum mechanics problem requires *all fields to be continuous*. This is a very stringent requirement. For example, the geometric domain of the reference configuration must have smooth boundaries; the continuum mechanics allows no sharp corners. At a sharp corner, the traction vector will be in general discontinuous, leading to a discontinuous stress field, which a continuum problem does not allow. Point loads, being similarly discontinuous, are likewise disallowed. In addition, sharp re-entrant corners permit infinite stresses and strains to appear, resulting in a non-quantitative problem solution.

Usually, engineers recognize these singularities as being spurious. For example, in a linear elastic finite element analysis of a steel connection with a reentrant corner, the engineer may rationalize that the non-convergent high stresses near the reentrant corner are not of concern because mild steel is "forgiving". This necessity for "engineering judgment" in combination with erroneous computational simulation results can lead to disaster, as for example in the North Ridge earthquake, in which supposedly ductile steel connections unexpectedly failed in a fracture mode.

Additionally, even if all of the fields associated with a continuum problem start out being continuous, it is possible for discontinuous fields to develop. For example, when the rate of change of stress with respect to strain becomes negative, instability occurs, in which infinite strain localizes to infinitesimal material volume, leading again to a singularity, which is not physically realistic.

Engineers have addressed the problem of fracture within the continuum mechanics framework by augmenting the theory with "localization limiters". This heaps complexity upon complexity, resulting in loss of understanding, and thus practicing engineers are rightly reluctant to use these models.

We prefer to go back to basics, with some slightly different assumptions about the material model, and see if we can develop a simpler theory for predicting the mechanics of deformable solids that allows cracks to initiate and grow unhindered by unwarranted assumptions of material continuity.

3.9 Exercises

3.1 Show that, according to Definition 3.2, the set of all polynomials $\boldsymbol{P} = \sum a_n x^n, n = 1,2,3,$ on the interval $0 \leq x \leq 1$, is a vector space P_n over the scalar field of real numbers. Find two different sets of basis vectors that can be used to describe all vectors in P_n.

3.2 Show that the vector space defined in Problem 3.1 is a valid inner product space if the inner product is defined as $\boldsymbol{P} \cdot \boldsymbol{Q} = \int_0^1 (p_n x^n)(q_m x^m) dx$, and the norm is defined as $\|\boldsymbol{P}\| = \boldsymbol{P} \cdot \boldsymbol{P}$.

3.3 Using a Euclidian space with the definition of the dot product being $\boldsymbol{a} \cdot \boldsymbol{b} = \|\boldsymbol{a}\| \|\boldsymbol{b}\| cos(\theta)$, prove that $\boldsymbol{a} \cdot \boldsymbol{b} = a_i b_i$, where a_i and b_i are the Cartesian components of $\boldsymbol{a}$ and $\boldsymbol{b}$.

3.4 Expand the following indicial expressions to as many terms as possible, and then express them as matrices. Indicate the rank of each matrix. The indices all vary from 1 to 3.

a. $A_i x_i$
b. $A_i x_j$
c. $A_{ij} x_j$
d. $A_{ijk} x_k$
e. $A_{ij} x_i x_j$

3.5 Show that the deformation gradient $\boldsymbol{F}$ is a second-order tensor.

3.6 Prove that $\frac{dV}{dV_0} = det\boldsymbol{F}$.

3.7 A body is rotated as a rigid body about the $\hat{\mathbf{e}}_3$ axis by the angle θ.
a) Determine the deformation function $\boldsymbol{x} = \boldsymbol{\varphi}(\boldsymbol{X})$ in terms of θ.
b) Determine the deformation gradient, $\boldsymbol{F}$.
c) Determine the Cauchy-Green deformation tensor $\boldsymbol{C}$.
d) Determine the Lagrangian strain tensor $\boldsymbol{E}$.
e) Determine the small-strain tensor $\boldsymbol{\epsilon}$.
f) Show that $\boldsymbol{\epsilon}$ and $\boldsymbol{E}$ are approximately the same for small θ.

3.8 Show that the Cauchy stress tensor is symmetric if body moment is absent: $\boldsymbol{\sigma} = \boldsymbol{\sigma}^T$.

3.9 In the reference configuration, the mass density of a body is given by the function $\rho_0(\boldsymbol{X})$. Show that the conservation of mass equation, Eq. 3.41, is satisfied for all deformations of the form $x_i = \alpha_i(t)X_i$, (no summation implied) where functions $\alpha_i(t)$ are differentiable functions of time. (Hint: $div\, \boldsymbol{v} = \frac{\frac{\partial}{\partial t}(det\boldsymbol{F})}{det\boldsymbol{F}}$).

3.10 Solve the problem specified in Fig. 3.10 in the time domain, assuming a uniform density $\rho_0 = 1.0\, kg/m^3$, $E = 1.0\, N/m^2$, $L = 1.0\, m$, $A = 1.0\, m^2$, $k = 0.001$, and that the applied tip displacement $\Delta = v_0 t$ is a ramp function, where $v_0 = 2kL/t_{max}\, m/s$ is the rate of end-displacement application, and $t_{max} = 25s$ is the total time of the simulation. Use small-displacement theory, as the strain ϵ is much smaller than unity. Use MatLab to solve the problem numerically using 10, 15 and 25 particles.

3.11 The cantilever beam shown in Fig. 3.11 has four corners at which the stress field is problematic. Explain why the stress field is problematic. Assume small deformation linear elastic theory.

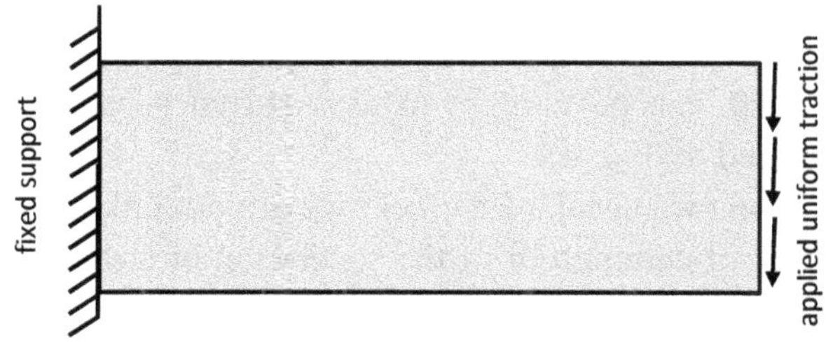

Figure 3.11. Cantilever beam for Problem 3.11.

Chapter 4

Fracture Mechanics

Chapter objectives are to:

- develop a familiarity with the concepts of "crack" and "fracture";
- review the history of the discipline of fracture mechanics;
- provide an overview of linear elastic fracture mechanics (LEFM);
- present the main ideas of nonlinear fracture mechanics;
- introduce computational fracture mechanics;
- describe the limitations of the current state of fracture mechanics.

The entire discipline of fracture mechanics is necessary because of limitations in the scope of applicability of continuum mechanics. Fracture is a mode of deformation not easily represented by continuum mechanics.

What is a "fracture"? English synonyms include "break", "crack", "split", and "rupture". Webster's New World Dictionary defines "crack" as "a break, usually without complete separation of parts; partial fracture; flaw".

These definitions are not very precise. Intuitively, perhaps, one can think of a fracture as a complete separation of a material body into two parts, and a crack can be thought of as a partial separation of a solid body, as illustrated by Fig. 4.1.

However, when we speak of a crack, we are not talking about pulling apart a soft stick of caramel, or of the separation of a drop of liquid into two smaller drops, in which each of the parts are grossly deformed after

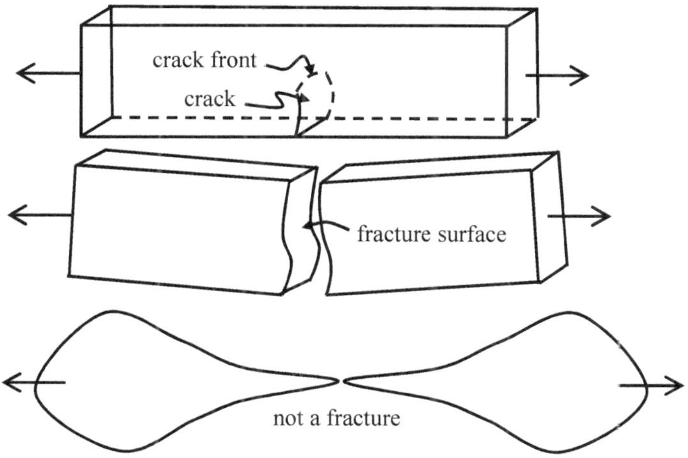

Figure 4.1. Crack, fracture, non-fracture.

the separation, as shown in the bottom case of Fig. 4.1. Thus, for the separation between parts to be considered a fracture, each of the parts must deform in an approximately elastic fashion, so that after the fracture occurs, the parts can be, at least approximately, fitted back together. If two parts separate, but not as a fracture, perhaps the separation could be termed a "rupture". Cracking and fracture are usually associated with brittle structures. A brittle structure is one that breaks "crisply", and whose broken parts fit sufficiently well that they can be glued back together.

From an engineering mechanics point of view, perhaps we can better think of a crack as "highly localized damage that occurs at a small scale, whose details cannot be readily predicted by the principles of continuum mechanics". The locus of the zone of separation (the crack) is approximately a two-dimensional surface (or more precisely, a pair of matching surfaces) within a three-dimensional material body.

These ideas lead to the concept of a *crack front*, which is the locus of the propagating boundary of a crack, as shown in the top sketch in Fig. 4.1. Apparently, something must be happening at the crack front, at some smaller scale, that allows a crack front to propagate, enabling the crack to grow and eventually become a fracture. The volume of material close to the crack front, within which damage takes place, we call the "fracture process zone (FPZ)".

Not all cracks, however, have a localized crack front. As a body is pulled apart, it is possible for a fracture to evolve simultaneously over its entire extent. In this case, the FPZ does not localize within a propagating crack front.

As the reader can discern, the concept of "crack" is vague. History shows that engineers' conception of "crack" has evolved over time. Let us next trace the history of the discipline of fracture mechanics.

4.1 Historical overview of fracture mechanics

The pioneers of elasticity theory, from Hooke and Newton on through the natural philosophers of the mid-1800's, largely ignored the problems of fracture. They were probably simply too involved with developing continuum theories of stress and strain to be concerned with the nasty problems involved with cracking and fracture. Additionally, the natural philosophers of this time were strongly attracted to the recently-developed calculus of Newton and Leibnitz, and this calculus was based upon continuous, smooth, functions. Cracks, involving sharp discontinuities, were not easily described by smooth functions. The natural philosophers possibly considered cracks "ugly", and unworthy of study.

With the industrial revolution and the advent of railroads in the mid-1800's, fracture of iron rails, wheels, and axles became a huge safety and economic problem. Railroad accidents due to fatigue occurred on an almost weekly basis in England and in the United States. Wilhelm Albert, in 1837, published the first paper describing fatigue of metals, followed in 1842 by a paper by William Rankine describing the danger of stress concentrations in railway axles (Rankine, 1842).

In the1860's, William Fairburn and August Wöhler independently developed fatigue data that were represented as S-N curves (cyclic stress amplitude versus number of cycles of applied stress to fracture), which in 1945 were unified for variable amplitude cyclic loadings in what later became known as Miner's Rule. We cannot consider this early work in fatigue failure as true fracture mechanics, because it does not say anything about the rate and direction of crack growth; only something about, for a body with no initial crack, the total number of cycles of applied loading necessary to cause complete fracture.

Only in 1961 was a model for the rate of growth of a crack subject to cyclic loading developed by Paul Paris in what we now call the Paris Law. Understanding of the rate of fatigue crack growth is crucial for modern aircraft, in which "crack control plans" allow fatigue cracks to be routinely identified, monitored, and accounted for in determining the remaining life (or required time to next repair) of an airplane.

The first analytical stress solution to a "crack-like" problem was published early in the twentieth century by Charles Edward Inglis (Inglis, 1913), in which he provided an analytical solution to the Navier-Cauchy equations of elasticity for an infinitely large plane-stress plate subject to uniaxial tension, with an elliptical hole whose major axis is transverse to the direction of applied load. Inglis showed that as the ellipse becomes increasingly narrow and elongated, the stress within the directly adjacent material on the major axis of the ellipse approaches infinity. Due to this singularity in stress, engineers could not use the traditional allowable-stress-based approach for predicting the failure load of a solid containing a sharp crack.

Griffith (Griffith, 1921) was the major pioneer of linear elastic fracture mechanics (LEFM). He explained, based upon experiment and theory, that because the stress at the tip of a sharp crack is theoretically infinite, one must look beyond stress, to energy, to develop a criterion for crack propagation. He developed what we now call the "Griffith energy criterion" for crack propagation.

The Griffith energy criterion was found to be incorrect when a significant amount of plasticity occurs within the fracture process zone (FPZ) at the crack tip (actually, in 3D, the crack front) of a propagating crack. To account for all dissipated energy, Irwin (Irwin, 1957), developed the concept of stress-intensity factor, K_I, and critical stress-intensity factor, also called fracture toughness, K_{IC}. Irwin modified Griffith's idea of LEFM so that it included a zone of plasticity close to the crack tip, and correspondingly assumed that all energy dissipated by this plastic zone, as well as the energy of fracture due to damage, was accounted for by the critical stress intensity factor, K_{Ic}, or equivalently, the critical energy release rate, G_{Ic}.

Westergaard (Westergaard, 1939), using an elegant complex analysis approach, presented a very important solution to the linear elastic boundary value problem of a straight crack in a plane-stress plate of infinite extent under uniaxial loading applied transverse to the crack. Confirming the work of Inglis, he showed that the stress at the crack tip is unbounded. This conclusion of course conflicts with what any real material is able to sustain, and hence there is a paradox. Elasticity theory predicts infinite stress at the crack tip, and yet no real material can sustain infinite stress. Thus, there must be a facture process zone at the crack tip, reducing the stress to a finite magnitude. However, explicitly modeling the FPZ, which itself may be full of discontinuities, is much too complicated, certainly for any closed-form solution. In LEFM, this difficulty is resolved by assuming that the FPZ is of negligible size compared to the size of the region where the leading term in elastic singular solution is dominant. Unfortunately, this situation hardly ever occurs in real structures. Thus, LEFM is a wonderful thought experiment, but except for a few cases such as fracture of glass and fatigue crack propagation in metals, LEFM is rarely applicable to real structures.

Another very important linear elastic crack solution is due to Williams (Williams, 1957). He developed this solution by looking at a set of analytical eigensolutions to the elastic boundary value problem near a re-entrant corner in an infinite, two-dimensional, plane stress domain. As the exterior angle of the re-entrant corner approaches zero, the near-tip stress solution approaches those found by Inglis and by Westergaard.

All of these solutions involve approximations, in which researchers cast out certain terms in the solution for dubious reasons. Indeed, as recently as 1995, well-regarded researchers have questioned the basic assumptions and applicability of LEFM (Hui, C.Y. and Ruina, A., 1995).

With the advent of computers, the boundary integral equation method, and the finite element method, it became possible to compute the stress intensity factors for a wide variety of problems. A "stress analysis of cracks handbook (Tada, Paris, and Irwin, 2000)" was published, containing a large number of stress-intensity factor solutions, almost all obtained computationally.

In the 1960s through the 1980s, it became evident that LEFM is not always applicable to real-world engineering problems, because the FPZ is usually too big to allow the singular term to dominate the solution surrounding the crack tip, especially when applied to mild steels and cementitious materials like concrete. Thus, researchers developed the discipline of nonlinear fracture mechanics over the course of the 1960s through the 1980s, first led by Dugdale and Barrenblatt, and later by Hillerborg.

In the 1980s through the present, the dominant model for fracture of concrete has become the "cohesive crack model", which assumes that concrete acts as a linear elastic continuum, except that there is a cohesive process zone of zero width extending in front of the traction-free portion of the crack. The cohesive crack model characterizes the FPZ using a unique cohesive-traction versus crack-opening-displacement relationship.

An alternative approach is "nonlocal damage mechanics", which allows continuum damage to develop within a continuous material body, subject to the restrictions of a localization limiter, which preserves the continuous nature of the strain and stress fields, so that continuum mechanics remains applicable.

In both linear elastic and nonlinear cohesive discrete fracture mechanics, one must discretely represent the crack as a pair of coincident boundaries within bulk of the solid continuum. Thus, to predict the crack surface locus, crack propagation modeling requires a method to predict the locus of the crack tip (or the crack front, in 3D) as the crack propagates. Researchers have made various assumptions of quasistatic crack propagation to facilitate prediction of crack locus as the crack propagates, but often cracks actually propagate dynamically. Additionally, the prevailing concept of "crack" assumes a pair of non-interacting surfaces behind a crack front that is at least piecewise continuous in character. Often however, FPZs are complex, with distributed microcracks, plasticity, dynamic effects, geometric nonlinearity, and material rate dependency all involved. Thus, the prediction of discrete crack development using linear elastic and nonlinear cohesive crack propagation methods is fraught with questionable assumptions. In essence, the entire discipline of fracture mechanics, while absolutely necessary and natural given the theory of continuum mechanics, would be difficult to defend as being of fundamental importance if a theory of solid mechanics existed that intrinsically permitted discontinuous as well as continuous deformation fields.

As of this writing, Google Scholar turns up 450,000 hits on the keyword "fracture mechanics", and the main reason for the whole discipline is the incompleteness of the discipline of continuum mechanics. The lattice models, which allow both continuum mechanics and fracture mechanics to emerge naturally, have the potential supersede both of these disciplines, especially as increasingly powerful computers are able to solve problems with ever more material particles.

4.2 Linear elastic fracture mechanics

Inglis (Inglis, 1913) found the stress solution to the Navier-Cauchy equations for an elliptical hole with major and minor axes a and b in an infinite plate, transversely loaded with applied stress σ_0 as indicated in Fig. 4.2. Inglis found that the maximum tensile stress σ_{max} located on the major axis, is given by

$$\sigma_{max} = \sigma_0 \left(1 + 2\sqrt{\frac{a}{\rho_c}}\right), \tag{4.1}$$

where $\rho_c = \frac{b^2}{a}$ is the radius of curvature of the ellipse at the end of the major axis. Eq. 4.1 shows that for a circular hole with $\rho = b = a$, the stress concentration factor is $\frac{\sigma_{max}}{\sigma_0} = 3$. However, as the radius of curvature becomes very small compared to the length of the major axis of the ellipse, $\frac{\sigma_{max}}{\sigma_0} \approx 2\sqrt{\frac{a}{\rho_c}}$, and for an infinitely sharp crack, $\frac{\sigma_{max}}{\sigma_0} = \infty$.

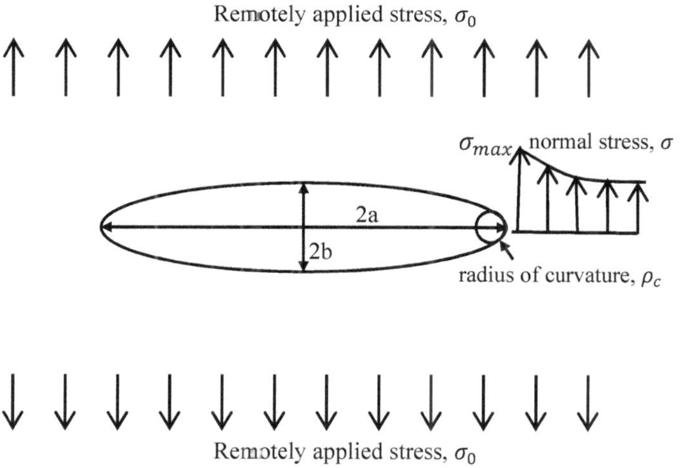

Figure 4.2. Inglis' infinite plate with elliptical hole.

How does the stress field vary near the crack tip? Westergaard (Westergaard, 1939) and Williams (Williams, 1957) found the stresses near the crack tip shown in Fig. 4.3, for plane stress and plane strain problems, regardless of the boundary conditions remote from the crack tip, in terms of r and θ, are given by:

$$\begin{aligned}
\sigma_{rr} &= \frac{1}{\sqrt{2\pi r}}\left[K_I\left(\frac{5}{4}cos\frac{\theta}{2}-\frac{1}{4}cos\frac{3\theta}{2}\right)+K_{II}\left(-\frac{5}{4}sin\frac{\theta}{2}+\frac{3}{4}sin\frac{3\theta}{2}\right)\right]+O(\sqrt{r})\,;\\
\sigma_{\theta\theta} &= \frac{1}{\sqrt{2\pi r}}\left[K_I\left(\frac{3}{4}cos\frac{\theta}{2}+\frac{1}{4}cos\frac{3\theta}{2}\right)-K_{II}\left(\frac{3}{4}sin\frac{\theta}{2}+\frac{3}{4}sin\frac{3\theta}{2}\right)\right]+O(\sqrt{r})\,;\\
\sigma_{r\theta} &= \frac{1}{\sqrt{2\pi r}}\left[K_I\left(\frac{1}{4}sin\frac{\theta}{2}+\frac{1}{4}sin\frac{3\theta}{2}\right)+K_{II}\left(\frac{1}{4}cos\frac{\theta}{2}+\frac{3}{4}cos\frac{3\theta}{2}\right)\right]+O(\sqrt{r})\,.
\end{aligned} \tag{4.2}$$

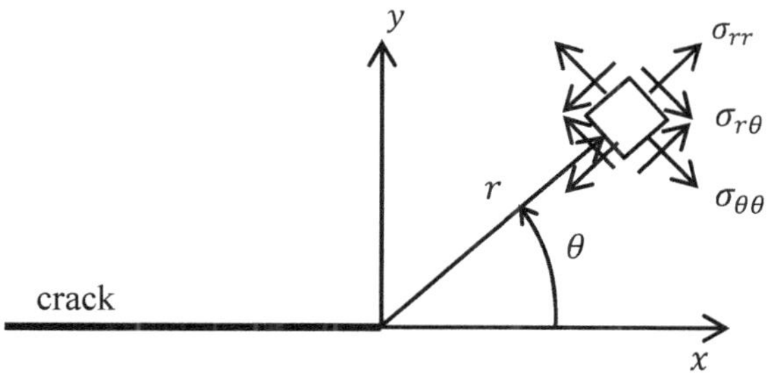

Figure 4.3. Stresses at crack tip in polar coordinates.

In Eq. 4.2, the term $O(\sqrt{r})$ (said in English as "order of $\sqrt{r}$") represents terms that are multiplied by $\sqrt{r} = r^{\frac{1}{2}}$ and all higher-order terms $r^{\frac{3}{2}}$, $r^{\frac{5}{2}}$, etc. As r goes to zero, these higher-order terms all go to zero, *regardless of remotely-applied boundary conditions*, while the first term, $r^{-\frac{1}{2}}$, goes to infinity. Thus, near the crack tip, the first term in the series solution dominates. The two undetermined constants K_I and K_{II}, called the Mode I and Mode II stress-intensity factors, depend upon the geometry and remotely-applied boundary conditions of each particular problem. A Mode III stress intensity factor K_{III} defines the stress field near the crack tip due to the so-called anti-plane shear, or tearing mode:

$$\sigma_{xz} = \frac{K_{III}}{\sqrt{2\pi r}}\left(sin\frac{\theta}{2}\right) + O(\sqrt{r});$$
$$\sigma_{yz} = \frac{K_{III}}{\sqrt{2\pi r}}\left(cos\frac{\theta}{2}\right) + O(\sqrt{r}); \tag{4.3}$$
$$\sigma_{zz}=0 \text{ (plane stress)}; \sigma_{zz}=\nu(\sigma_{xx} + \sigma_{yy}) \text{ (plane strain and 3D)}.$$

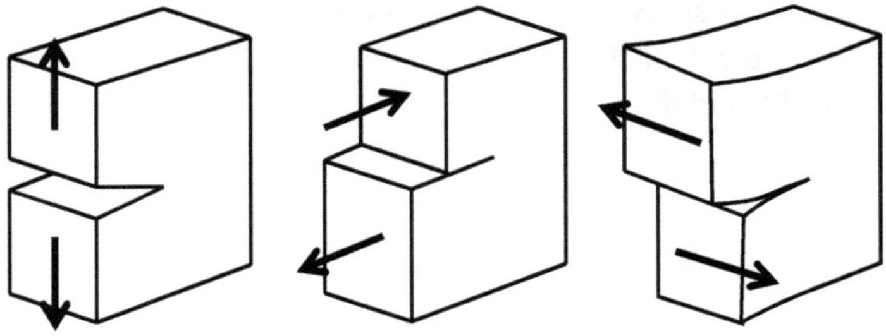

Mode I
(Opening Mode)

Mode II
(Sliding Mode)

Mode III
(Tearing Mode)

Figure 4.4. Modes of deformation near crack tip.

Fig. 4.4 shows the deformations associated with the three modes of stress intensity. The equations for deformation within the crack tip singular elastic zone are simple trigonometric functions, each multiplied by $\sqrt{r}$.

To summarize, with LEFM, it is possible to describe the stress field at points within a region, called the singular elastic zone, located "close" to the crack tip but "far" outside the FPZ using just three stress-intensity factors: K_I, K_{II}, and K_{III}. We give, in rectangular coordinates, the near-tip LEFM stress and displacement fields in Table 4.1.

Table 4.1. Linear elastic fracture mechanics near-crack-tip stress and displacement fields, with reference to Fig. 4.3.

Near-Crack-Tip Stress Components
$\sigma_{xx} = \frac{1}{\sqrt{2\pi r}}\left\{K_I cos\frac{\theta}{2}\left[1 - sin\frac{\theta}{2}sin\frac{3\theta}{2}\right] - K_{II}sin\frac{\theta}{2}\left[2 + cos\frac{\theta}{2}cos\frac{3\theta}{2}\right]\right\}$
$\sigma_{yy} = \frac{1}{\sqrt{2\pi r}}\left\{K_I cos\frac{\theta}{2}\left[1 + sin\frac{\theta}{2}sin\frac{3\theta}{2}\right] + K_{II}sin\frac{\theta}{2}cos\frac{\theta}{2}cos\frac{3\theta}{2}\right\}$
$\tau_{xy} = \frac{1}{\sqrt{2\pi r}}\left\{K_I cos\frac{\theta}{2}sin\frac{\theta}{2}cos\frac{3\theta}{2} + K_{II}cos\frac{\theta}{2}\left[1 - sin\frac{\theta}{2}sin\frac{3\theta}{2}\right]\right\}$
$\sigma_{zz} = 0$ (plane stress); $\sigma_{zz} = \nu(\sigma_{xx} + \sigma_{yy})$ (plane strain and 3D)
$\tau_{yz} = \frac{K_{III}}{\sqrt{2\pi r}}cos\frac{\theta}{2}$
$\tau_{xz} = \frac{K_{III}}{\sqrt{2\pi r}}sin\frac{\theta}{2}$
Near-Crack-Tip Displacement Components
$u = \frac{1}{2\mu}\sqrt{\frac{r}{2\pi}}\left\{K_I cos\frac{\theta}{2}\left[\kappa - 1 + 2sin^2\frac{\theta}{2}\right] + K_{II}sin\frac{\theta}{2}\left[\kappa + 1 + 2cos^2\frac{\theta}{2}\right]\right\}$
$v = \frac{1}{2\mu}\sqrt{\frac{r}{2\pi}}\left\{K_I sin\frac{\theta}{2}\left[\kappa + 1 - 2cos^2\frac{\theta}{2}\right] - K_{II}cos\frac{\theta}{2}\left[\kappa - 1 - 2sin^2\frac{\theta}{2}\right]\right\}$
$w = z\epsilon_{zz} + \frac{2K_{III}}{\mu}\sqrt{\frac{r}{2\pi}}sin\frac{\theta}{2}$
Notes: μ is the shear modulus, $\mu = E/2(1+\nu)$; E is Young's modulus ν is Poisson's ratio; $\kappa = \frac{3-\nu}{1+\nu}$ for plane stress; $\kappa = 3 - 4\nu$ for plane strain.

The LEFM singular solutions are only valid if the three following conditions are satisfied.

1) All applied loads and boundaries of the cracked structure (other than the crack itself) must be remote from the crack tip singular elastic zone. In other words, the crack tip singular solution provided by Eqs. 4.2 and 4.3 is only valid within the region that looks, locally, like an infinite cracked plate statically loaded at infinite distance from the crack tip. In practice, the region within

which the stress singularity dominates the stress solution has perhaps a diameter of one percent of the least characteristic dimension of the cracked body. (The least characteristic dimension is often the length of the crack or the length of the remaining ligament in front of the crack.)

2) The singular elastic stress field predicted by Eqs. 4.2 and 4.3 is invalid very close to the crack tip, where inelastic and nonlinear material processes within the FPZ occur. The size of the FPZ is material-dependent, and can be approximated, for ductile materials having yield strength σ_{yield} and fracture toughness K_{IC} approximately as $r_p = \left(\frac{K_{IC}}{\sigma_{yield}}\right)^2$. ($K_{IC}$, the fracture toughness, is defined in the next section.) In quasi-brittle materials like concrete, the FPZ size is perhaps 100 times the average aggregate size.
3) The conditions must be quasistatic, so that inertial effects are negligible. In addition, the material must be isotropic and homogeneous.

It has been repeatedly stated in the literature that for LEFM to be applicable, the relevant dimensions, such as the length, thickness, and width of the structure, should be larger than $2.5r_p$. In light of the previous discussion, this restriction seems much too weak, and the lower limit on structure size should instead be more like $1000r_p$ if the theory of LEFM is to be reasonably reliable.

Thus, the applicability of LEFM to real problems is quite limited (Hui, C.Y. and Ruina, A., 1995). Nonetheless, LEFM provides a useful "limiting-case thought experiment" that helps in understanding the mechanics of fracture. To develop this comprehension further, in the following two sections we describe how linear elastic cracks propagate under quasistatic conditions.

4.3 Griffith energy criterion

Griffith (Griffith, 1921) brought an energy perspective to LEFM that allows us to predict crack propagation behavior despite the infinite stresses at the tip of a sharp crack predicted by Inglis, Westergaard, and Williams.

Griffith performed mechanical tests on glass rods and fibers, and found that the glass rods failed at average tensile stresses (100 MPa or 15,000 PSI) that were one-thousand times lower than would be expected based upon the theoretical stress (10,000 MPa or 1,500,000 PSI) needed to break the atomic chemical bonds of the silicon oxide molecules of which glass consists.

Griffith found that the strength of the glass rods was highly sensitive to both the diameter of the rod and to the presence of tiny scratches existing upon the surfaces of the rods. He hypothesized that the presence of flaws, or cracks, in the bulk material as well as on the surface of the glass strongly influenced its macroscopic strength.

This observation led him to consider the energy exchange that takes place as a crack propagates. Regardless of the particulars of the strain and stress field, as a crack propagates quasi-statically, the potential energy (which depends upon both the stored strain energy as well as upon the potential energy of the applied loads) decreases, and as the total energy must remain constant, energy becomes available to break bonds at the tip of the crack.[a]

Chemical surface energy is the energy needed to separate atoms, thus creating new surfaces, within a solid material. This energy is, roughly, equal to the integral of the atomic bond force over the distance needed completely to separate the two surfaces on opposite sides of the crack, times the number of bonds per unit cracked area.

Thus, Griffith developed the very important fracture propagation criterion that when the rate of change (with respect to newly-formed crack area) of potential energy G exceeds a critical value G_F, called the critical energy release rate, the crack has the necessary potential energy to propagate:

$$\text{Griffith fracture criterion:} \quad G \geq G_F. \tag{4.4}$$

However, having the necessary energy to propagate the crack is not the same as having sufficient conditions to propagate the crack. A further criterion for crack propagation is that the conditions at the crack tip be

[a] In addition, energy is made available for dynamic motion (sound!).

severe enough to break atomic bonds. Thus, for example, a sharp crack and a crack with a smooth rounded tip in a large body both have approximately the same energy release rates. However, the crack with the rounded crack tip may not have sufficient conditions to propagate if the concentrated stress at its rounded crack tip is below the tensile strength of the material.[b]

Griffith found that for glass, his fracture criterion worked very well. However, when U.S. naval "liberty" ships, made of steel, began cracking in half during the Second World War, the U.S. Navy found that steel is much more resistant to crack growth than would be predicted by the Griffith criterion.

Irwin's contribution was to recognize that surface energy is not the only type of energy that is absorbed as a crack in steel propagates. Indeed, he discovered that the FPZ in steel includes a plastic zone that absorbs much more energy per unit crack propagation (and releases this energy primarily as heat) than would be expected if the only energy absorbed were the surface energy associated with breaking atomic bonds between iron atoms. So Irwin proposed that the critical fracture energy, G_F, must include not only surface energy, but also the energy of plastic work:

$$G_F = G_{surface} + G_{FPZplasticity} \,. \tag{4.5}$$

It is important to understand that the nature of the plastic portion of the FPZ is sensitive to the problem geometry and loading. However, in the case where the FPZ is tiny compared to all other dimensions characterizing the problem, and only in this case, the FPZ is entirely contained by a stress field that is sensitive only to three parameters determined from the applied loading: the Mode I, Mode II, and Mode III stress intensity factors K_I, K_{II}, and K_{III}. In other words, as far as the FPZ is concerned, it sees only the surrounding LEFM crack-tip stress field. If the FPZ happens to be very small compared to other geometric dimensions of a pure Mode I crack problem, then $G_{FPZplasticity}$ is a

[b] Hence, a circular hole drilled at the tip of a slowly growing crack in a glass windshield can arrest further crack growth.

material property, unaffected by the geometry and boundary conditions of the problem. One can calculate the size of the steady-state FPZ in an infinitely large material specimen if one has an elasto-plastic model for the material. Alternately, one can experimentally measure the size of the FPZ if one has a sufficiently large test sample and testing machine.

Consider a symmetrically loaded crack. The elastic stress field in the vicinity of the crack tip (far from problem boundaries, but also far outside the vicinity of the FPZ) has Cauchy stress components $\sigma_{ij}(r,\theta)$, where r is the distance from the crack tip, and θ is the angle with respect to the plane of the crack, as shown in Fig. 4.3. The elastic stress field is uniquely determined by one parameter, called the Mode I stress-intensity factor, K_I:

$$\sigma_{ij(r,\theta)} = \frac{K_I}{\sqrt{2\pi r}} f_{ij}(\theta)\,, \tag{4.6}$$

where f_{ij} are simple trigonometric functions of θ, given, for stress components in radial coordinates, by Eq. 4.2 and in Cartesian coordinates, by Table 4.1.

More generally, the near-tip elastic crack-tip stress field is completely determined by the Mode I, Mode II, and Mode III stress intensity factors K_I, K_{II}, and K_{III} as shown by Eqs. 4.2 and 4.3:

$$\sigma_{ij}(r,\theta) = \frac{1}{\sqrt{2\pi r}}\left[K_I f_{ij}(\theta) + K_{II} g_{ij}(\theta) + K_{III} h_{ij}(\theta)\right], \tag{4.7}$$

where f_{ij} , g_{ij}, and h_{ij} are the simple trigonometric functions of θ given in Table 4.1. Again, this solution is only valid within the domain that is sufficiently far from boundaries of the problem and sufficiently far from the FPZ.

For planar crack growth, there is a direct relationship between the energy release rate G and the stress-intensity factors:

$$G = \frac{{K_I}^2}{E'} + \frac{{K_{II}}^2}{E'} + \frac{{K_{III}}^2}{2\mu}\,, \tag{4.8}$$

where $E' = E$ for plane stress, $E' = \frac{E}{1-\nu^2}$ for plane strain, $\mu = \frac{E}{2(1+\nu)}$ is the shear modulus, E is Young's modulus, and ν is Poisson's ratio. Note that for Mode II crack growth, the crack immediately changes direction as it propagates, and therefore Eq. 4.8 would subsequently not be valid. In addition, for three-dimensional crack problems, in which the three stress-intensity factors may vary along the crack front, Eq. 4.8 is not meaningful.

Irwin called the stress intensity factor at which a plane-strain Mode I crack begins to propagate the *critical Mode I stress intensity factor*, or *fracture toughness* K_{IC}. Note that if a plane-stress problem has an out-of-plane thickness smaller than, or of the same order of magnitude as, the FPZ size, this out-of-plane thickness will affect the characteristics of the FPZ. In this case, the plane strain fracture toughness K_{IC} would be inapplicable, although perhaps a plane-stress fracture toughness K_{IC}' (which is dependent upon plate thickness, and thus not really a material property) could perhaps be used instead.

Thus, we see that there are so many caveats to the theory of LEFM that it is applicable only in very special cases. The theory is more of an academic exercise (and a very helpful teaching exercise) than a general tool for engineering analysis. However, there are specific situations where LEFM is appropriate for problem solving in engineering practice.

4.4 Mixed-mode fracture propagation criteria

If a crack satisfies the conditions of LEFM, then the FPZ at a point on the crack front is entirely surrounded by an elastic zone whose stress state is fully defined by the stress intensity factors, K_I, K_{II}, and K_{III}. We can determine these stress intensity factors by a linear elastic stress analysis, and they are completely insensitive to the nature of the FPZ, which is assumed to be highly localized at the crack tip and of negligible size compared to the size of the elastic zone dominated by the singular terms. The FPZ behavior depends upon the stress intensity factors, but the stress intensity factors do not depend upon the FPZ characteristics, except that if the crack propagates as a consequence of FPZ behavior, thus changing the problem geometry, the stress-intensity factors will change accordingly. Thus, the mesomechanical nonlinear mechanisms active within the FPZ are only weakly coupled with the macro-scale problem, solely through the stress-intensity factors.

Researchers have proposed several criteria to predict crack propagation, most prominently:

1) the maximum circumferential tensile stress criterion;
2) the maximum energy release rate criterion; and
3) the minimum strain energy density criterion.

Erdogan and Sih (Erdogan, F., and Sih, G. C., 1963) proposed the maximum circumferential tensile stress criterion for mixed-mode fracture in planar problems. This criterion assumes that (a) the crack propagates radially from the crack tip in the direction θ_0 (measured with respect to the previously-existing crack, as shown in Fig. 4.3) of the maximum circumferential tensile stress $\sigma_{\theta\theta max}$; and (b) crack propagation occurs when $\sigma_{\theta\theta max}$, at a given small distance r_c from the crack tip, is equal to the maximum circumferential tensile stress that leads to Mode I fracture. Thus, for plane strain mixed-mode problems, if the stress intensity factors K_I and K_{II} are known, it can be shown that the crack propagates in the direction θ_0 which solves the following equation:

$$K_I sin\theta_0 + K_{II}(3cos\theta_0 - 1) = 0\,. \tag{4.9}$$

With the direction of crack propagation θ_0 known, the condition, in terms of stress intensity factors, for the crack to propagate, occur when K_I and K_{II} satisfy the following equation:

$$K_I cos^2\frac{\theta_0}{2} - \frac{3}{2}K_{II} sin\theta_0 \geq \frac{K_{Ic}}{cos\frac{\theta_0}{2}}, \tag{4.10}$$

where K_{IC} is the critical stress intensity factor for Mode I plane strain crack propagation, considered to be a material property. Note that this theory is limited to plane stress and plane strain problems.

The theory of *maximum energy release rate* (Hussain, M. A., Pu, S. I., and Underwood, J., 1974) gives similar results as the maximum circumferential tensile stress theory. A third theory, called the *minimum strain energy density criterion* also was proposed by (Sih, 1974). We do not present these latter two theories in detail here. All three crack propagation "theories" (actually models) mentioned here predict very similar loads and directions for crack propagation.

4.5 Fatigue crack propagation

One of the most practical applications of LEFM is in the prediction of fatigue crack behavior. With fatigue crack propagation, the FPZ is often very small, and therefore LEFM is more frequently applicable than in the case of monotonically applied loading.

When a cracked metal structure is subject to cyclically applied loads, the stresses surrounding the FPZ are strongly concentrated, leading to highly localized plastic deformations of large magnitude. The large-magnitude plastic deformations at the crack tip lead to localized bond breakage due to work hardening and rupture. The repeated loading causes the crack tip grow slowly, but in a predictable manner. The crack grows by a very small increment in each loading cycle, even under very low loadings and at low stress-intensity factors. Paris et al. proposed, in the early 1960's, what we know call the Paris model for fatigue crack propagation:

$$\frac{da}{dN} = C\Delta K_I{}^m, \tag{4.11}$$

where a is the crack length, N is the number of cycles of applied load, ΔK_I is the variation in the Mode I stress-intensity factor per cycle of applied loading, and C and m are material properties. This model works approximately for most metallic materials, although there are many caveats, such as a lower threshold in ΔK_I below which fatigue crack propagation is null. In addition, of course, if $K_I > K_{IC}$ then the crack will propagate regardless of any cyclically applied loading. We refer the reader to (Anderson, 2005) for more detail.

4.6 Nonlinear fracture mechanics

When engineering researchers attempted to apply LEFM to concrete, they found that the observed fracture toughness K_{IC} was strongly dependent upon specimen shape and size for a given type of concrete. In retrospect, this would be expected, as the size of the FPZ in normal concrete is perhaps 0.5 m (1.5') long – much too big for LEFM to be applicable in structural elements of typical size. Even defining the "crack length" is impossible when the FPZ is diffuse, and the size of the FPZ is perhaps even larger than the traction-free portion of the crack. In concrete, the very concept of "discrete crack" is unclear at best.

Researchers have proposed various models for the FPZ in concrete over the years since 1960. If LEFM conditions hold, then no model for the FPZ is required if crack propagation behavior is sought as K_{IC} is sufficient to predict crack direction and the applied load necessary to propagate the crack, at least according to the LEFM crack propagation theories presented in Section 4.4. (In dynamic situations, LEFM is not well developed; we refrain from commenting further on the state of dynamic LEFM, other than to say that we do not understand dynamic crack propagation very well, despite the fact that researchers have published many papers on the topic.)

The Dugdale (Dugdale, 1960), Barenblatt (Barenblatt, 1962), and the Hillerborg fictitious crack (Hillerborg, A., Modeer, M., and Petersson, P. E., 1976) models are all cohesive strip models of the FPZ, in which all nonlinear behavior is considered as closing tractions acting on a plane preceding the traction-free portion of the crack, with the surrounding material being modeled as linearly elastic. Researchers developed all of these models with Mode I crack problems in mind.

In the Dugdale model, the cohesive zone is of an assumed constant length, with a closing traction equal to the uniaxial yield strength of the ductile material.

Barenblatt, recognizing that the closing traction may be decreasing as the crack opens, extended Dugdale's model to develop the "cohesive crack model", although this was not its name at the time it was published, in which the closing traction is assumed to be a function of crack opening displacement, $\sigma = \tilde{\sigma}(COD)$, as shown in Fig. 4.5. At the time when Barenblatt proposed the cohesive crack model, it was completely impractical, as the following passage from his paper indicates:

> "... The following is possible in principle for solving the problem of cracks. The distance between the opposing crack faces at each point on its surface is determined as a function of the unknown distribution of cohesive forces along the surface. Assuming an assigned relationship f(y) [$\sigma(COD)$]

expressing the intensity of cohesive forces as a function of distance, we may find from it an expression determining the distribution of cohesive forces along the surface of the crack.

This approach to the problem of cracks cannot be carried out in practice. First of all, the function f(y) is not known to a sufficient extent for any real material. Even if this function were known, the problem would reduce to a very complex nonlinear integral equation whose effective solution presents great difficulty even in the simplest cases."[c]

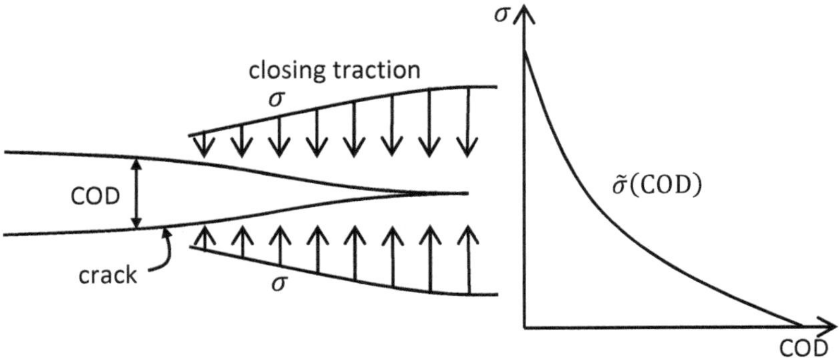

Figure 4.5. Cohesive crack model.

The Hillerborg fictitious crack model assumes the closing traction is a unique function of (fictitious) crack opening displacement, as shown in Fig. 4.5. Hillerborg essentially applied the Barenblatt model to concrete structures, so the model hardly merits a new name. However, Hillerborg showed that: (1) it is indeed possible to obtain a cohesive force law by using sufficiently stiff tensile testing machine; and (2) it is indeed possible (approximately) to solve a cohesive crack problem by using a nonlinear finite element method implemented on a computer. The Hillerborg fictitious crack model has been in recent years renamed the "cohesive crack model".

Thus, we see that although in 1962 Barrenblatt considered his own cohesive crack model to be entirely impractical, by 1976 Hillerborg and his students were able to solve the problem numerically using a very stiff

---

[c] The expression $[\sigma(COD)]$ has been added for clarity.

tensile test machine to obtain the cohesive crack relation, and a computer to solve the resulting nonlinear problems.

None of these detailed models of the FPZ is necessary unless the FPZ is of significant size compared to the size of the structure. If the FPZ is of significant size compared to the structure, so that LEFM does not apply, then the model is termed a "nonlinear fracture mechanics" model.

There are many types of nonlinear fracture mechanics models. The simplest is perhaps the cohesive crack model, in which we consider the domain as linear elastic, and the only nonlinearity comes from the cohesive crack model depicted in Fig. 4.5. It is also possible, although difficult, to marry the cohesive crack model with any number of nonlinear and inelastic continuum models for the surrounding domain.

Finally, the researchers have modeled the FPZ as part of the material continuum with "nonlocal continuum damage mechanics". Bazant (Bazant, Z. P. and Jirasek, M., 2002) and others have found that strain-softening continuum damage mechanics yields spurious results unless a "localization limiter" of some sort accompanies it. Unless the continuum damage model provides a localization limiter, damage immediately localizes to a band of zero width, resulting in infinite strain across the crack band, and any theory that includes infinity as a solution is not a quantitative theory.

Almost none of these nonlinear fracture mechanics models are possible to solve analytically, and thus computational models are required to provide approximate solutions. The recent availability of powerful computers indicates that the traditional linearized and analytical methods of modeling of deformation and fracture of solids should be updated.

4.7 Computational fracture mechanics

A state-of-the-art report entitled "Computational Analysis of Fracture in Concrete Structures" traces the history of computational fracture mechanics, at least as it relates to the analysis of concrete structures (ACI Committee 446, 2009).

Roughly speaking, the approaches to computational fracture mechanics include three methods:

(1) discrete crack models;

(2) smeared crack (or continuum damage mechanics) models; and

(3) lattice and particle models.

With discrete crack models, the model represents the crack as a geometric feature, as shown in Fig. 4.6. The difficulty with the discrete crack modeling approach is that it is a very challenging problem to represent a propagating crack using a CAD geometric modeler. Even if one could conquer the problem of representing a propagating crack within a CAD model, the problem remains of automatically creating a new finite element mesh for computational stress analysis at each crack propagation increment. With each successive crack increment, the program must transfer inelastic responses from the previous mesh to the new mesh. Thus, most discrete crack propagation models do not attempt to update the CAD geometry, but rather update only a finite element representation of the geometric domain. Notable finite element programs that implement fracture propagation algorithms are FRANC2D and FRANC3D (Bittencourt, T.N., Wawrzynek, P.A., Ingraffea, A.R. and Sousa, J.L., 1996).

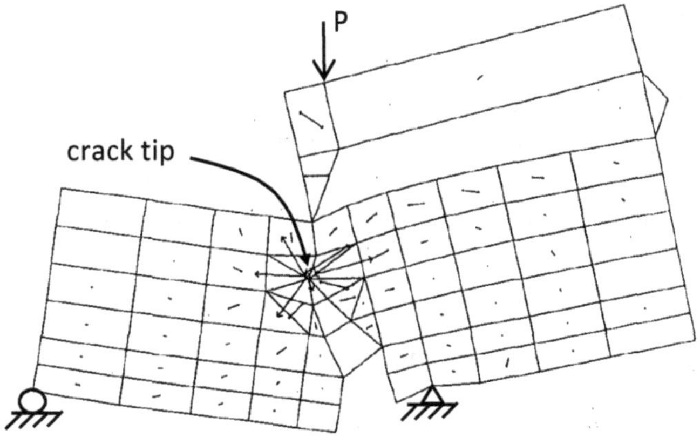

Figure 4.6. Discrete crack model of a cracked beam (Xie, M. and Gerstle, W., 1995).

The basic approach to modeling of LEFM fracture propagation using finite elements is to:

- Perform a linear elastic stress analysis of a problem with one or more cracks.
- Compute the mixed-mode stress-intensity factors at each point along each crack front.
- Use the fracture propagation criteria such as those given in Section 4.4 to predict crack propagation direction and distance.
- Modify the mesh to represent the propagated cracks.
- Repeat.

It is possible to compute the mixed-mode stress intensity factors in a number of different ways, but the most accurate methods use energy approaches.

Others have used the boundary element method to represent cracked domains. Boundary element methods ease the meshing problem, but are essentially limited to linear elastic fracture problems (Gerstle, W., Ingraffea, A.R., and Perucchio, R., 1988).

Ingraffea and his coworkers have developed methods for automatically representing propagating discrete cracks in 2D and 3D finite element meshes. However, these methods have limited capability: for example, multiple propagating and interacting discrete cracks are difficult to model. Additionally, mesh convergence studies to assure objective results are problematic.

Although discrete crack propagation methods are certainly preferable in situations where LEFM is applicable, such as fatigue crack propagation in metals like aluminum, there are many engineering applications where the conditions for LEFM is simply do not exist.

In recent years, researchers have renamed the *smeared crack model* the *continuum damage mechanics model* (Jirásek, 1998). This type of model does not represent the crack explicitly; rather, the material stiffness is decreased as the material is damaged due to excessive stress or strain. A strain-softening material model is one in which stress decreases with increasing strain, as shown in Fig. 4.7. Strain-softening materials are unstable, as described in Section 3.7, and dynamically the strain distribution localizes to a band of zero width, resulting in indeterminate fracture energy. In a quasistatic process, as the finite element mesh is refined, the energy dissipated by the damage decreases without limit. Thus, researchers have found the need to regularize the model.

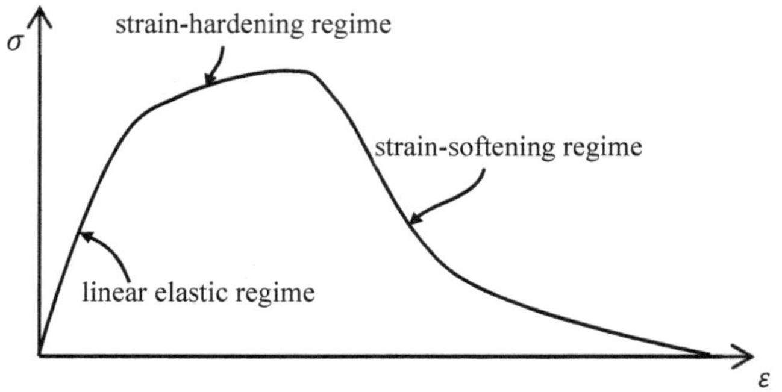

Figure 4.7. Stress-strain relationship including strain-softening regime.

To regularize continuum damage models, researchers have introduced a "localization limiter", producing what is known as the discipline of "nonlocal continuum damage mechanics" (Jirásek, 1998). With a localization limiter, one usually assumes that the stress at a material point is a function not of the strain at that point, but rather of the average strain over a finite neighborhood of the point, producing a nonlocal stress-strain model. To determine this average strain over a finite neighborhood, at least in the neighborhood of a damaged location, requires very fine finite element meshes, and thus convergence studies can be problematic. Smeared crack models have been widely used to predict the nonlinear behavior of concrete structures, as shown in Fig. 4.8.

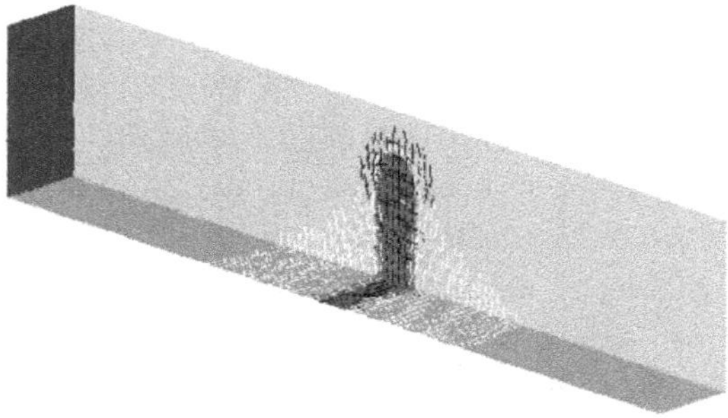

Figure 4.8. Smeared crack model of a cracked beam (Jirásek, 1998).

Lattice and particle models are the last category of model that researchers have used computationally to simulate fracture – mostly in cementitious materials.

Particle models have typically assumed particles of finite size, and have assumed particle interactions of a local nature, in the form of nonlinear contact forces between rigid bodies of finite size.

Physicists have developed particle models to simulate the behavior of crystals at an atomic level, and most of these models have been linear elastic.

Lattice models, on the other hand, have mostly assumed that the lattice objects are linear elastic truss or frame elements connected at nodes. When the force becomes sufficiently high, the model automatically removes the truss or frame element from the mesh.

Most of the particle and lattice models to date have been used to model materials at the meso-mechanical level, rather than at the macroscale.

4.8 Conclusions regarding fracture mechanics

In this chapter, we have reviewed the history of fracture mechanics and the main ideas involved. Early attempts to model cracks were linear elastic discrete crack models – and researchers mostly assumed the cracks were planar.

Recently, the limitations of linear elastic fracture mechanics (LEFM) have become apparent, and this has led to the development of nonlinear fracture models. Nonlinear models are not amenable to analytical solution, and researchers usually solve them in an approximate way using computational techniques. These computational techniques (for example using the finite element method) usually require automatic meshing methods and associated convergence studies.

Computational fracture mechanics has historically required the solution of continuum mechanics problems, augmented by fracture mechanics theories.

The particle methods have been ad-hoc, primarily aimed at understanding mesomechanical behavior, and more complex than required if we seek to simulate macro-scale behavior.

The peridynamic model proposed by Silling in 2000 showed that a more fundamental and rigorous approach to fracture mechanics, and indeed solid mechanics in general, is possible. We present Silling's original bond-based peridynamic model in the next chapter.

4.9 Exercises

(Note: the following exercises are challenging, and may require further study of the fracture mechanics literature and perhaps some help from your professor. This is simply the nature of fracture mechanics: nothing comes easily. The problems perhaps best serve as examples of the type of work that we can avoid by using alternate computational approaches like the SPLM.)

4.1 Refer to the paper (Westergaard, 1939) for this problem. Using the Westergaard semi-inverse complex analysis approach, determine the near-tip stress field for an infinite plate with a straight crack of length $2a$, loaded in pure in-plane shear as shown in Fig. 4.9. Assume an Airy stress function $\Phi = -yRe\left(\overline{Z(z)}\right)$, satisfying the biharmonic equation $\nabla^4\Phi = 0$, with $Z(z) = \frac{\tau_0}{\sqrt{1-\frac{a^2}{z^2}}}$.

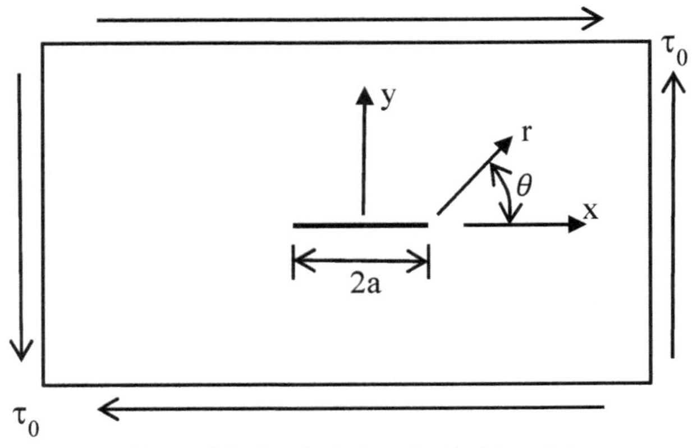

Figure 4.9. Cracked plate for Problem 4.1.

a. Derive the expressions

$$\sigma_x = 2Im(Z) + yRe(Z')$$
$$\sigma_y = -yRe(Z')$$
$$\tau_{xy} = Re(Z) - yIm(Z')$$

b. Show that all of the required boundary conditions, both on the crack surface and far from the crack, are satisfied by this choice of Airy stress function.

c. Defining the Mode II stress-intensity factor as $K_{II} \equiv \tau_0\sqrt{\pi a}$, develop the near-tip stress components:

$$\sigma_x = \frac{-K_{II}}{\sqrt{2\pi r}} sin\left(\frac{\theta}{2}\right)\left(2 + cos\left(\frac{\theta}{2}\right) cos\left(\frac{3\theta}{2}\right)\right)$$
$$\sigma_y = \frac{K_{II}}{\sqrt{2\pi r}} sin\left(\frac{\theta}{2}\right) cos\left(\frac{\theta}{2}\right) cos\left(\frac{3\theta}{2}\right)$$
$$\tau_{xy} = \frac{K_{II}}{\sqrt{2\pi r}} cos\left(\frac{\theta}{2}\right)\left(1 - sin\left(\frac{\theta}{2}\right) sin\left(\frac{3\theta}{2}\right)\right)$$

d. Derive the Mode II radial and tangential stress components σ_r, σ_θ and $\tau_{r\theta}$ in terms of r and θ.

4.2 Refer to the papers: Williams, M.L., "Stress Singularities Resulting From Various Boundary Conditions in Angular Corners of Plates in Extension", J. Appl. Mech., Vol. 19, 1952., and Williams, M.L., "On the Stress Distribution at the Base of a Stationary Crack", J. Appl. Mech., Vol. 24, 1957.

Following the Williams approach, derive, at the level of detail that satisfies you, the singular stress field, in radial coordinates, at the tip of a perfectly sharp crack in a plane stress linear elastic solid, as shown below.

For Mode I:

$$\sigma_{rr} = \frac{K_I}{\sqrt{2\pi r}}\left(\frac{5}{4}cos\frac{\theta}{2} - \frac{1}{4}cos\frac{3\theta}{2}\right) = \frac{K_I}{\sqrt{2\pi r}}cos\frac{\theta}{2}\left(1 + sin^2\frac{\theta}{2}\right)$$
$$\sigma_{\theta\theta} = \frac{K_I}{\sqrt{2\pi r}}\left(\frac{3}{4}cos\frac{\theta}{2} + \frac{1}{4}cos\frac{3\theta}{2}\right) = \frac{K_I}{\sqrt{2\pi r}}cos\frac{\theta}{2}\left(1 - sin^2\frac{\theta}{2}\right)$$
$$\tau_{r\theta} = \frac{K_I}{\sqrt{2\pi r}}\left(\frac{1}{4}sin\frac{\theta}{2} + \frac{1}{4}sin\frac{3\theta}{2}\right) = \frac{K_I}{\sqrt{2\pi r}}sin\frac{\theta}{2}cos^2\frac{\theta}{2}$$

For Mode II:

$$\sigma_{rr} = \frac{K_{II}}{\sqrt{2\pi r}}\left(\frac{-5}{4} sin\frac{\theta}{2} + \frac{3}{4} sin\frac{3\theta}{2}\right)$$

$$\sigma_{\theta\theta} = \frac{K_{II}}{\sqrt{2\pi r}}\left(\frac{-3}{4} sin\frac{\theta}{2} - \frac{3}{4} sin\frac{3\theta}{2}\right)$$

$$\tau_{r\theta} = \frac{K_{II}}{\sqrt{2\pi r}}\left(\frac{1}{4} cos\frac{\theta}{2} + \frac{3}{4} cos\frac{3\theta}{2}\right)$$

Using MatLab, in (xy) space, make the following plots of the stress components at the tip of a crack:

For Mode I: $\frac{\sigma_{rr}}{K_I}, \frac{\sigma_{\theta\theta}}{K_I}, \frac{\tau_{r\theta}}{K_I}$, and

For Mode II: $\frac{\sigma_{rr}}{K_{II}}, \frac{\sigma_{\theta\theta}}{K_{II}}, \frac{\tau_{r\theta}}{K_{II}}$.

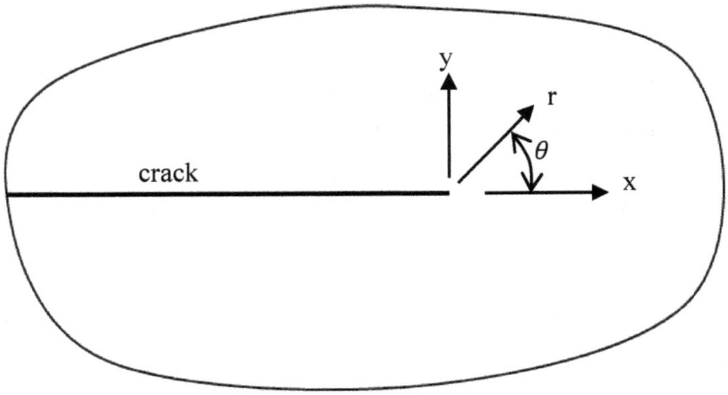

Figure 4.10. Cracked plate for Problem 4.2.

4.3 Read the paper: Griffith, A. A., "The Phenomena of Rupture and Flow in Solids", Phil. Trans. Royal Soc., Series A., Vol. 221, 1920, pp. 163-198. In this problem, derive the equation

$G = G_I + G_{II} + G_{III} = \frac{K_I^2}{E'} + \frac{K_{II}^2}{E'} + \frac{K_{III}^2}{2\mu}$ where

$\mu = \frac{E}{2(1+v)}$ is the shear modulus,

$E' = E$ for plane stress problems,

$E' = \frac{E}{1-\nu^2}$ for plane strain problems,

E = Young's modulus, and

ν = Poisson's ratio.

a) Using the near-tip stress solutions, write an expression for the non-zero stress components directly in front of the crack ($\theta = 0$).
b) Using the near-tip stress solutions, write an expression for the non-zero displacement components directly behind the crack tip ($\theta = \pm\pi$).
c) Explain why the potential energy of a linear elastic body is equal to one-half the potential of the applied loads.
d) Explain why as the crack extends by a small increment, Δa, the potential energy due to loads that are remote from the crack tip does not change.
e) Develop an integral expression for the change in potential energy, $\Delta\Pi$, due to a small change in crack length, Δa, in terms of all three stress intensity factors, the shear modulus μ, the factor $\kappa = 3 - 4\nu$ (plane strain) $\kappa = {(3-\nu)}/{(1+\nu)}$ (plane stress), and Δa. Use crack tip displacement fields for Mode I and Mode II from Table 4.1.
f) Evaluate the integral in part (e).
g) Finally, do the algebra to show that $G = \frac{K_I^2}{E'} + \frac{K_{II}^2}{E'} + \frac{K_{III}^2}{2\mu}$.

4.4 Under what conditions are the assumptions of linear elastic fracture mechanics (LEFM) applicable? In this problem we investigate this question using a case study. Assume that the hull of a ship is constructed with plates of 1-inch-thick American Bureau of Shipping (ABS) Grade A steel, so that one side of the hull can be considered to be a flat plate 100 ft × 100 ft × 1 inch in size.

ABS Grade A steel has the following mechanical properties at room temperature: yield strength, F_y = 34 KSI; ultimate strength, F_u = 58 KSI; fracture toughness, K_{IC} = 114 KSI-in$^{1/2}$. We consider a straight crack of length 2a = 6 ft in the center of one side of the hull.

a) Assuming LEFM conditions, what is the level of far-field stress, $\sigma_0 = \sigma_{xx0} = \sigma_{yy0}$, in tension, with the principal directions, x and y, assumed parallel and perpendicular to the crack plane, respectively, that will cause the crack to propagate?

b) Using MatLab, make a contour plot of the elastic stress field surrounding the crack, assuming that the far-field stress is $\sigma_0 = \sigma_{xx0} = \sigma_{yy0}$ from Problem 1. (Plot the Westergaard solution, $\sigma_{VMWestergaard}$, for Von Mises stress for a crack in an infinite plate, including all terms. You can make use of bilateral symmetry: thus, the plot should have extents $0 < x < 10ft$; $0 < y < 10ft$, with the coordinate origin located at the center of the crack. Include 100 contour lines and a color bar in your plot. Also, provide a copy of your MatLab code.)

c) Make a contour plot of the elastic stress field surrounding the crack, again assuming that the far-field stress is $\sigma_{xx0} = \sigma_{yy0} = \sigma_0$ from Problem 1. This time, include only the singular term of the Von Mises stress in your solution, $\sigma_{VMSingular}$, again assuming the crack is in a plate of infinite size. (Use the same contour plot parameters as in Problem 2.) Provide a copy of your MatLab code.

d) Make a contour plot of the error in the Von Mises stress in assuming that the stress near the crack tip is given by the singular term, $\sigma_{VMSingular}$, rather than the full Westergaard solution, $\sigma_{VMWestergaard}$. Define the error as $e = \frac{\sigma_{VMSingular} - \sigma_{VMWestergaard}}{\sigma_{VMWestergaard}}$.

e) Within what radius surrounding the crack tip is it acceptable to use $\sigma_{VMSingular}$, rather than the complete elastic solution, $\sigma_{VMWestergaard}$, if 10% error in the computed elastic Von Mises stress is acceptable?

f) Make a contour plot of a first approximation of the boundary between elastic and plastic behavior, assuming $\sigma_{VMWestergaard} = F_Y$. Provide a copy of your MatLab code.

g) Is the LEFM prediction of strength expected to be accurate for the current situation of a 6 ft crack in the side of a ship? Discuss.

h) If the applied stress, σ_0, is sufficient to cause the crack to propagate, will the propagation be stable or unstable, if σ_0 is held constant?

4.5 A vertical-axis cylindrical water tower is 120 ft high and 50 ft in diameter. It is constructed of 0.5-inch-thick 4340 high-strength steel (F_y = 214 KSI, plane strain K_{Ic} = 90 K-in$^{-3/2}$). Water weighs 62.4 lb/ft^3.

(a) Is it valid to use the plane strain (rather than plane stress) fracture toughness?
(b) How large a vertical crack, a_{cr}, at the base of the tower can be tolerated if crack growth is to be prevented?
(c) Once the crack reaches this critical size, a_{cr}, will failure be slow or will it be catastrophic?
(d) Does LEFM apply to this problem?

4.6 The principal stress trajectories in Fig. 4.11(a) show that a crack in an infinite plate causes stress relief in the shaded triangular regions next to the crack. As an approximation, assume the stress relief region to be limited by lines of constant slope, k, as shown in Fig. 4.11(b), and assume the stresses inside the stress relief region are zero while remaining unchanged outside.

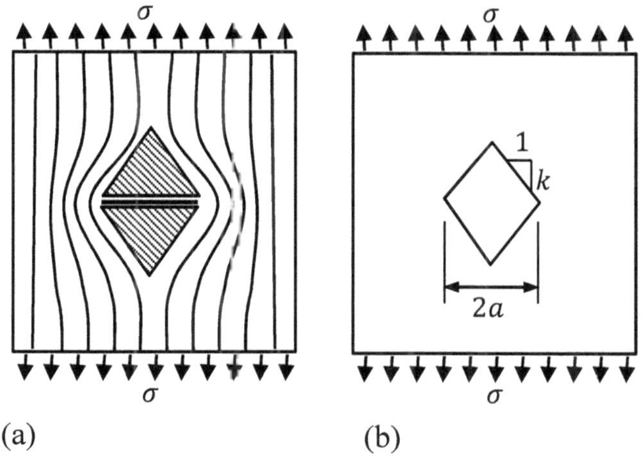

Figure 4.11. Infinite plate with crack for Problem 4.6.

(a) Use the Griffith criterion to evaluate G_I and K_I as functions of σ, E, k, and a.

(b) Show that the energy release rate depends upon whether the specimen is in plane stress or plane strain, while the stress-intensity factor is the same for both cases.
(c) What value of k gives the exact solution to this problem?

4.7 Assume simple Euler-Bernoulli bending theory and plane stress conditions for the problem shown in the Fig. 4.12.

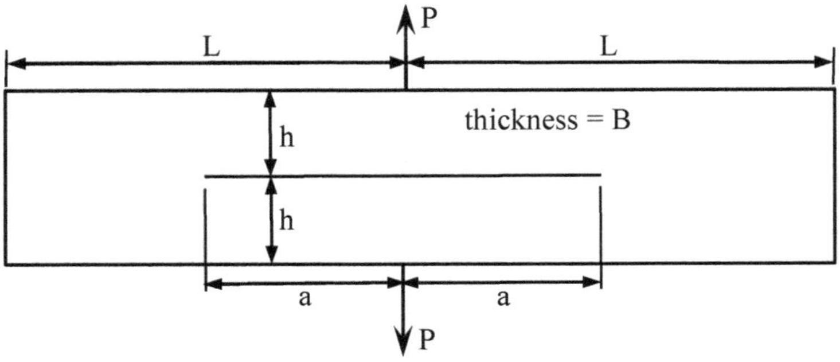

Figure 4.12. Beam with longitudinal central crack for Problem 4.7.

(a) Derive the formula that relates G and the rate of change of compliance. (Following Anderson, $C \equiv \frac{\Delta}{P}$, where Δ is the crack opening displacement at the location where P is applied.)
(b) Use the compliance derivative approach to calculate G_I and K_I.
(c) Choosing an appropriate path, Γ, use the J-integral approach to calculate G_I and K_I.
(d) If the structure has an initial crack length of a_0, then what is the maximum load P which the structure will sustain, if the fracture toughness is K_{IC}? Will crack growth be stable?
(e) If the structure is loaded with a wedge, which causes an opening displacement of Δ, what is the maximum value of Δ which can be sustained, if the fracture toughness is K_{IC}? Will crack growth be stable?

4.8 An offshore oilrig is subjected to a complete cycle of wave loading twice per minute. One of its components is a sheet of ¼" thick steel plate. During each cycle of wave loading, the plate is subject to a maximum average uniaxial tensile stress of 10 KSI and a minimum average uniaxial tensile stress of 2 KSI. In the first inspection, after 278 days of service, a pair of cracks, each of length 0.10 inch is discovered emanating in opposite directions (co-linearly) from a pre-existing ¼ inch – long slot. After another 70 days of service, a second inspection reveals that each of the cracks has grown to a length of 0.2 inch. Finally, after another 70 days of service, a third inspection shows that each of the cracks has grown to a length of 0.40 inch. The plate is known to have a plane strain fracture toughness of 50 ksi-in$^{1/2}$ and yield strength of 100 KSI.

a. Determine the material parameters, C and m, in the Paris Law from the available data.
b. Predict the number of days (after the third inspection) to catastrophic failure of the plate.
c. How long will the cracks be just prior to catastrophic failure?

Chapter 5

Bond-Based Continuum Peridynamics

Chapter objectives are to:

- introduce the bond-based continuum peridynamic model;
- present the kinematics of the model;
- discuss the kinetics of the model;
- investigate restrictions on the pairwise force function;
- explore peridynamic material isotropy;
- learn about elastic bond-based peridynamic models;
- investigate the relationship to Navier-Cauchy theory;
- consider the application of boundary conditions;
- explore modeling of fracture using the bond-based model;
- show how the bond-based peridynamic theory is simultaneously less general and more general than conventional theory;
- present a micropolar bond-based peridynamic theory;
- show that even micropolar bond-based peridynamic theory is inadequate to represent some features of real materials.

Silling (Silling, 2000) wrote the seminal journal paper describing the first theory of peridynamics. In a generalization of Navier's approach, this paper assumes that the force between any two particles in a material body is a function only of the states of the two particles – and the inter-particle force is insensitive to the states of other nearby particles. This original theory of peridynamics was later named the "bond-based peridynamic theory", to distinguish it from the more general "state-based peridynamic theory" published in (Silling, S.A., Epton, M., Weckner, O., Xu, J., and Askari, E., 2007). The bond-based theory seems reasonable, but was found by Silling to be insufficiently general to capture some known behaviors of real materials. We present the bond-based theory in this chapter as a way to motivate an understanding of the more general state-based theory of peridynamics, which we defer to Chapter 8.

Silling, retaining some of the essential aspects of continuum mechanics (such as mass density) has presented both of his peridynamic models as revised versions of continuum mechanics. This is a bit different from the approach taken in this book, which is to dispense entirely with assumptions of spatial material continuity, while still holding on to the concept of "material body".

In this chapter on bond-based peridynamics, we adhere closely to the continuum approach taken by (Silling, 2000). When we present the state-based peridynamic model in Chapter 8, we abandon the continuum approach in favor of a discrete lattice-based approach.

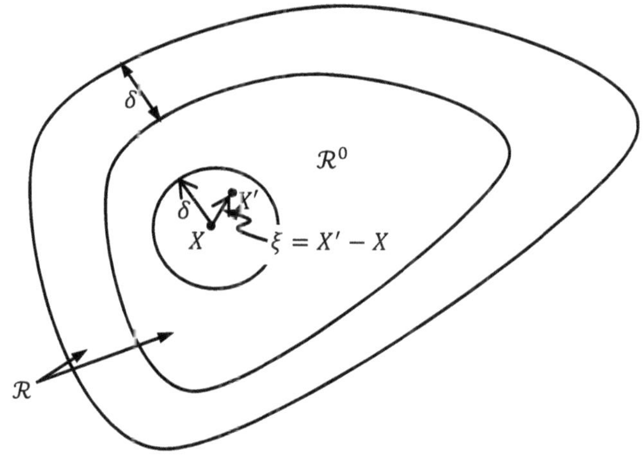

Figure 5.1. Schematic of regions $\mathcal{R}$ and $\mathcal{R}^0$ of a body, reference points $\boldsymbol{X}$ and $\boldsymbol{X}'$, and horizon δ in the reference configuration.

5.1 Introduction to the bond-based peridynamic theory

A solid body is assumed to occupy an initial undeformed reference configuration in an N_R-dimensional spatial region $\mathcal{R} \in \mathbb{R}^{N_R}$, as shown in Fig. 5.1. The dimensionality N_R of the physical space is not explicitly defined in (Silling, 2000), but a three-dimensional real Euclidian space, $\mathbb{R}^3$, is implied.

Truesdell (Truesdell, 1960) points out that a material point P should be simply thought of as a label for a particular material particle, rather than as a physical location in space. The vector $\boldsymbol{X}(P) \in \mathcal{R}^0 \subset \mathbb{R}^3$ gives the initial undeformed location in $\mathbb{R}^3$ of point P.[a] Later in this book, for convenience in computation, we will consider the material configuration the be the set of material points $\mathcal{P}$, as distinct from their undeformed configuration $\mathcal{R}$, to be the primitive elements of the set of material points in a lattice body, $\mathcal{L}$, residing within an N_D-dimensional Cartesian integer space, $\mathbb{Z}^{N_D}$, which would be stated mathematically as $\mathcal{P} = \{P \in \mathcal{L} \subset \mathbb{Z}^{N_D}\}$.[b] Essentially, we will discretely number the material particles with integers as a means of uniquely identifying them. However, for now, we stick with Silling's (and Truesdell's) definition of material point as residing within a real Cartesian space; in other words, within a continuum.

5.2 Kinematics of the bond-based peridynamic theory

Each particle P at initial reference, undeformed, location $\boldsymbol{X}$ deforms to a deformed location $\boldsymbol{x}$ due to applied loads, as shown in Fig. 5.2. The *displacement* of particle P is the vector $\boldsymbol{u} = \boldsymbol{x} - \boldsymbol{X}$. Similarly, a particle P', with reference location $\boldsymbol{X}'$, deforms to the deformed location $\boldsymbol{x}'$. The vector from $\boldsymbol{X}$ to $\boldsymbol{X}'$ is called a "reference bond" $\boldsymbol{\xi}$. The displacement of particle P' is the vector $\boldsymbol{u}' = \boldsymbol{x}' - \boldsymbol{X}'$. The vector labelled $\boldsymbol{\xi} + \boldsymbol{\eta}$ is the deformed configuration of reference bond $\boldsymbol{\xi}$. The vector $\boldsymbol{\xi} + \boldsymbol{\eta}$ is called the image of the reference bond $\boldsymbol{\xi}$ under deformation, or simply the "deformed bond".[c]

---

[a] $\boldsymbol{X}(P) \in \mathcal{R}^0 \subset \mathbb{R}^3$ means: the position vector $\boldsymbol{X}$, a function of material particle P, is an element of the set of spatial points $\mathcal{R}^0$, which is a subset of $\mathbb{R}^3$.

[b] "{ }" means "set of", "∈" means "element of", and "⊂" means "subset of".

[c] ξ is the Greek letter for x, pronounced "xi". η is the Greek letter for h, pronounced "eta". δ is the Greek letter for d, pronounced "delta".

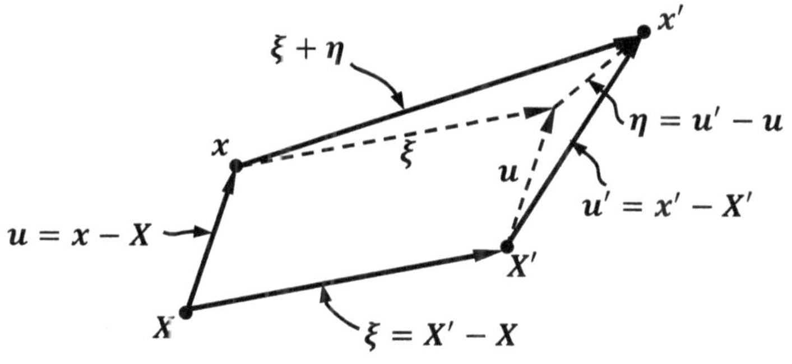

Figure 5.2. Kinematics of two nearby particles, $\boldsymbol{X}$ and $\boldsymbol{X}'$, within a material body.

In bond-based peridynamics, the assumption is that the force acting between two particles is dependent only upon the reference $\boldsymbol{\xi}$ and deformed $\boldsymbol{\xi} + \boldsymbol{\eta}$ bond configurations. There is no need for the concepts of "stress" and "strain", because the force acting between each pair of particles, associated with each bond, and its deformed configuration, is considered explicitly, as described in the next section.

5.3 Kinetics of the bond-based peridynamic theory

In the bond-based theory, each pair of infinitesimal particles, with reference positions $\boldsymbol{X}$ and $\boldsymbol{X}'$, as shown in Fig. 5.2, interact though a vector-valued function $\boldsymbol{f}$, which is assumed to be a function of the undeformed bond $\boldsymbol{\xi} \equiv \boldsymbol{X}' - \boldsymbol{X}$, and the deformed bond $\boldsymbol{\xi} + \boldsymbol{\eta} \equiv (\boldsymbol{X}' + \boldsymbol{u}') - (\boldsymbol{X} + \boldsymbol{u}) = (\boldsymbol{X}' - \boldsymbol{X}) + (\boldsymbol{u}' - \boldsymbol{u})$, of the two reference particles. Thus, we define the pairwise force function $\boldsymbol{f} = \boldsymbol{f}(\boldsymbol{\eta}, \boldsymbol{\xi})$.[d]

[d] When we say $\boldsymbol{f} = \boldsymbol{f}(\boldsymbol{\eta}, \boldsymbol{\xi})$, we mean that $\boldsymbol{f}$ is a function of $\boldsymbol{\eta}$ and $\boldsymbol{\xi}$, and *nothing else*. Of course, $\boldsymbol{f}$ also depends upon material type, temperature, etc., but these other arguments of $\boldsymbol{f}$ are assumed fixed, and so are not explicitly declared.

On physical grounds, two particles $\boldsymbol{X}'$ and $\boldsymbol{X}$ in a material body do not interact unless they are sufficiently close together in the undeformed configuration. Therefore, we make the assumption that

$$\boldsymbol{f}(\boldsymbol{\eta},\boldsymbol{\xi}) = \boldsymbol{0}\,,\ \forall\ \boldsymbol{\eta},\ \forall\ \|\boldsymbol{\xi}\| > \delta \tag{5.1}$$

where δ is some finite, pre-specified distance.[e] We call δ the *material horizon radius*.

Thus, at any time t, the force per unit reference volume $\boldsymbol{L_u}$ acting upon reference particle P originally located at $\boldsymbol{X}$ is the integral (infinite vector sum) of all pairwise infinitesimal forces acting on particle $\boldsymbol{X}$:

$$\boldsymbol{L_u}(\boldsymbol{X},t) = \int_{\mathcal{R}} \boldsymbol{f}(\boldsymbol{u}(\boldsymbol{X}',t) - \boldsymbol{u}(\boldsymbol{X},t), \boldsymbol{X}' - \boldsymbol{X})\, dV_{\boldsymbol{X}'}, \quad \forall \boldsymbol{X} \in \mathcal{R},\ \ t \geq 0. \tag{5.2}$$

Defining $\boldsymbol{u}' \equiv \boldsymbol{u}(\boldsymbol{X}',t)$, $\boldsymbol{u} \equiv \boldsymbol{u}(\boldsymbol{X},t)$, and $dV' = dV_{\boldsymbol{X}'}$,

$$\boldsymbol{L_u}(\boldsymbol{X}) = \int_{\mathcal{R}} \boldsymbol{f}(\boldsymbol{u}' - \boldsymbol{u},\ \boldsymbol{X}' - \boldsymbol{X})\, dV', \quad \text{on } \mathcal{R}. \tag{5.3}$$

Thus, the internal force density per unit volume acting upon particle $\boldsymbol{X}$ is a *functional* of the displacement field $\boldsymbol{u}$.[f] Note that $\boldsymbol{L_u}(\boldsymbol{X})$ is not a function of $\boldsymbol{X}'$, because $\boldsymbol{X}'$ disappears from Eq. 5.3 after integration.

Applying Newton's second law to particle P with differential mass $dm = \rho_0 dV \equiv \rho_0 dV(\boldsymbol{X})$, considered as a free body, with mass-per-unit-reference-volume, ρ_0, and with externally applied body force-per-unit-reference-volume $\boldsymbol{b}$,

$$dm\ddot{\boldsymbol{u}} = \rho_0 dV \ddot{\boldsymbol{u}} = (\boldsymbol{L_u} + \boldsymbol{b})dV \quad \text{on } \mathcal{R}\,, \tag{5.4}$$

---

[e] "$\forall$" means "for all"; $\|\boldsymbol{v}\|$ means "magnitude of vector $\boldsymbol{v}$".

[f] A functional can be thought of as a function of a vector field. As an example of a functional, the potential energy of a deformable body is a scalar function of a vector displacement field. $\boldsymbol{L_u}$ in Eq. 5.2 is a vector functional of the displacement vector field.

and, dividing Eq. 5.4 by dV, along with Eq. 5.3,

$$\rho_0 \ddot{\boldsymbol{u}} = \int_{\mathcal{R}} \boldsymbol{f}(\boldsymbol{u}' - \boldsymbol{u},\ \boldsymbol{X}' - \boldsymbol{X})\, dV' + \boldsymbol{b} \quad \text{on } \mathcal{R}\,. \tag{5.5}$$

The function $\boldsymbol{f}(\boldsymbol{u}' - \boldsymbol{u},\ \boldsymbol{X}' - \boldsymbol{X}) = \boldsymbol{f}(\boldsymbol{\eta}, \boldsymbol{\xi})$ is called the *pairwise force function*, and has units of force-per-(volume squared). Specifying this pairwise force function defines the "constitutive model" for the material.

For static problems, the acceleration term in Eq. 5.5 is zero, so that

$$0 = \int_{\mathcal{R}} \boldsymbol{f}(\boldsymbol{\eta}, \boldsymbol{\xi})\, dV' + \boldsymbol{b} \quad \text{on } \mathcal{R}\,. \tag{5.6}$$

Notice that this formulation depends upon two hold-over continuum concepts: density $\rho_0 \equiv \lim_{\Delta V \to 0} \left(\frac{\Delta m}{\Delta V}\right)$, and loading force density, $\boldsymbol{b} \equiv \lim_{\Delta V \to 0} \left(\frac{\Delta \boldsymbol{F}}{\Delta V}\right)$. Both of these concepts (of density and of body force) depend upon continuity of the mass distribution and of the reference space, $\mathcal{R}$; the limits do not exist unless the reference space $\mathcal{R}$ is assumed to be a continuous open[g] subset of $\mathbb{R}^3$. This assumption was not stated explicitly in (Silling, 2000). The assumption that $\mathcal{R}$ is a continuous domain in $\mathbb{R}^3$ is in fact unnecessary, and is perhaps even counterproductive, as will be discussed further in Chapter 6.

5.4 Restrictions on the pairwise force function, *f*

The pairwise force function, $\boldsymbol{f}(\boldsymbol{\eta}, \boldsymbol{\xi})$ in Eqs. 5.5 and 5.6 contains the constitutive information for a given, homogeneous, material. Of course, the pairwise force function would be different for steel than for aluminum. On the other hand, if P happens to be of one material and P' is of another material, then the pairwise force function would need additional arguments to be meaningful: $\boldsymbol{f}(\boldsymbol{\xi}, \boldsymbol{\eta}, \boldsymbol{X}, \boldsymbol{X}')$. More precisely, we would need to specify $\boldsymbol{f}(\boldsymbol{\xi}, \boldsymbol{\eta}, P, P')$ as well as a function designating the type of material associated with each point. Indeed, because the pairwise

[g] "open" in this context means $\mathcal{R}$ excludes its boundary.

force function acts between particles that are a finite distance apart, the very notion of "material" must be clarified; this we do in Chapter 6. To avoid this complexity, we assume in the remainder of this chapter that the body $\mathcal{R}$ consists of a single, homogeneous, material.

The reference (undeformed) configuration is termed *equilibrated* if Eq. 5.6 is satisfied with null body force field $\boldsymbol{b}(\boldsymbol{X}) = \boldsymbol{0}$, and displacement field $\boldsymbol{u}(\boldsymbol{X}) = \boldsymbol{0}$ for all $\boldsymbol{X}$. Similarly, if $\boldsymbol{f}(\boldsymbol{\eta} = \boldsymbol{0}, \boldsymbol{\xi}) = \boldsymbol{0}$ for all bonds $\boldsymbol{\xi} \neq \boldsymbol{0}$ we call the reference configuration *pairwise equilibrated*. Note that it is possible for an equilibrated body to be not pairwise-equilibrated.

The pairwise function $\boldsymbol{f}(\boldsymbol{\eta}, \boldsymbol{\xi})$ need not be a smooth function with respect to either of its arguments; however, it must be integrable in the Riemann sense.[h]

As defined so far, the pairwise force function $\boldsymbol{f}(\boldsymbol{\eta}, \boldsymbol{\xi})$ does not contain history-dependent state variables such as plastic stretch or damage; therefore this is a material *without memory*. Other forms of the pairwise force function are certainly possible. For example, if the pairwise force function were made a function only of the current particle positions, $\boldsymbol{f} = \boldsymbol{f}(\boldsymbol{x}', \boldsymbol{x})$, with $\boldsymbol{x} \equiv \boldsymbol{X} + \boldsymbol{u}$ and $\boldsymbol{x}' \equiv \boldsymbol{X}' + \boldsymbol{u}'$, the represented material would be a *material without structure*, because the reference (undeformed) configuration $\boldsymbol{X}$ is not considered in determining $\boldsymbol{f}$. Molecular dynamics simulations, also called *atomistics*, represent materials without structure, because in such models the forces acting between atoms depend only upon the current atomic positions.

There is a restriction on the pairwise force function: it must satisfy Newton's third law, given in Section 3.1. Thus, for a given material, we require that

$$\boldsymbol{f}(-\boldsymbol{\eta}, -\boldsymbol{\xi}) = -\boldsymbol{f}(\boldsymbol{\eta}, \boldsymbol{\xi}) \qquad \forall \boldsymbol{\eta},\ \boldsymbol{\xi}\,. \tag{5.7}$$

---

[h] Riemann integration is defined for "nice-enough" functions. More involved definitions of integration such as the Riemann–Stieltjes integral and the Lebesgue integral apply to more complicated types of functions.

Silling [(Silling, 2000) , Eq. 7], also states that another restriction of $\boldsymbol{f}$ arises from conservation of angular momentum:

$$(\boldsymbol{\xi}+\boldsymbol{\eta})\times\boldsymbol{f}(\boldsymbol{\eta},\boldsymbol{\xi})=\mathbf{0} \qquad \forall\boldsymbol{\eta},\boldsymbol{\xi}.^{\mathrm{i}} \tag{5.8}$$

This equation states that the peridynamic force between two particles must align with the current positions of the particles, in what we call a "central force model". However, this assertion is not true if particles can sustain moments, as can be shown by the reader through counter example (Problem 5.4). Nonetheless, throughout the rest of this chapter, for simplicity, we will assume that particles sustain no moment, and that Eq. 5.8 is satisfied; thus the pairwise force function $\boldsymbol{f}(\boldsymbol{\eta},\boldsymbol{\xi})$ acts centrally between the deformed particle locations $\boldsymbol{x}$ and $\boldsymbol{x}'$, as shown in Fig. 5.3. This type of force interaction, in which the forces acting between each pair of particles are functions only of the reference and deformed locations of those two particles (i.e. *bond-based*), and in which the forces are equal, opposite, and collinear with the deformed locations of the two particles, (i.e. *ordinary*) is what is called the "ordinary bond-based peridynamic model".

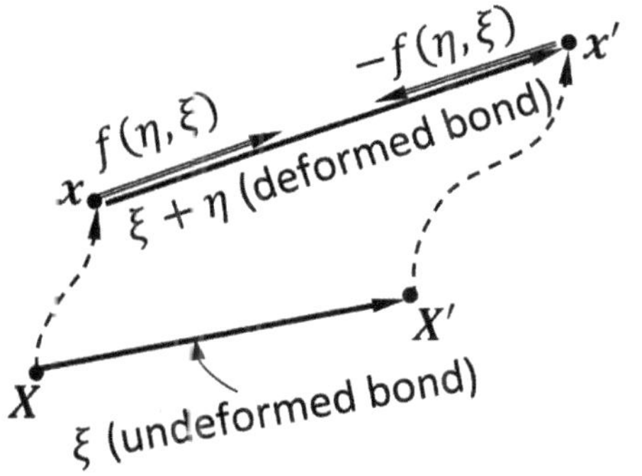

Figure 5.3. The ordinary bond-based peridynamic pairwise force function, $\boldsymbol{f}$.

Perhaps arguably, Silling (Silling, 2000) continues with the assertion that the most general form of the pairwise force function $\boldsymbol{f}$ is

---

[i] × is the vector cross product.

$$f(\eta, \xi) = F(\eta, \xi)(\xi + \eta), \qquad \forall \eta,\ \xi\,, \tag{5.9}$$

where $F(\eta, \xi)$ is a scalar-valued function such that

$$F(-\eta, -\xi) = F(\eta, \xi), \qquad \forall \eta,\ \xi\,. \tag{5.10}$$

Together, Eqs. (5.9) and (5.10) say that the pairwise forces between any two particles must be equal, opposite, and collinear with the pair of particles in their current configuration. Crucially, Eqs. (5.9) and (5.10) indicate that the pairwise peridynamic force function $\boldsymbol{f}$ is a function only of $\boldsymbol{\xi}$ and $\boldsymbol{\eta}$ (which depend only upon the reference and current locations of both particles). This assertion is later found to be insufficiently general for many materials, and it is corrected in a later paper (Silling, S.A., Epton, M., Weckner, O., Xu, J., and Askari, E., 2007). However, in keeping with the bond-based theory, we will nonetheless tentatively assume that all pairwise force functions are equal, opposite, and collinear with the deformed positions of the particles, as expressed by Eqs. 5.9 and 5.10. Again, the pairwise force functions depend only upon the two interacting particles' reference and deformed positions, and not upon the states of any other nearby particles. This assumption will however be relaxed in the state-based theory presented in Chapter 8.

5.5 Isotropic ordinary bond-based peridynamic model

A material is *isotropic* if its essential behavior does not change with material orientation with respect to a specific coordinate system. To put a mathematical handle on "material orientation", we need the concept of "proper orthogonal tensor".

As described in Section 3.2, a proper orthonormal second-order tensor $\boldsymbol{Q}$ is a tensor whose transpose is its inverse, and whose determinant is unity:

$$\boldsymbol{Q}^T = \boldsymbol{Q}^{-1} \text{ and } \det(\boldsymbol{Q}) = 1 \tag{5.11}$$

Thought of as a linear transformation, a proper orthonormal second-order tensor is a rigid-body rotation.[j] We designate the set of all proper orthonormal tensors as $\mathcal{O}^+$.

A peridynamic material as *isotropic* if

$$\boldsymbol{f}(\boldsymbol{Q\eta}, \boldsymbol{Q\xi}) = \boldsymbol{Qf}(\boldsymbol{\eta}, \boldsymbol{\xi}) \quad \forall \boldsymbol{\eta}, \boldsymbol{\xi}, \ \forall \boldsymbol{Q} \in \mathcal{O}^+, \tag{5.12}$$

which means that the response of the pairwise force function is independent of the orientation of the material with respect to a reference coordinate system. Note that in an isotropic material, if the bond $\boldsymbol{\xi}$ and the relative displacement vector $\boldsymbol{\eta}$ are rotated by $\boldsymbol{Q}$, then the pairwise force vector $\boldsymbol{f}$ is also rotated by $\boldsymbol{Q}$. Additionally, with this definition, note that a particle within the horizon distance δ of the boundary of $\mathcal{R}$, i.e. outside of $\mathcal{R}^0$ in Fig. 5.1, cannot be isotropic. The proof is left to the reader in Problem 5.6.

In Chapter 8, we distinguish between "micro-isotropic" and "macro-isotropic" materials. We show that a material that is macro-isotropic does not necessarily need to be micro-isotropic.

5.6 Elastic bond-based peridynamic models

A set of material particles (a body) is said to be *conservative* if the work done on the body vanishes for all deformed configurations that carry the system around any closed path in the deformed configuration space.

A body is *elastic* if its stored mechanical energy is a function only of deformation, and if the stored mechanical energy is independent of the deformation path through which the deformation was achieved. Thus, a

[j] Thus, all vectors $\boldsymbol{\xi}$ are rotated by a specific angle about a specific axis using the linear transformation $\boldsymbol{\xi}^* = \boldsymbol{Q\xi}$ to image vectors $\boldsymbol{\xi}^*$ with $|\boldsymbol{\xi}^*| = |\boldsymbol{\xi}|$. The magnitude of the vector is left unchanged by the transformation.

conservative system is elastic (but an elastic body is not always part of a conservative system).[k]

Elastic bodies have *energy potential functions*. An energy potential function might for example, be the strain energy density $\mathfrak{U}$. If we assume that the strain energy density is a function of several kinematic variables, say, the Cauchy small strain components ϵ_{ij}, then $\mathfrak{U} = \mathfrak{U}(\epsilon_{ij})$ and by the chain rule,

$$d\mathfrak{U}(\epsilon_{ij}) = \frac{\partial \mathfrak{U}}{\partial \epsilon_{ij}} d\epsilon_{ij} = \frac{\partial \mathfrak{U}}{\partial \epsilon_{11}} d\epsilon_{11} + \frac{\partial \mathfrak{U}}{\partial \epsilon_{12}} d\epsilon_{12} + \cdots + \frac{\partial \mathfrak{U}}{\partial \epsilon_{33}} d\epsilon_{33}. \tag{5.13}$$

Considering the physical definition of differential change in strain energy density,

$$d\mathfrak{U}(\epsilon_{ij}) = \sigma_{ij} d\epsilon_{ij} = \sigma_{11} d\epsilon_{11} + \sigma_{12} d\epsilon_{12} + \cdots + \sigma_{33} d\epsilon_{33}, \tag{5.14}$$

where σ_{ij} are the Cauchy stress components. Because Equations 5.13 and 5.14 must hold for arbitrary sets of strain components $d\epsilon_{ij}$, we see that $\sigma_{ij} = \frac{\partial \mathfrak{U}}{\partial \epsilon_{ij}}$. So we say that the Cauchy stress and the small strain components are *work-conjugate*. We can also assume that the strain energy density $\mathfrak{U}$ is a function of any other set of kinematic variables, say q_{ij}, and automatically generate the set of work-conjugate components p_{ij} as the set of partial derivatives $p_{ij} = \frac{\partial \mathfrak{U}}{\partial q_{ij}}$.

Mathematically, a differentiable vector field $\boldsymbol{F}$ is conservative if it satisfies any of the three following equivalent conditions:

(1) the curl of $\boldsymbol{F}$ is zero[l]:

---

[k] An elastic body can be part of a nonconservative system if the applied loads are nonconservative. For example, an elastic body subject to frictional applied forces is a nonconservative system.

[l] The curl of a vector field is a vector representing the angular rotation of the field at a point. In Cartesian (x, y, z) coordinates, with $\boldsymbol{F} = F_x\boldsymbol{i} + F_y\boldsymbol{j} + F_z\boldsymbol{k}$, $\nabla \times \boldsymbol{F} = det \begin{vmatrix} \boldsymbol{i} & \boldsymbol{j} & \boldsymbol{k} \\ \frac{\partial}{\partial x} & \frac{\partial}{\partial y} & \frac{\partial}{\partial z} \\ F_x & F_y & F_z \end{vmatrix}$. The curl is only defined for vectors in 3D Euclidean space.

$$\nabla \times \boldsymbol{F} = \boldsymbol{0}\ ; \tag{5.15}$$

(2) the work, W, done on traversing any closed path in configuration space is zero[m]:

$$W = \oint_{\Gamma} \boldsymbol{F} \cdot \boldsymbol{dr} = 0\ ;\ \text{or} \tag{5.16}$$

(3) the force can be obtained as the gradient[n] of a scalar potential function W:

$$\boldsymbol{F} = \nabla W\ . \tag{5.17}$$

A peridynamic material is *microelastic* if it is conservative; that is if the work done on any material point $\boldsymbol{X}'$ due to interaction with another particle $\boldsymbol{X}$ as $\boldsymbol{x}'$ moves along any closed path Γ is zero:

$$\oint_{\Gamma} \boldsymbol{f}(\boldsymbol{\eta},\boldsymbol{\xi}) \cdot d\boldsymbol{\eta} = 0 \quad \forall \text{ closed paths } \Gamma,\ \ \forall \boldsymbol{\xi} \neq \boldsymbol{0} \tag{5.18}$$

where $d\boldsymbol{\eta}$ represents the differential path length along Γ. Equivalently, by Stokes' theorem,

$$\nabla_{\eta} \times \boldsymbol{f}(\boldsymbol{\eta},\boldsymbol{\xi}) = \boldsymbol{0}\,, \qquad \forall \boldsymbol{\xi} \neq \boldsymbol{0}\ .^{\text{o}} \tag{5.19}$$

---

[m] The work done by a force $\boldsymbol{F}$ moving along a path $\boldsymbol{r}$ is a scalar function defined as $dW = \int \boldsymbol{F} \cdot \boldsymbol{dr}$, where $\cdot$ is the vector dot product.

[n] In a Cartesian coordinate system, the gradient operator, $\nabla\Phi = \frac{\partial \Phi}{\partial x}\boldsymbol{i} + \frac{\partial \Phi}{\partial y}\boldsymbol{j} + \frac{\partial \Phi}{\partial z}\boldsymbol{k}$ gives the vector whose components are the partial derivatives of the potential function $\Phi(x,y,z)$. More generally, the gradient of a scalar field gives the magnitude and direction of the maximum slope of the scalar field.

[o] The curl of a vector field is a vector representing the rotation of the field at a point. In Cartesian (x,y,z) coordinates, with $\boldsymbol{f} = f_x\boldsymbol{i} + f_y\boldsymbol{j} + f_z\boldsymbol{k}$, $\nabla \times \boldsymbol{f} = det\begin{vmatrix} \boldsymbol{i} & \boldsymbol{j} & \boldsymbol{k} \\ \frac{\partial}{\partial x} & \frac{\partial}{\partial y} & \frac{\partial}{\partial z} \\ f_x & f_y & f_z \end{vmatrix}$. $\nabla_{\eta} \times \boldsymbol{f}(\boldsymbol{\eta},\boldsymbol{\xi})$ means that the vector curl operator is evaluated with respect to the (perhaps local) coordinates of $\boldsymbol{\eta}$.

Another consequence of Stokes' theorem is that a necessary and sufficient condition for an ordinary bond-based peridynamic material to be microelastic is the existence of a differentiable, scalar-valued function $w(\boldsymbol{\eta}, \boldsymbol{\xi})$, called the *pairwise potential function*, (w has units of energy per [volume squared]), such that

$$\boldsymbol{f}(\boldsymbol{\eta}, \boldsymbol{\xi}) = \frac{\partial w}{\partial \eta}(\boldsymbol{\eta}, \boldsymbol{\xi}) \,. \tag{5.20}$$

(Silling, 2000) shows that for an ordinary bond-based peridynamic material, the pairwise potential function $w(\boldsymbol{\eta}, \boldsymbol{\xi})$ can depend only upon the initial and the deformed lengths of the bond, and not upon the orientation of either of these vectors:

$$w(\boldsymbol{\eta}, \boldsymbol{\xi}) = \widetilde{w}(|\boldsymbol{\xi} + \boldsymbol{\eta}|, |\boldsymbol{\xi}|) \,. \tag{5.21}$$

However, Eq. 5.21 is actually not fully general, as in fact the pairwise potential function could also actually depend upon rotations of bonds and even upon the deformations of other bonds within the peridynamic horizon distance δ. The state-based theory presented in the Chapter 8 generalizes the behavior of the pairwise potential function.

So far we have modeled the material from a *microelastic* viewpoint. The *macroelastic energy density* functional $W_u(\boldsymbol{X})$ is obtained by summing up the one-half of the differential energy stored in each bond $\boldsymbol{\xi}$ attached to particle $\boldsymbol{X}$:

$$W_u(\boldsymbol{X}) = \frac{1}{2} \int_{\mathcal{R}} w(\boldsymbol{\eta}, \boldsymbol{\xi}) \, dV' \,, \tag{5.22}$$

where dV' is the differential volume in the undeformed configuration $\mathcal{R}$ associated with particle $\boldsymbol{X}'$. The factor of ½ is included because the energy stored in each bond $\boldsymbol{\xi}$ is associated equally with the two particles $\boldsymbol{X}$ and $\boldsymbol{X}'$.

The *total macroelastic energy* functional Φ_u for an entire material body $\mathcal{R}$ is obtained by integrating the macroelastic energy density W_u over the body:

$$\Phi_u = \int_{\mathcal{R}} W_u(\boldsymbol{X})\, dV\,, \tag{5.23}$$

where dV is the differential volume in the undeformed configuration $\mathcal{R}$ associated with particle $\boldsymbol{X}$.

By assuming that the displacement field, $\boldsymbol{u}$ is time-dependent, Silling demonstrated that the time rate of change of the total macro elastic energy $\dot{\Phi}_u$, the time rate of change of the total kinetic energy $\dot{T}_u$, and the time rate of change of the work of the externally applied body forces $\dot{V}_u$ are related. Inserting Eq. 5.22 into Eq. 5.23 and taking the derivative of both sides with respect to time:

$$\begin{aligned} \dot{\Phi}_u &= \frac{\partial}{\partial t}\left(\int_{\mathcal{R}} \left(\frac{1}{2}\int_{\mathcal{R}} w(\boldsymbol{\eta},\boldsymbol{\xi})\, dV'\right) dV\right) \\ &= \frac{1}{2}\int_{\mathcal{R}}\int_{\mathcal{R}} \frac{\partial}{\partial t} w(\boldsymbol{\eta},\boldsymbol{\xi})\, dV'\, dV \\ &= \frac{1}{2}\int_{\mathcal{R}}\int_{\mathcal{R}} \frac{\partial w}{\partial \eta}(\boldsymbol{u}'-\boldsymbol{u},\, \boldsymbol{x}'-\boldsymbol{x})\cdot(\dot{\boldsymbol{u}}'-\dot{\boldsymbol{u}})\, dV'\, dV\,. \end{aligned} \tag{5.24}$$

By recognizing that bond forces come in pairs and using Eq. 5.7, and by interchanging the variables $\boldsymbol{x}'$ and $\boldsymbol{x}$, Eq. 5.24 is simplified to

$$\dot{\Phi}_u = -\int_{\mathcal{R}}\int_{\mathcal{R}} \frac{\partial w}{\partial \eta}(\boldsymbol{u}'-\boldsymbol{u},\, \boldsymbol{x}'-\boldsymbol{x})\cdot\dot{\boldsymbol{u}}\, dV'\, dV\,, \tag{5.25}$$

and using Eq. 5.20,

$$\dot{\Phi}_u = -\int_{\mathcal{R}}\int_{\mathcal{R}} \boldsymbol{f}(\boldsymbol{\eta},\boldsymbol{\xi})\cdot\dot{\boldsymbol{u}}\, dV'\, dV = -\int_{\mathcal{R}}\int_{\mathcal{R}} \boldsymbol{f}(\boldsymbol{\eta},\boldsymbol{\xi})\, dV'\cdot\dot{\boldsymbol{u}}\, dV, \tag{5.26}$$

or, using Eq. 5.5,

$$\dot{\Phi}_u = -\int_{\mathcal{R}} (\rho_0 \ddot{\boldsymbol{u}} - \boldsymbol{b}) \cdot \dot{\boldsymbol{u}}\, dV\,, \tag{5.27}$$

or

$$\dot{\Phi}_u = -\int_{\mathcal{R}} \left[\frac{\partial}{\partial t}\left(\frac{\rho_0}{2}\dot{\boldsymbol{u}} \cdot \dot{\boldsymbol{u}}\right) - \boldsymbol{b} \cdot \dot{\boldsymbol{u}}\right] dV\,, \tag{5.28}$$

or

$$\dot{\Phi}_u + \frac{\partial}{\partial t}\left[\int_{\mathcal{R}} \left(\frac{\rho_0}{2}\dot{\boldsymbol{u}} \cdot \dot{\boldsymbol{u}}\right) dV\right] = \int_{\mathcal{R}} [\boldsymbol{b} \cdot \dot{\boldsymbol{u}}]\, dV. \tag{5.29}$$

Recognizing that the kinetic energy of a particle is $\mathrm{T}_u = \frac{1}{2} m\dot{\boldsymbol{u}}^{\mathbf{2}}$, we finally arrive at the equation

$$\dot{\Phi}_u + \dot{\mathrm{T}}_u = \dot{\mathrm{V}}_u\,, \tag{5.30}$$

and integrating with respect to time, we obtain

$$\Phi_u + \mathrm{T}_u = \mathrm{V}_u + C\,, \tag{5.31}$$

where C is an arbitrary constant of integration.

This last equation is expected for a conservative system, expressing macro elastic conservation of energy distributed between the internally stored elastic energy Φ_u, the kinetic energy T_u, and the potential energy of the applied forces V_u. It is a statement of conservation of mechanical energy, with the assumption that there is no energy dissipation due to friction or internal generation of heat. In addition, this model does not allow for internal chemical reactions.

The fact that the micro elastic peridynamic model results in a traditional macro elastic conservation of mechanical energy model is no surprise. If every subdomain of a conservative system is conservative, then the collection of these systems should also be conservative.

5.7 Relationship between peridynamic traction and stress

Even though the peridynamic theory is more general than the conventional continuum theory, in some cases it is possible to relate the peridynamic bond forces and their distributions, to the notion of "stress" in the continuum theory, as shown next.

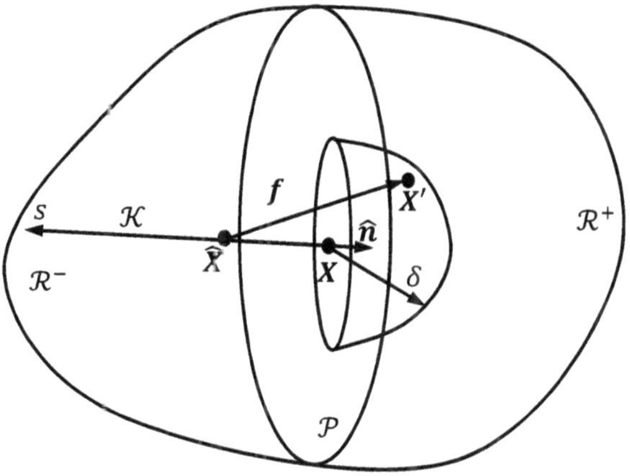

Figure 5.4. Relationship between pairwise force $\boldsymbol{f}$ and conventional stress, $\boldsymbol{\sigma}$.

Consider an equilibrated peridynamic body $\mathcal{R}$, as shown in Fig 5.4. Suppose we want to find the "peridynamic traction" $\boldsymbol{\tau}$ with respect to plane $\mathcal{P}$, with outward pointing unit normal $\hat{\boldsymbol{n}}$ at reference point $\boldsymbol{X}$. Plane $\mathcal{P}$ divides body $\mathcal{R}$ into two subregions, $\mathcal{R}^+$ and $\mathcal{R}^-$, as shown in the figure. $\mathcal{R}^+$ exerts a force on $\mathcal{R}^-$ through nonlocal peridynamic bond forces $\boldsymbol{f}$.

With $\boldsymbol{X}$ and $\hat{\boldsymbol{n}}$ given, let us define the following point sets

$$\begin{aligned} \mathcal{R}^+ &= \{\boldsymbol{X}' \in \mathcal{R}\colon (\boldsymbol{X}' - \boldsymbol{X}) \cdot \hat{\boldsymbol{n}} \geq 0\}, \\ \mathcal{R}^- &= \{\boldsymbol{X}' \in \mathcal{R}\colon (\boldsymbol{X}' - \boldsymbol{X}) \cdot \hat{\boldsymbol{n}} < 0\}. \end{aligned} \tag{5.32}$$

Also, let $\mathcal{K}$ be the following set of collinear points:

$$\mathcal{K} = \{\hat{\boldsymbol{X}} \in \mathcal{R}^-\colon (\hat{\boldsymbol{X}} = \boldsymbol{X} - s\,\hat{\boldsymbol{n}}),\ 0 \leq s < \infty\}, \tag{5.33}$$

We define the peridynamic traction, or *areal force density*, $\boldsymbol{\tau}(\boldsymbol{X},\widehat{\boldsymbol{n}})$ at point $\boldsymbol{X}$ in body $\mathcal{R}$ in the direction of unit vector $\widehat{\boldsymbol{n}}$ as

$$\boldsymbol{\tau}(\boldsymbol{X},\widehat{\boldsymbol{n}}) = \int_{\mathcal{K}} \int_{\mathcal{R}^+} \boldsymbol{f}\left(\boldsymbol{u}' - \widehat{\boldsymbol{u}}, \boldsymbol{X}' - \widehat{\boldsymbol{X}}\right) dV_{X'}\, ds\,, \tag{5.34}$$

where ds represents the differential path length over $\mathcal{K}$. Equivalently, the integration domain can be further restricted by recognizing that the integrand is null outside the material horizon δ of point $\boldsymbol{X}$:

$$\begin{aligned} \mathcal{K}^\delta &= \left\{\widehat{\boldsymbol{X}} \in \mathcal{R}^-\colon \left(\widehat{\boldsymbol{X}} = \boldsymbol{X} - s\,\widehat{\boldsymbol{n}}\right),\ 0 \le s \le \delta\right\}, \\ \mathcal{H} &= \{\boldsymbol{X}' \in \mathcal{R}\colon |\boldsymbol{X}' - \boldsymbol{X}| \le \delta\,\}, \\ \mathcal{H}^+ &= \{\boldsymbol{X}' \in \mathcal{H}\colon (\boldsymbol{X}' - \boldsymbol{X}) \cdot \widehat{\boldsymbol{n}} \ge 0\,\}, \text{ and} \\ \boldsymbol{\tau}(\boldsymbol{X},\widehat{\boldsymbol{n}}) &= \int_{\mathcal{K}^\delta} \int_{\mathcal{H}^+} \boldsymbol{f}\left(\boldsymbol{u}' - \widehat{\boldsymbol{u}}, \boldsymbol{X}' - \widehat{\boldsymbol{X}}\right) dV_{X'}\, ds\,. \end{aligned} \tag{5.35}$$

$\boldsymbol{\tau}$ has units of force per unit area, like stress, and represents the nonlocal force per unit area acting between $\mathcal{R}^+$ and $\mathcal{R}^-$. If $\mathcal{H}^+$ and $\mathcal{K}^\delta$ are both subsets of $\mathcal{R}$, and if the peridynamic force function $\boldsymbol{f}(\boldsymbol{\xi},\boldsymbol{\eta})$ is invariant with respect to $\boldsymbol{X}$, then areal force density is the classical stress vector (or traction) $\boldsymbol{P} \cdot \widehat{\boldsymbol{n}}$ where $\boldsymbol{P}$ is the first Piola-Kirchoff stress tensor, defined in Section 3.5, because the areal force density $\boldsymbol{\tau}$ represents force with respect to unit area in the reference configuration. This is similar to the concept for stress that Cauchy describes, although in the case of the peridynamic model we do not take the limit as the material horizon δ goes to zero, as Cauchy does. On the other hand, if $\boldsymbol{f}(\boldsymbol{\xi},\boldsymbol{\eta})$ varies with respect to $\boldsymbol{X}$, then the areal force density $\boldsymbol{\tau}$ cannot be cleanly identified with a stress $\boldsymbol{P}$.

A "peridynamic stress tensor" is also defined by (Lehoucq, R.B., and Silling, S.A., 2007).

5.8 Microelasticity under homogeneous deformation

As discussed in Section 3.6, in continuum elasticity, a *hyperelastic* material is defined as a material whose stress-strain relationship is derived from a strain energy density function $\widehat{W}$, which depends only upon the deformation gradient tensor $\boldsymbol{F}$. We would like to compare the microelastic bond-based peridynamic model to the classical continuum model for a hyperelastic material. In a hyperelastic material, the strain energy density is a function $\widehat{W}$ of the deformation gradient: $\widehat{W} = \widehat{W}(\boldsymbol{F})$.

First, let us define a "spatially-homogeneous deformation field":

Definition 5.1 In a peridynamic material with material horizon $\mathcal{H}$, a *spatially-homogeneous deformation field* within a domain $\mathcal{B} \subset \mathcal{R}$ satisfies

$$\boldsymbol{\eta}\langle\boldsymbol{\xi}\rangle[\boldsymbol{X}] = \boldsymbol{\eta}\langle\boldsymbol{\xi}\rangle[\boldsymbol{X}^*] \,, \forall\boldsymbol{\xi} \in \mathcal{H}, \ \forall \boldsymbol{X}, \boldsymbol{X}^* \in \mathcal{B} \subset \mathcal{R} \,. \tag{5.36}$$

This means that the deformed configuration $\boldsymbol{\eta} + \boldsymbol{\xi}$ of a particular bond $\boldsymbol{\xi}$ is invariant with respect to particle $\boldsymbol{X}$ within domain $\mathcal{B}$. The kinematic variables are defined in Figs. 5.1 and 5.2. Apparently, $\mathcal{B}$ must be a subset of $\mathcal{R}^0$ for the deformation field on $\mathcal{B}$ to be considered spatially homogeneous.

On the other hand, in the continuum model, the deformation field is spatially homogeneous if the deformation gradient $\boldsymbol{F}$ is invariant with respect to $\boldsymbol{X}$. Note that the peridynamic definition of spatially-homogeneous deformation is more general, as it does not require that the deformation field $\boldsymbol{x}$ be differentiable.

Definition 5.2 A classical material point $\boldsymbol{X}$ and a peridynamic material point $\boldsymbol{X}^*$ with material horizon $\mathcal{H}$ have *equivalent spatially-homogeneous deformation fields* within peridynamic horizon $\mathcal{H}$ if

$$\boldsymbol{F}[\boldsymbol{X}] \cdot \boldsymbol{\xi} = (\boldsymbol{\xi} + \boldsymbol{\eta})\langle\boldsymbol{\xi}\rangle \,, \forall\boldsymbol{\xi} \in \mathcal{H}[\boldsymbol{X}^*] \,. \tag{5.37}$$

This means that corresponding deformed bonds in both the classical and the peridynamic body undergo equivalent stretches and rotations under deformation.

Also, let us define when materials are *macroscopically energy equivalent under homogeneous deformation*:

Definition 5.3 Two hyperelastic materials are *macroscopically energy equivalent under homogeneous deformation* if, for equivalent spatially homogeneous deformation fields, the materials store the same strain energy per unit reference volume.

We defined the peridynamic macroelastic energy density W_u in Eq. 5.22. To consider the peridynamic and the classical models as "macroscopically-energy-equivalent under homogeneous deformation", we require that $\widehat{W} = W_u$ in the material neighborhood $\mathcal{H}(\boldsymbol{X})$ of reference point $\boldsymbol{X}$ for "equivalent spatially-homogeneous deformation fields".[p] For nonhomogeneous deformation fields, and also for particles closer than the material horizon, δ, of the domain boundary of $\mathcal{R}$, the continuum and the peridynamic models are too different to be directly compared.

In the peridynamic model, the strain energy density is given by Eq. 5.22 as

$$W_u(\boldsymbol{\eta}) = \frac{1}{2}\int_{\mathcal{R}} w(\boldsymbol{\eta}, \boldsymbol{\xi})\, dV' , \tag{5.38}$$

and with $\boldsymbol{\eta} = \boldsymbol{u}' - \boldsymbol{u} = (\boldsymbol{F} - \mathbf{1})(\boldsymbol{X}' - \boldsymbol{X}) = (\boldsymbol{F} - \mathbf{1})\boldsymbol{\xi}$,

$$W_u(\boldsymbol{F}) = \frac{1}{2}\int_{\mathcal{H}} w\big((\boldsymbol{F} - \mathbf{1})\boldsymbol{\xi}, \boldsymbol{\xi}\big)\, dV' , \tag{5.39}$$

which shows that this is indeed a hyperelastic material, with the strain energy density function W_u depending only upon $\boldsymbol{F}$ (for a given pairwise microelastic potential function, w).

Notice that the macroelastic strain energy density $W_u(\boldsymbol{F})$ in Eq. 5.39 is dependent upon the spatial integral over the material horizon $\mathcal{H}$ of the pairwise microelastic potential function w (points $\boldsymbol{\xi}$ in $\mathcal{R}$ but outside of $\mathcal{H}$ add no bond energy w). Many different forms of the microelastic potential function w can produce the same macroelastic potential W_u;

[p] A "homogeneous deformation field" produces a strain state that is invariant spatially. This is guaranteed if the deformation gradient, $\boldsymbol{F}$, is spatially constant, so $\nabla \boldsymbol{F} = \mathbf{0}$.

thus the bond-based peridynamic approach is in a certain sense more general than the classical model.

When the deformation field is nonhomogeneous, "strain energy density $\widehat{W}$" is not defined in the peridynamic model; the peridynamic model is simply different than the continuum model, and no direct comparison can be made.

5.9 Classical elastic moduli

Without going into too much mathematical detail, it is sufficient here to say that the microelastic model can be linearized with respect to relative deformation $\boldsymbol{\eta}$ by assuming small deformations, so that $|\boldsymbol{\eta}| \ll |\boldsymbol{\xi}|$. In this case the pairwise force function can be expressed as

$$\boldsymbol{f}(\boldsymbol{\eta},\boldsymbol{\xi}) = \boldsymbol{C}(\boldsymbol{\xi})\boldsymbol{\eta} + \boldsymbol{f}(\boldsymbol{0},\boldsymbol{\xi}) \qquad \forall \boldsymbol{\eta},\boldsymbol{\xi}, \tag{5.40}$$

where, according to the Taylor series expansion, ignoring higher-order terms, $\boldsymbol{C}$ is the second-order *micromodulus* tensor given by

$$\boldsymbol{C}(\boldsymbol{\xi}) = \frac{\partial \boldsymbol{f}}{\partial \boldsymbol{\eta}}(\boldsymbol{0},\boldsymbol{\xi}) \ . \tag{5.41}$$

For homogeneous deformation states, this linearization of the bond-based peridynamic model gives macro-elastic Young's modulus E and Poisson's ratio ν of

$$E = \frac{\pi}{3}\int_0^\infty \lambda(r) r^5\, dr \quad \text{and } \nu = \frac{1}{4} \ , \tag{5.42}$$

$$\text{where } \lambda(r) = \frac{1}{\|\boldsymbol{\xi}\|^2}\boldsymbol{\xi} \cdot \frac{\partial \boldsymbol{F}}{\partial \boldsymbol{\eta}}(\boldsymbol{0},\boldsymbol{\xi}) \ . \tag{5.43}$$

Details of this derivation are given in (Silling, 2000). As with Navier's paper in Section 2.2, Poisson's ratios other than ¼ cannot be modeled using the ordinary bond-based peridynamic model.

Thus, the ordinary bond-based model is insufficiently general to model all of the materials that Navier-Cauchy theory of elasticity can model. Nonetheless, the bond-based model may be sufficiently general for engineering purposes, as most engineering materials have Poisson's ratio close to ¼, and in many cases it is unnecessary to simulate Poisson's contraction effect with a high degree of accuracy. The state-based peridynamic theory, described in Chapter 8, although more complicated, corrects this shortcoming.

5.10 Boundary conditions

In the conventional continuum mechanics theory, we pose and then solve the governing differential equations subject to the condition that the solution must satisfy the essential boundary conditions (applied displacements and velocities) and the natural boundary conditions (applied tractions) on surfaces of the computational domain. These are known as *boundary value problems*.

One usually poses the essential boundary conditions as specified smoothly-spatially-varying displacements and velocities applied to smooth segments of the domain boundaries (or perhaps applied even to single points of the domain, which produce strain and strain singularities). The natural boundary conditions are applied as surface tractions (forces per unit area), or as point loads (which again produce singularities).

The situation is different with peridynamics, in which the domain boundaries do not have to be piecewise smooth, as would be the case for the conventional continuum theory.

So how are loads and displacements applied? In peridynamics, loads are applied to finite subdomains of body $\mathcal{R}$ as body forces, and displacements are also applied to finite regions within the domain $\mathcal{R}$. In continuum peridynamics, concentrated point, line, and surface forces can produce infinite responses. The same can be said for point, line, and surface applications of specified displacements, which also yield infinities in the response.

To obtain reasonable behavior at the "macro" level, we simulate tractions by applying loads on a boundary layer that is at least as wide as the peridynamic horizon. Similarly, displacements are applied to boundary layers at least as wide as the peridynamic horizon. Although this need to apply loads and displacements to finite volumes of the domain might seem disconcerting at first, it is really a feature, not a defect, of the peridynamic model, in that this practice prevents solution singularities. Arguably, applied loads are in reality applied to finite material subdomains; not to discrete points or to smooth surfaces. In fact, one could argue that the concepts of "material point" and "smooth bounding surface" are idealizations that do not occur in the physical world of material bodies. These idealizations are nonetheless useful for some problems, however, and it is still possible to approximate "material points" and "smooth bounding surfaces" using finite volumes of the domain.

5.11 Fracture

Continuum mechanics has no means of allowing discontinuities in the deformation field to form. Thus the auxiliary discipline of fracture mechanics, presented in Chapter 4, was developed over the course of the twentieth century. Starting particularly in the 1970's, with the rise of nonlinear computational methods in solid mechanics, the struggle between the competing approaches of discrete fracture mechanics, continuum damage mechanics, and particle methods remains unresolved.

The main advantage of the peridynamic model is its ability to simulate continuous behavior, damage, and fracture in a unified computational framework. If the value of the pairwise peridynamic force $\boldsymbol{f}$ (or the value of the bond stretch, bond energy, etc.) exceeds a certain limiting value, then peridynamic micro-damage ω can occur, causing the value of the pairwise force function to decrease with increasing bond stretch. If micro-damage occurs, discontinuities in the deformation field may form. These discontinuities usually start as distributed damage, and ultimately localize to form what has been called a "crack" or a "fracture" when the micro-damage is complete and consequently the areal force density, $\boldsymbol{\tau}$, across a plane becomes zero.

It is necessary to define the damage function in such a way that the correct fracture energy is replicated as damage progresses, allowing a crack to grow.

5.12 Generality of the bond-based peridynamic model

The ordinary bond-based model assumes that the pairwise force $\boldsymbol{f}(\boldsymbol{\eta}, \boldsymbol{\xi})$ between two particles $\boldsymbol{X}$ and $\boldsymbol{X}'$ is (1) a function only of the undeformed relative position $\boldsymbol{\xi}$ and relative displacement $\boldsymbol{\eta}$ between the two particles; and (2) is collinear with a line through the current positions of the two particles. As already pointed out in (Silling, 2000), these two assumptions produce a pairwise force function that is insufficiently general to model materials with Poisson's ratio other than one-quarter, and this force function is also insufficiently general to model plastic materials, in which the bond forces should depend upon collective bond deformations in order to prevent plastic volume changes.

One problem with the bond-based peridynamic model is that it is posed assuming that the material is a continuum $\mathcal{R}$, which contains an infinite number of material points. Thus, Eq. 5.5 represents an infinite number of equations, and the integral in Eq. 5.5 is over an infinite number of points. Thus, except for very special cases, the spatial integrals can be only approximately computed. The problem needs to be discretized. In Chapter 6, we propose that the material domain be idealized as a discrete lattice, so that it will be better suited for discrete computational solution.

The question also arises: Should peridynamic bonds be capable of applying moments to their associated particles? This question is addressed in the next section.

5.13 Bond-based micropolar peridynamic model

Even though the differential particles dV' in Eq. 5.5 are infinitesimally small, it would seem reasonable (and perhaps in some cases even necessary) to assume that internal peridynamic micro-moments as well as internal peridynamic forces can act upon the particles. In this case, in addition to possessing differential masses, the particles should also possess differential mass moments of inertia. The particles' translational degrees of freedom can be augmented to include rotational degrees of freedom. Thus the original bond-based peridynamic equations of motion are augmented with moment equations, to become:

$$\begin{aligned} \rho\ddot{\boldsymbol{u}} &= \int_{\mathcal{R}} \boldsymbol{f}(\boldsymbol{u},\ \boldsymbol{u}',\ \boldsymbol{X},\ \boldsymbol{X}')\, dV' + \boldsymbol{b} \qquad \text{on } \mathcal{R} \text{ and} \\ \boldsymbol{\mathcal{J}}\ddot{\boldsymbol{\theta}} &= \int_{\mathcal{R}} \boldsymbol{m}(\boldsymbol{u},\ \boldsymbol{u}',\ \boldsymbol{X},\ \boldsymbol{X}')\, dV' + \boldsymbol{m}_{ext} \qquad \text{on } \mathcal{R}\,, \end{aligned} \tag{5.44}$$

where $\boldsymbol{X}$ is now represents not only the translational degrees of freedom but also the rotational degrees of freedom of the particle in its reference configuration. In addition, $\boldsymbol{u}$ represents not only the translational displacement components, but also the rotational components of the particle. $\boldsymbol{\mathcal{J}}$ is the mass moment of inertia per unit volume (a tensor), $\boldsymbol{\theta}$ is the angular position, $\ddot{\boldsymbol{\theta}}$ is the angular acceleration vector, $\boldsymbol{m}$ is the peridynamic moment vector applied to particle $\boldsymbol{X}$ (with units of moment per unit volume squared), and $\boldsymbol{m}_{ext}$ is the applied moment vector per unit volume. This model, which includes peridynamic particle moments and rotations, is called the micropolar peridynamic theory (Gerstle, W., and Sau, N., and Silling, S., 2007).

For small displacements and rotations, $\boldsymbol{u}$, the micropolar model is relatively simple, because small rotations possess the nice properties of vectors. However, for large rotations, the kinematics is quite complicated.

Several advantages of the bond-based micropolar peridynamic model are that materials with variable Poisson's ratio can be modeled, and more complex material behaviors can be modeled with the limited numbers of particles. For example, a one-dimensional strand of particles can model beam bending using micropolar peridynamics, while the original bond-based model requires at least a two-dimensional lattice of particles to model the same beam.

Computational simulation of composite structures, such as reinforced concrete beams, may benefit from the micropolar model. Using the micropolar model, we can simulate reinforcing bars, including their bending and shear behavior, using a single line of peridynamic particles. The concrete particles may be of the ordinary bond-based peridynamic type, the bond-based micropolar peridynamic type, or state-based peridynamic type.

5.14 Summary

The ordinary bond-based peridynamic model is essentially a generalization of Navier's theory of elasticity (Navier, 1827). With the peridynamic theory, no assumption is made about continuity of deformation field $\boldsymbol{x}$, and no assumption is made about the linearity of the force versus stretch pairwise function $\boldsymbol{f}$. However, as with Navier's theory, the particles remain infinitesimal in size.

However, the bond-based peridynamic theory is insufficiently general for the same reason that Navier's theory was insufficiently general: both theories assume that the central force acting between two particles is a function only of the positions of those two particles. Real materials are more complex than this.

Thus, Silling (Silling, S.A., Epton, M., Weckner, O., Xu, J., and Askari, E., 2007) generalized his own peridynamic theory by introducing peridynamic states, making the state-based peridynamic theory much more general than the theory of continuum mechanics.

The bond-based peridynamic model is extremely simple, and its essence can be fully described in a page or two. Much of the difficulty in reading and understanding Silling's original paper is due to the admirable effort to compare the bond-based peridynamic theory to the theory of continuum mechanics. This is laudable, but actually it is not necessary to view continuum mechanics as "the true" theory that peridynamics needs to emulate.

Silling based the bond-based peridynamic model upon a continuous material space, and thus the model still needs to be in some way discretized if we want to implement it on a digital computer.

As is pointed out in Silling's original peridynamic paper (Silling, 2000), many forms of the peridynamic micro model can lead to the same macro behavior. This leads to the idea that perhaps we can simplify the material space itself, thus avoiding an assumption of continuity of the reference body. Additionally, the bond-based theory does not lend itself to analytical solution except in the very simplest cases.

This leads us to the approach taken by the state-based peridynamic lattice model (SPLM), in which we model the material as a finite number of interacting lattice particles, rather than as an infinite number of continuum particles. This latter approach is even more fundamental than Navier's theory, and branches off directly from Newtonian mechanics.

However, before we proceed with the SPLM, we present a particle lattice model for solids in the next chapter.

5.15 Exercises

5.1 Show, using sketches, that the definition of density, $\rho_0 \equiv \lim_{\Delta V \to 0} \left(\frac{\Delta m}{\Delta V}\right)$, fails at points on the boundary of a closed continuous domain $\mathcal{R}$, but is well-defined at all points within an open continuous domain $\mathcal{R}$.

5.2 Show, using sketches, that the definition of density, $\rho_0 \equiv \lim_{\Delta V \to 0} \left(\frac{\Delta m}{\Delta V}\right)$, fails if ρ_0 is spatially discontinuous.

5.3 Give an example of a body, containing just four particles, which is equilibrated, but not pairwise equilibrated.

5.4 Provide a counterexample to the assertion that $(\boldsymbol{\xi} + \boldsymbol{\eta}) \times \boldsymbol{f}(\boldsymbol{\eta}, \boldsymbol{\xi}) = \mathbf{0}$ for a pairwise force function $\boldsymbol{f}(\boldsymbol{\eta}, \boldsymbol{\xi})$ connecting two particles with relative deformed position $\boldsymbol{\xi} + \boldsymbol{\eta}$.

5.5 Prove that the pairwise force function $\boldsymbol{f}(\boldsymbol{\eta}, \boldsymbol{\xi}) \equiv F(\|\boldsymbol{\xi}+\boldsymbol{\eta}\|)(\boldsymbol{\xi}+\boldsymbol{\eta})$ is isotropic for any point $\boldsymbol{x} \in \mathcal{R}^{\mathbf{0}}$, where $\mathcal{R}^0$ is defined in Fig. 5.1.

5.6 Prove that the material represented by particle, $\boldsymbol{x}$, located less than the horizon distance, δ, of the boundary of $\mathcal{R}$ cannot be isotropic.

5.7 Derive Equation 5.14, $d\mathfrak{U}(\epsilon_{ij}) = \sigma_{ij} d\epsilon_{ij} = \sigma_{11} d\epsilon_{11} + \sigma_{12} d\epsilon_{12} + \cdots + \sigma_{33} d\epsilon_{33}$, where $d\mathfrak{U}$ is the total differential of the strain energy density $\mathfrak{U}$, ϵ_{ij} are the small strain components, and σ_{ij} are the Cauchy stress components.

5.8 Why is a pendulum supported on a frictionless bearing in an evacuated glass chamber a conservative system?

5.9 Why is a pendulum in a glass chamber filled with air nonconservative?

5.10 Derive Eq. 5.25 from Eq. 5.24.

5.11 A linear-elastic uniaxial bar is stretched axially. The Young's modulus E and cross-sectional area A are constant with respect to the axial coordinate X. Assuming a material space $X \in \mathcal{R}^0 \in \mathbb{R}^1$, determine the pairwise force function f in terms of E, A, and the peridynamic horizon δ. Assume $f = k\frac{|\xi+\eta|-|\xi|}{|\xi|}$, where k is a constant, to be expressed in terms of E, A, and L.

5.12 A linear-elastic uniaxial bar is stretched axially. The Young's modulus E and cross-sectional area A are constant with respect to the axial coordinate X. Assuming a material lattice $X \subset \mathcal{R}^0 \subset \mathbb{Z}^1$, with initial reference lattice spacing L, determine the pairwise force function, f, in terms of E, A, and L, if (a) the peridynamic horizon is $\delta = L$; (b) $\delta = 2L$; (c) $\delta = 3L$; (d) $\delta = nL$. In all cases, assume that $f = k\frac{|\xi+\eta|-|\xi|}{|\xi|}$. k is a constant, to be expressed in terms of E, A, and L.

5.13 Repeat Problem 5.12, this time assuming that pairwise force function f varies as $f = k\frac{|\xi+\eta|-|\xi|}{|\xi|}\left(1-\left|\frac{\xi}{\delta}\right|\right) \forall\ \xi \leq \delta$, and $f = 0 \forall\ \xi > \delta$.

5.14 In a planar ordinary continuum bond-based peridynamic material body with thickness t, the magnitude of the pairwise force function is $f = kt\frac{\|\xi+\eta\|-\|\xi\|}{\|\xi\|} = ktS$, where S is the bond stretch, for all bonds within the peridynamic horizon δ. Assuming homogeneous deformation, determine the Young's modulus E and Poisson's ratio ν of the equivalent linear elastic classical material, assuming (a) plane stress conditions; (b) plane strain conditions.

5.15 In a three-dimensional ordinary continuum bond-based peridynamic material body, the magnitude of the pairwise force function is $f = k\frac{\|\xi+\eta\|-\|\xi\|}{\|\xi\|} = kS$, where S is the bond stretch for all bonds within the peridynamic horizon δ. Assuming homogeneous deformation, determine the Young's modulus E and Poisson's ratio ν of the equivalent linear elastic classical material.

Chapter 6

Particle Lattice Model for Solids

Chapter objectives are to:

- describe current computational solid modeling practices;
- appreciate the need for a lattice model for solid bodies;
- understand the differences between material, referential, spatial, and relative descriptions of solid bodies;
- develop a framework for lattice modeling of solid bodies;
- propose a model for lattice topology.

In Chapter 5, following (Silling, 2000), we presented the bond-based continuum peridynamic model. From Eq. 5.5, we see that this model is represented by an integro-differential equation that must be satisfied at each of an infinite number of points $\boldsymbol{X}$. In addition, the spatial integral involves an infinite number of terms in the summation. Discretization of the system of equations is necessary for computational solution.

One way to discretize the model is to relax the requirement that the solid body's geometry be represented as an analytic continuum, and instead to represent the solid body as a lattice of material particles.

We propose to discretize the material body as a lattice *prior* to creating the peridynamic model, rather than *subsequent* to developing the model. This approach has the triple benefit of being physically reasonable, conceptually simple, and computationally efficacious.

We begin by reviewing the idea of *continuity*, and then we discuss the current practices used in *solid modeling*. Then we go on to introduce lattice solid models.

6.1 Real numbers and continuous functions

A three-dimensional continuum model of a solid body is usually defined with respect to a three-dimensional Euclidian space of real numbers $\mathbb{R}^3$. Indeed, the concept of "real number" originates in the efforts of the early Greek philosophers to "measure the world": geometry. Real numbers were invented only relatively recently, beginning with work by Descartes in the seventeenth century.

Real numbers are necessary for describing physical space. For example, the Pythagorean theorem, $a^2 + b^2 = c^2$, where a and b are the lengths of the two shortest sides of a right triangle, and c is the length of the hypotenuse, does not always (although it may sometimes) have an integer or a rational (integer ratio) solution. For example, if a and b are each of unit length, then $c = \sqrt{2}$, which is an irrational number: it cannot be expressed as a ratio of two integers. Similarly, the circumference, $C = \pi D$, of a circle of diameter D, if D has unit length, is the transcendental number, π. Neither $\sqrt{2}$ nor π can be exactly represented as a ratio of integers.[a]

A real number is the name of one of an ordered set of points along a continuous line, perhaps of finite length, containing an infinite number of points. The real numbers include the integers, the rational numbers (expressible as ratios of integers), the irrational numbers, such as $\sqrt{2}$, and the transcendental numbers (not being a root of a polynomial equation with rational coefficients), such as π. Each real number can be approximated to any degree of precision by a decimal representation, such as $\pi = 3.1415926 \ldots$. The real number line is thus an infinite continuum of points, each with a name, represented by the real numbers.

The real numbers are uncountable; that is, there is no one-to-one mapping from the real numbers to the natural numbers 1, 2, 3,... . Calculus and analytical methods (involving limits) depend critically upon the continuous nature of the real number line. However, computers represent the set of real numbers only approximately, using the mechanism of the *floating-point number* (the set of which are countable).

---

[a] The ancient Egyptians represented π as $256/81 \approx 3.1605$, with an error of about 0.6 percent.

As far as we know, physical space is continuous, and to be perfectly accurate, must be measured using the set of real numbers – even though most of the real numbers can never be exactly represented numerically on a digital computer.

A function is continuous if infinitesimally small changes in input result in infinitesimally small changes in output, as is indicated by Definition 3.1 and Fig. 3.3. Fig. 6.1 shows an example of a discontinuous function. Classical physics usually assumes that physical fields in nature are continuous. However, this is an assumption that may be detrimental to the solution of problems in solid mechanics, in which discontinuities, or fractures, are often an important aspect of mechanical behavior.

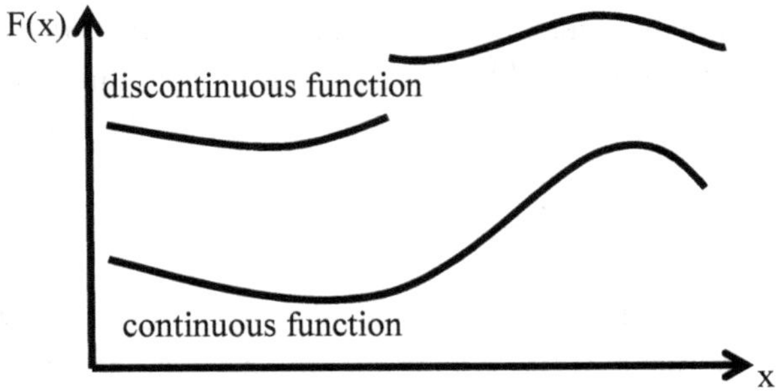

Figure 6.1. Continuous and discontinuous functions.

Mathematically, one sees by Definition 3.1 that continuity is fundamentally dependent upon the continuous nature of the real numbers. For example, Cauchy defined the continuity of a function as follows: a function $y = F(x)$ is continuous if an infinitely small increment α of the independent variable x always produces an infinitely small change $F(x + \alpha) - F(x)$ of the dependent variable y. Only the real numbers allow for these infinitely small increments in α.

6.2 Foundations of solid modeling

According to Truesdell (Truesdell, 1965), there are four "methods" for describing the motion of a body: the *material*, the *referential*, the *spatial*, and the *relative* (shown in Fig. 6.2). The differences lie in choosing what one considers as the independent variables describing the various physical fields.

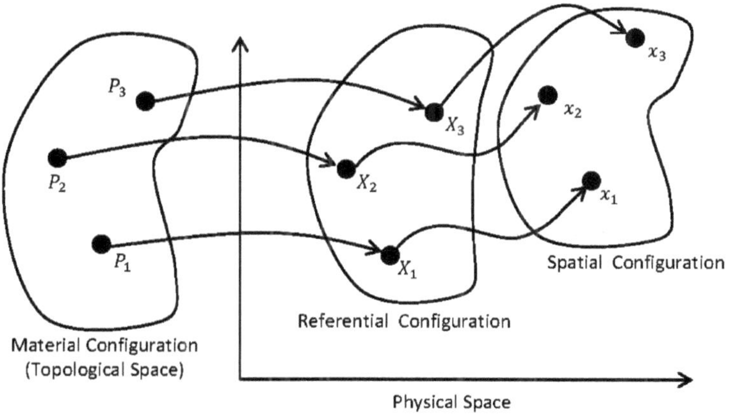

Figure 6.2. Material, referential, and spatial configurations.

In the *material* description, abstract particles P and time t are the independent variables describing the domain of the position and velocity functions; thus position $\boldsymbol{x} = \boldsymbol{x}(P, t)$ and velocity $\boldsymbol{v} = \boldsymbol{v}(P, t)$, etc. This is the method that is most natural both for the application of Newton's laws and for implementation on digital computers, and which we propose to employ in this book. The symbol P represents the name of a particle and not a position in space. Truesdell implies that P is a real number or a set of real numbers; however, for ease in computer modeling of solid bodies, we propose to name particle P using either a single integer or a list of integers.

In contrast, in continuum mechanics, rather than the material description, one typically uses the referential, the spatial, or the relative description as the independent variable. Fig. 6.2 illustrates the material, referential, and spatial notions.

In the *referential* description, the current physical position $\boldsymbol{x}$ as well as other attributes of each particle like velocity, density, and acceleration, are described with respect to an assumed initial position $\boldsymbol{X}$. This is often called the *total Lagrangian* description of motion, and is the most common method for describing the motion of solids. The notion of "solid" fundamentally assumes that there exists an initial, undeformed reference configuration in space. Thus, the current position $\boldsymbol{x}$ and the current velocity $\dot{\boldsymbol{x}}$ are described in terms of the independent variables $\boldsymbol{X}$ and t: $\boldsymbol{x} = \boldsymbol{x}(\boldsymbol{X},\ t)$, $\dot{\boldsymbol{x}} = \dot{\boldsymbol{x}}(\boldsymbol{X},\ t)$, etc.

On the other hand, in the *spatial* description, most often used in fluid mechanics, in which the initial, undeformed configuration is both unknown and unimportant, the location $\boldsymbol{x}$ and time t are the independent variables. With the spatial description, $\dot{\boldsymbol{x}} = 0$, because both $\boldsymbol{x}$ and time t are independent variables. One cannot take a partial derivative of one independent variable with respect to another independent variable, as the operation is not mathematically defined. Thus, for example, the velocity $\boldsymbol{v}$ of the particle currently occupying location $\boldsymbol{x}$ at time t is expressed as $\boldsymbol{v} = \boldsymbol{v}(\boldsymbol{x}, t)$. The spatial description presents a challenge when applying Newton's laws, which one must apply to a particular particle considered as a free body, and not to a particular location in space. Also, if the material body has moving boundaries, then portions of the spatial domain may contain no mass after some time has elapsed, rendering Newton's laws inapplicable.

Finally, in the *relative* description of motion, one can take any homeomorph[b] of the body as the reference configuration. This is the approach taken in *updated Lagrangian* finite element programs, in which the program periodically updates the reference configuration as the deformation progresses. In this case, a function at time t is expressed in terms of particle positions $\boldsymbol{x}^*$ at some earlier time t^*: $\boldsymbol{x} = \boldsymbol{x}(\boldsymbol{x}^*,\ t)$, $\dot{\boldsymbol{x}} = \dot{\boldsymbol{x}}(\boldsymbol{x}^*,\ t)$, etc.

In small-deformation analysis, there is essentially no difference between the material, referential, the spatial, and the relative descriptions of motion, because the geometric change in particle positions as a function of time is assumed to be negligible (even though the displacements are nonzero). Traditionally, undergraduate courses in mechanics of deformable solids and structural analysis tacitly assume small deformations; thus, many students are unaware of these important

[b] A homeomorph of a body is any configuration that one can obtain via continuous deformation of the body.

differences, which only become apparent when considering large deformations.

Most books on continuum mechanics start out by assuming that the material reference domain is a continuous, open subset[c] of $\mathbb{R}^3$, and furthermore that all future deformations are homeomorphs (continuous mappings) of the reference configuration. Thus, the material is described by a set of points in $\mathbb{R}^3$, in which each material point at time zero is labeled by its reference, undeformed, spatial position, $\boldsymbol{X} = (X_1, X_2, X_3)$. The material body is thus initially continuous, due to the inherited properties of real numbers. One thus conflates the material and the referential configurations, and induces an unintended, non-physical topology upon the solid body. The essential assumptions of continuum mechanics prohibit discontinuities and cracks from ever forming.

Topology is the study of connectivity in metric spaces.[d] Topology is the study of the properties of subsets of spaces that remain invariant under continuous deformation. Because $\mathbb{R}^3$ is continuous, the topology of open sets within $\mathbb{R}^3$ possess topological properties such as homotopy.[e] Topology is an important mathematical subject, and is significant in our study of peridynamic lattices. The topology of an integer metric space $\mathbb{Z}^3$ is much simpler than that of a real metric space, $\mathbb{R}^3$, in the sense that, given a lattice point and a distance metric, a finite, countable number of nearest neighbors in the lattice exists. On the other hand, with $\mathbb{R}^3$ metric spaces, it is impossible to identify the "nearest neighbor" to a point, as there are always an infinite number of points between any two distinct points on the real number line.

Computationally, it is desirable to define the set of particles with which each material particle within a solid body potentially interacts. We call this set of points its material neighborhood. While it is possible to compute the force interactions between all pairs of particles within a

---

[c] An open subset does not include its bounding points.

[d] Roughly speaking, a metric space is a set of points, together with a definition of the distances between all pairs of points within the set.

[e] Two objects are *homotopic* if one can be continuously deformed into the other. The concept of homotopy was first formulated by Poincaré around 1900. Homotopic and homeomorphic are related, but distinct concepts.

solid body, even if most such interactions are null, this is prohibitively expensive because on digital computers null floating point computations require as much time as non-null computations. For a domain with N_P particles, the number of force interactions per time step is $N_F = \frac{N_P(N_P-1)}{2}$, which is approximately $\frac{N_P{}^2}{2}$ for large numbers of particles. To limit the necessary number of force computations, several topological ideas can be employed.

The most common approach to describe, at least approximately, topological information amongst particles is to decompose the domain into an array of cubical cells of size somewhat larger than the peridynamic horizon δ (the distance beyond which forces between particles are guaranteed to be null). Using this domain decomposition approach, each particle need only check for interactions with particles located within its own cell and within directly adjacent cells. Thus the number of force interactions is now $N_F \approx \frac{N_P \times N_A \times N_Q}{2}$, where N_Q is the approximate number of particles per cell ($N_Q \approx 18$ particles per cell for 3D close-packed lattices) and N_A is the number of adjacent cells ($N_A = 27$ cells for 3D cell arrays). Therefore, for N_P equal to one million particles, without cell decomposition, there are about $\frac{N_P{}^2}{2} = 5 \times 10^{11}$ force computations per time step, while with cell decomposition there are only about $\frac{N_P \times N_A \times N_Q}{2} = 243 \times 10^6$ force computations per time step (two thousand times less than without cell decomposition). For some types of problems, particles are sufficiently mobile that they can interact with new neighbors (and discontinue interactions with old neighbors) as the simulation progresses; thus the topology changes. For such problems, it is advantageous to update particle cell addresses periodically. We call this "particle shuffling", described in detail in Chapter 13.

In many problems in solid mechanics, however, particularly with solids that do not deform too much, one can assume that particles will not change neighbors during the course of the simulation. In such cases, one can realize computational savings by precomputing a *neighbor list* for each particle before entering the time integration loop. If we can precompute a neighbor list before entering the time integration loop, then a decomposition of the domain into cells may be unnecessary.[f] The neighbor list defines the topology of a discrete set of particles.

The appropriate choice of computational data structure depends upon the class of particle dynamics problem. As we shall see in Chapters 10 and 11, when modeling plasticity or damage, plastic and damage state variables may be associated with the bonds between particles, rather than with the particle itself. In such cases, one may partition and store bond information with each of the interacting particles. Thus, rather than to require a separate data structure defining the bond information, each particle may have an ordered list of bonds, each of which contains the necessary plasticity and damage variables.

Both Newtonian and Einsteinian physics posit that physical space is continuous[g], but neither variety of physics says anything about *material bodies* being continuous. In fact, atomic theory and quantum mechanics state that matter is essentially discontinuous. The assumption that solid materials are continuous crept into the engineering literature with the assumptions of Euler, the Bernoulli brothers, Navier, and Cauchy. They did their work before atomic theory was established fact. However, in material bodies, cracks and fractures often occur. The assumption of continuity in modeling of solid and fluid bodies was a convenient, and even necessary, modeling assumption in the pre-computer age, but one that may, at least where appropriate, be relaxed in the age of fast computers with large memories.

It is not possible to represent a discretely cracked continuous body simply by defining its point set. One also needs topology. While exterior bounding surfaces emerge through the geometric description, cracks, usually characterized by a pair of identically located crack surfaces (with opposing outward-pointing normal vectors to the solid domain) and a crack front, must be defined separately and in addition to the geometric description of the locus of material points. In other words, simply knowing the locus of material points does not guarantee that one knows the bounding surfaces (and particularly the crack surfaces) of the

---

[f] Although a cell-based domain decomposition may still be advantageous for parallelization of the simulation.

[g] Except at black holes.

domain. This presents a problem for solid models based upon the $\mathbb{R}^3$ reference domain, in which cracking may be initially present or may emerge during the simulation.

While modeling of solid bodies using a Euclidian material space was necessary before computers became available, a lattice is more convenient if powerful computers are available. With continuum models, with infinitely many infinitesimally small continuum particles, it is necessary to define a mass density function. Then, it is necessary to satisfy explicitly the conservation of mass, momentum, and energy through appropriate differential equations. On the other hand, with a finite-mass- particle-based model, mass is automatically conserved by virtue of the simple assumption that particle mass does not vary with time. Momentum is conserved, as long as Newton's second and third laws are appropriately applied to the particles. If we select appropriate peridynamic force and thermodynamic interactions between the particles, then energy will also be conserved automatically.

The advantage of using a lattice, rather than a random distribution of particles, is that the lattice provides more structure and symmetry. With a lattice, for example, the particle mass is easily computed from the material mass density, as, by symmetry, each particle within a lattice occupies identical volume. Such is not the case with a random distribution of particles.

Thus, we propose to model solids using a material defined by a countable number of identical material particles within a reference lattice. This approach avoids any unintended assumption of continuity between material particles. The topology can be stated explicitly using cellular decompositions and neighbor lists. Additionally, and importantly, on computers, solid bodies can be more efficaciously represented using integers, rather than with real numbers, to identify the material particles.[h]

---

[h] With the finite element method, the discretization process requires interpolation, using analytical functions, between discrete nodal points.

6.3 Solid modeling in Euclidean space

As we have seen, solid modeling in Euclidean $\mathbb{R}^3$ space carries with it unwarranted topology. Let us review the methods for solid modelling using a Euclidian material space. In the continuum point-set solid modelling, all material points P in a body $\mathcal{B}$ that are elements of a subset $\mathcal{R}$ of $\mathbb{R}^3$ (expressed mathematically as $\mathcal{B} = \{P \in \mathcal{R} \subset \mathbb{R}^3\})$[i] must be classified according to their neighborhoods as interior, exterior, or boundary points. Interior material points P possess open balls (point set spaces that are homeomorphic with $\mathbb{R}^3$). For $\mathcal{B}$ to be considered solid, every neighborhood of $P \in \mathcal{R}$ must take the form of an open ball in $\mathbb{R}^3$. Boundary points can have a number of different topologies. But let us not get sidetracked by the mathematical discipline of geometric point set topology. Suffice it to say, that with $\mathbb{R}^3$ descriptions of geometry, the associated boundary topology can be very complex.

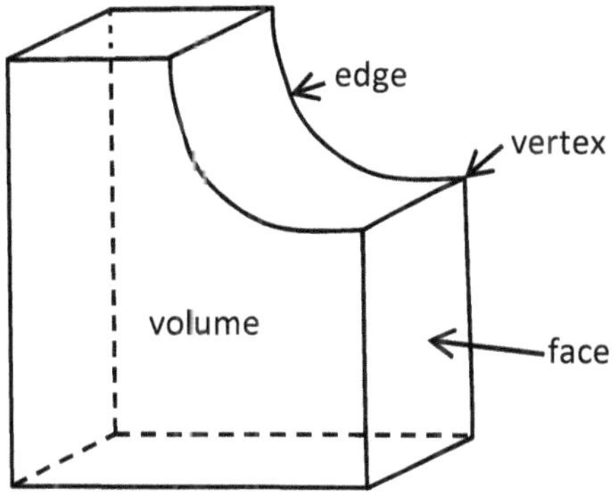

Figure 6.3. Vertex, edge, face, and volume topological elements in a solid model.

There is no standard way to represent solid geometry in $\mathbb{R}^3$.[j] An identical geometry can be expressed in many different ways. In general, real geometries cannot be evaluated exactly numerically, because they contain an infinite number of points. The surfaces, being analytical

---

[i] { } means "the set of"; ∈ means "element of"; ⊂ means "subset of"

[j] Despite efforts such as the STEP (STandard for the Exchange of Product model data) and IGES (Initial Graphics Exchange Specification) solid modeling standards.

functions, are not generally susceptible to exact numerical evaluation. Because there are an infinite number of points to check, this renders impossible even the seemingly simple chore of determining whether or not two geometric models in $\mathbb{R}^3$ are equivalent.

Usually, geometric models in $\mathbb{R}^3$ involve computational data structures containing representations of primitive geometric objects with analytically-defined surfaces, edges, and vertices, as shown in Fig. 6.3. Some of the primitive geometric objects include cuboids, tetrahedrons, spheroids, and NURBS (nonuniform rational B-splines). These primitive geometric objects are usually then combined to define a more complex solid object using Boolean point-set constructive solid geometry (CSG) operations.

The most common methods for solid modeling include parameterized primitive instancing, spatial occupancy enumeration (voxels), cell decomposition (finite element meshes), boundary representations, constructive solid geometry, sweeping, implicit representation using algebraic functions, and parametric feature-based modeling. If this sounds like a jumble of different methods, it certainly is. Because different engineering applications require quite different methods for defining the solid geometry, standardization does not appear to be in the offing.

Rendering a geometric model for the purpose of visualization always involves discretization and associated approximations: decomposing the object into a faceted finite element mesh, which can then be graphically drawn on a display screen using computer graphics. Similarly, simulating the deformation of a solid model involves decomposition into a finite element mesh, or something similar, to discretize the model.

For some materials, either due to the coarseness of their material fabric, or due to the limitations of computational resources, the extreme topological and geometric detail permitted by modeling in $\mathbb{R}^3$ is actually counterproductive to representing a solid model with an appropriate level of fidelity. Thus, "defeaturing" tools have arisen. Defeaturing tools automatically recognize and remove small features that analysts typically want to eliminate from a geometric model prior to meshing. Defeaturing simplifies the model by removing unnecessary detail.

For peridynamic simulations, we propose representing the material body using a particle lattice, as described further in the next section.

6.4 Solid bodies modeled as particle lattices

In engineering, "solid material body" implies some degree of regularity. While we recognize that all materials are very complex at a small-enough scale, consisting of atoms, molecules, crystallites, etc., the mechanics of deformable solids calls for regularization of properties at the macroscale.

However, we are not willing to assume continuity, for the many reasons described in the introduction to this book, nor are we willing to assume that the particles are completely random, as for example in molecular dynamics (MD) or in the discrete element method (DEM) used in soil mechanics.

There is a middle ground between the wilderness of DEM or MD, and the finely manicured grounds of continuum mechanics. A regular lattice of particles – in the initial reference configuration – provides a useful compromise that is well suited to the computational power of digital computers. Thus, a particle lattice seems to provide a reasonable representation of solid bodies upon which to build peridynamic models. In this section, we provide a mathematical description of a particle lattice model.

Let $\mathbb{Z}$ be the set of integers and let $\mathbb{R}^{N_R}$ be an N_R-dimensional real Euclidian space. We identify the initial reference position $\boldsymbol{X}$ of each particle P within $\mathbb{R}^{N_R}$ by the position vector $\boldsymbol{X} = \sum_{i=1}^{N_R} X_i \hat{\boldsymbol{e}}_i$, where $\hat{\boldsymbol{e}}_i$ are unit basis vectors in each of the physical coordinate directions X_i.

Definition 6.1 A *lattice basis* B^{N_R} is a set of N_R N_R-dimensional basis vectors $\boldsymbol{b}_i \in \mathbb{R}^{N_R}$ spanning $\mathbb{R}^{N_R}$. Thus $B^{N_R} = \{\boldsymbol{b}_1 \cdots \boldsymbol{b}_{N_R}\}$.

The lattice basis vectors $\boldsymbol{b}_i$ define the strides between the positions of lattice particles in each of the directions defined by $\boldsymbol{b}_i$.

Definition 6.2 A *lattice basis matrix* $[B^{N_R}]$ is a matrix whose columns are the components of the lattice basis vectors $\boldsymbol{b}_i$ with respect to the physical Cartesian basis vectors $\hat{\boldsymbol{e}}_j$. Thus $[B^{N_R}] = \begin{bmatrix} b_{11} & \cdots & b_{N_R 1} \\ \vdots & \ddots & \vdots \\ b_{1N_R} & \cdots & b_{N_R N_R} \end{bmatrix}$, where b_{ij} is the jth component of basis vector $\boldsymbol{b}_i$.

Definition 6.3 An N_R*-dimensional lattice* Λ^{N_R} in $\mathbb{R}^{N_R}$ is the set of points given by

$$\Lambda^{N_R} = \left\{ \boldsymbol{X} = \textstyle\sum_{i=1}^{N_R} a_i \boldsymbol{b}_i + \boldsymbol{X}_0 = \left[\lfloor a_i \rfloor [B^{N_R}{}_{ij}]^T + \lfloor X_{0j} \rfloor \right] \{\hat{\boldsymbol{e}}_j\} \;\middle|\; a_i \in \mathbb{Z}, \quad \boldsymbol{b}_i \in B^{N_R} \right\}.$$ [k]

The lattice Λ^{N_R} is thus unbounded, and contains an infinite number of lattice points represented by position vectors $\boldsymbol{X}$. When the dimensionality N_R of the lattice is clear or unimportant, we drop the superscript and represent a lattice simply as Λ. The lattice coordinates of a point P are given by the $N_R \times 1$ array of integer components $\{a_i\}$ with respect to lattice basis B^{N_R}. The reference (initial, undeformed) coordinates of the same point P are given by the array of components $\{X_i\}$ with respect to the basis vectors $\hat{\boldsymbol{e}}_i$. The origin of the lattice coordinates $\{a_i\} = \{0\}$ is located at point $\boldsymbol{X}_0$ in $\mathbb{R}^{N_R}$.

Let $\mathcal{R}$ be an N_R-dimensional solid body in $\mathbb{R}^{N_R}$.

Definition 6.4 A *lattice body* $\mathcal{L}_{\mathcal{R}}$ representing a region $\mathcal{R} \subset \mathbb{R}^{N_R}$ is the null particle P_ϕ together with the set of particles P whose positions $\boldsymbol{X}(P)$ in the lattice Λ^{N_R} are also elements of $\mathcal{R}$: $\mathcal{L}_{\mathcal{R}} = \{\Lambda^{N_R} \cap \mathcal{R},\ P_\phi\}$.

The lattice body $\mathcal{L}_{\mathcal{R}}$ thus contains a subset of the particles within the region $\mathcal{R}$, as well as a null particle P_ϕ. If $\mathcal{R}$ has finite volume, then it can be shown that $\mathcal{L}_{\mathcal{R}}$ contains a finite number of lattice points. The null point P_ϕ is included in $\mathcal{L}_{\mathcal{R}}$ as a mechanism to aid in describing the topology, described in the next section.

A homogeneous solid body can be partitioned, using a regular tiling, into regularly-spaced representative volume elements (RVE's) of

---

[k] Λ is the capital Greek letter lambda.

uniform shape and size. Each RVE is represented by a single lattice particle located at its centroid. The volume of a lattice particle is called the covolume of the lattice:

Definition 6.5 The *covolume* $d(\Lambda^{N_R})$ of a lattice with lattice basis B^{N_R} is equal to the determinant of the lattice basis matrix $d(\Lambda^{N_R}) = det[B^{N_R}]$.

The covolume is the N_R-dimensional volume of the parallepiped spanned by the lattice basis vectors.

Lattices Λ^{N_L} of dimension N_L less than N_R can be associated with N_L-dimensional subspaces $\mathbb{R}^{N_L}$ embedded within $\mathbb{R}^{N_R}$. For example, consider a one-dimensional real Euclidian space $\mathbb{R}^1$ defined as $\boldsymbol{X} = s\hat{\boldsymbol{e}}_s + \boldsymbol{X}_0$, where s is a distance coordinate, $\hat{\boldsymbol{e}}_s$ is a unit vector in $\mathbb{R}^3$, and $\boldsymbol{X}_0$ is the location of the origin of Λ^1 in $\mathbb{R}^3$. Then, Λ^1 defined with respect to $\mathbb{R}^1$ defines a one-dimensional lattice oriented in the direction of $\hat{\boldsymbol{e}}_s$ whose origin is at $\boldsymbol{X}_0$ in $\mathbb{R}^3$.

Similarly, planar lattices Λ^2 can be embedded in physical $\mathbb{R}^3$ space.

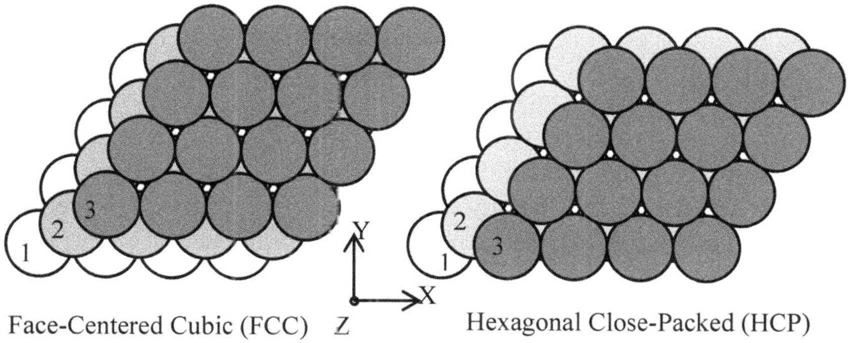

Figure 6.4. Face-centered cubic and hexagonal close-packed configurations.

Three-dimensional close-packed spheres come in two main forms: hexagonal close-packed (HCP) and face-centered cubic (FCC). These types of particle packings are depicted in Fig. 6.4. Notice that the stride between successive layers labeled 1, 2, and 3 is constant for the FCC packing, but not for the HPC packing. Thus, the FCC packing is a lattice, but the HCP is not a lattice, according to Definitions 6.2 and 6.3.

Close-packed FCC lattice basis matrices $[B^{N_R}]$ with respect to the coordinate system shown in Fig 6.4, for 1D, 2D, and 3D close-packed particle lattices are represented as

$$\begin{aligned}&\text{one-dimensional lattice: } [B^1] = L[1];\\ &\text{two-dimensional hexagonal lattice: } [B^2] = L[Q]\begin{bmatrix}1 & \frac{1}{2}\\ 0 & \frac{\sqrt{3}}{2}\end{bmatrix};\\ &\text{three-dimensional FCC lattice: } [B^3] = L[Q]\begin{bmatrix}1 & \frac{1}{2} & \frac{1}{2}\\ 0 & \frac{\sqrt{3}}{2} & \frac{1}{2\sqrt{3}}\\ 0 & 0 & \sqrt{\frac{2}{3}}\end{bmatrix},\end{aligned} \tag{6.1}$$

where L is the lattice particle spacing, and $[Q]$ is a proper (rigid body) rotation matrix.

One can construct a particle lattice body $\mathcal{L}_{\mathcal{R}}$ in many different ways; this is an issue of computational geometric modelling. For example, given a CAD geometry $\mathcal{R}$ and an associated geometry engine, and given the lattice Λ, it is possible to determine an associated particle lattice body $\mathcal{L}_{\mathcal{R}}$ by scanning over a domain containing the geometry, saving only those lattice points P that lie within $\mathcal{R}$. Alternately, if no CAD geometry model is given, specifying the points within $\mathcal{L}_{\mathcal{R}}$ can be the primary means by which one constructs a solid geometry.

Once the particle lattice body $\mathcal{L}_{\mathcal{R}}$ is determined, it is useful to rename its lattice particles P using a more compact naming method than using the vector of lattice coordinates $\{a_i\}$. For example, particles $\mathbb{P}$ constituting the lattice body might be numbered as integers sequentially from one to the number of particles in the body N_P: $\mathbb{P} = \{P \in \mathbb{Z} \mid 1 \leq P \leq N_P\}$. Or, if the reference space is decomposed into an N_R-dimensional cell array, each particle within a cell $C = (cell_1, \cdots, cell_{N_R})$ is given a sequential integer P_i relative to its cell, so that the name (or address) of a particle in a 3D cell array would be the integer address $P = (P_i, cell_1, \cdots, cell_{N_R})$. Regardless of how the particle is addressed, we identify a lattice particle using the symbol P, and we identify its reference position using the symbol $\boldsymbol{X}$.

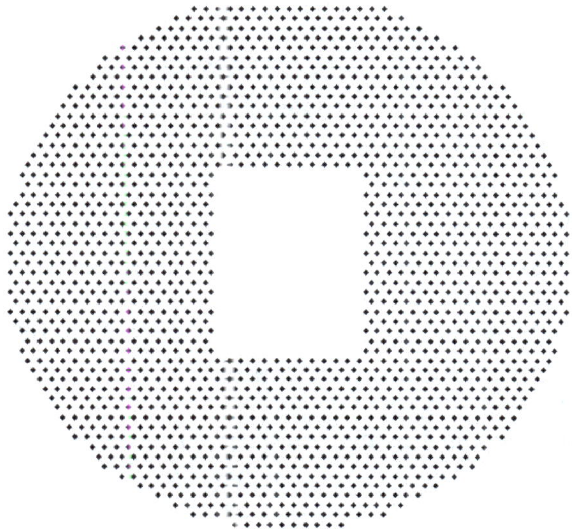

Figure 6.5. Planar hexagonal lattice model of a disk with a rectangular cutout.

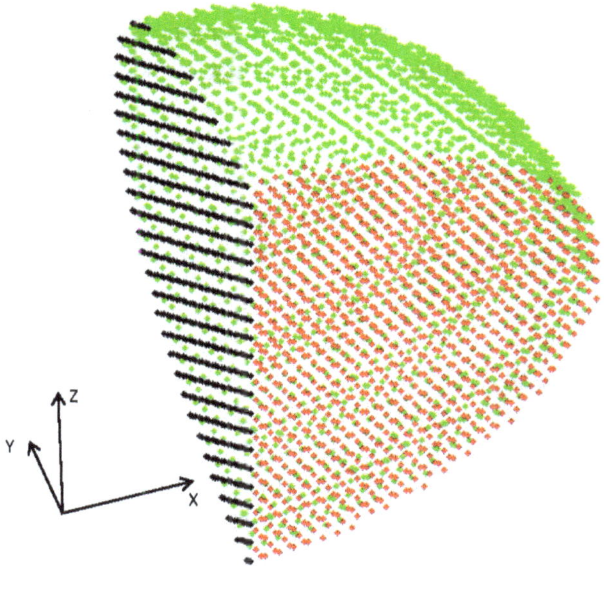

Figure 6.6. FCC lattice model of quarter-sphere.

As examples of particle lattice bodies, Fig. 6.5 shows a hexagonal lattice model of a two-dimensional disk with a rectangular cut-out, and Fig. 6.6 shows a three-dimensional FCC model of a quarter sphere.

Many material bodies of varying dimensionality, represented by different particle lattice bodies $\mathcal{L}_{\mathcal{R}}$, may be included within a single structural model. The computational peridynamic simulation then produces the deformed configuration $\boldsymbol{x}$ of each particle, at each time step.

As the simulation of the solid progresses, damage can occur, and as damage progresses, it can localize into lines and surfaces, forming cracks and fractures. Through the mechanism of damage, particles can become completely physically detached from the main body. Ultimately, the originally solid body can become completely fragmented, in which case particles move independently of one another, like a fluid. It is possible to supply short-range peridynamic force functions in such a case that prevent two particles from occupying the same location in space. At this point, the simulation is more like a molecular dynamics simulation, in which the initial configuration is no longer directly relevant. The simulation is also much more expensive, because it must conduct more checking to determine interactions between particles as they interact with new neighbors. Thus, we see that the lattice model has value for solids, but becomes less essential as the simulation converts the solid to a rubbelized state, or to a fluid, through the mechanism of damage. However, the initial discretization as a lattice with a finite number of particles of identical mass is very helpful, and the solid model, if indicated, can later transition to a rubbelized model as the simulation progresses.

6.5 Lattice topology

With solid models, if the deformation is not too large, particles maintain their neighbors as the deformation progresses. Therefore, it is helpful to develop the concept of *bond* and a *particle bond list*. The bond describes a potential interaction between two particles. These concepts are helpful in developing material models that depend upon the relative positions of a set of neighboring lattice particles.

We call the description of lattice particles adjacent to a particular particle the *lattice topology*. For a given particle P_i lattice topology is used to answer questions like: which particles are P_i's nearest neighbors, and which particles are P_i's second-nearest neighbors, etc.? Also, lattice topology organizes the neighboring particles to P_i in a structured manner so that the bond structure can be easily represented in the constitutive model. We can use lattice topology to answer questions such as, for a given bond, "which is my opposing bond?" and "which are my neighboring bonds?". Additionally, the lattice topology should be able to answer the question as to whether a particular particle is on the boundary of the lattice body $\mathcal{L}_\mathcal{R}$ or not.

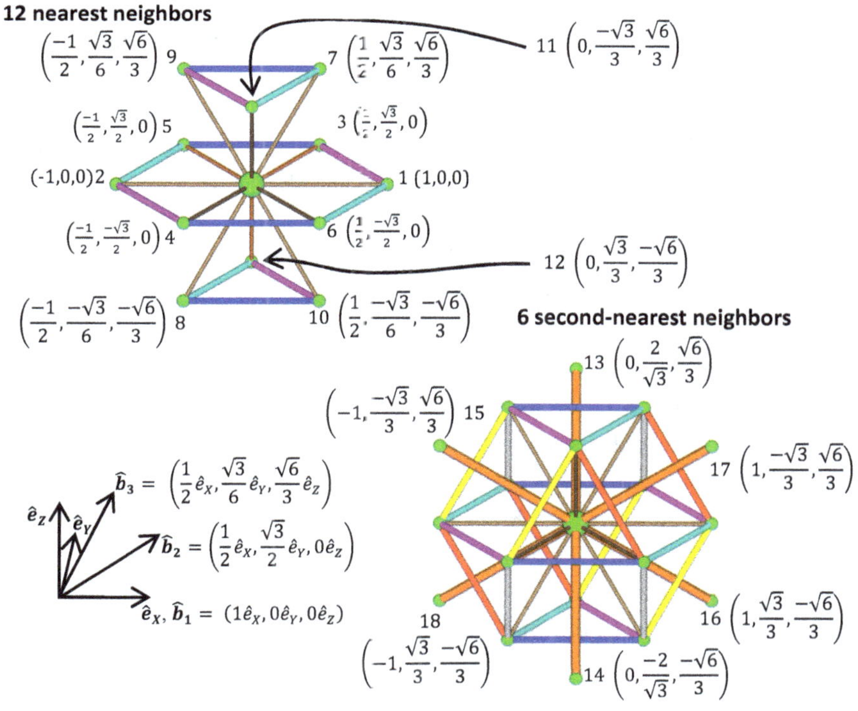

Figure 6.7. Bond numbering order and reference coordinates $\boldsymbol{X}$ of first- and second-nearest neighboring particles for the FCC lattice defined in Eq. 6.1 (assuming no lattice rotation; $\boldsymbol{Q} = \boldsymbol{I}$) in units of lattice spacing, L. Adapted from (Rahman, 2011).

For a given particle $P_i \in \Lambda$, potential interactions with other particles are called the particle's *bonds*. In a lattice Λ, each particle has a specified number of nearest neighbors, second-nearest neighbors, etc. The bonds for a single typical particle in a FCC particle lattice, defined by Eq. 6.1, are numbered and physically positioned as shown in Fig. 6.7. While the numbering order of the bonds is arbitrary, the numbering order must be the same for all particles in a given lattice Λ.

Definition 6.6 A lattice *particle bond list* $\mathbb{B}$ contains a sequential list of N_B bonds, B_i describing the topological connection between a given lattice particle P_i and N_B potentially interacting lattice particles in lattice Λ. Bonds B_j are numbered as sequential integers from 1 to the number of bonds per particle, N_B. Thus, $\mathbb{B} = \{B_j \in \mathbb{Z} \mid 1 \leq B_j \leq N_B\}$.

In Fig. 6.7, the twelve nearest neighbors, of undeformed length L, as well as the six second-nearest neighbors, of undeformed length $\sqrt{2}L$, are shown. Third-nearest neighbors, and so on, could also be included. Notice that we have chosen to number opposing bonds in odd/even sequential pairs. In addition, the bonds are numbered in a specific order spatially, starting with the two bonds on the X axis, then the remaining four bonds in the XY plane, next the six shorter out-of-plane the bonds, and finally the six longer out-of-plane bonds.

The particle bond list is a bond numbering system local to each particle. Each particle within a particle lattice Λ has an identical particle bond list $\mathbb{B}$. For the FCC lattice shown in Fig. 6.7, including first- and second-nearest neighbors, the number of bonds $N_B = 18$.

With reference to Fig. 6.7, in a one-dimensional lattice, only the first two bonds are included, so $\mathbb{B} = \{1,2\}$ and in a planar hexagonal lattice, only the first six bonds are included, so $\mathbb{B} = \{1, \cdots, 6\}$.

For a given particle P_i within a lattice body $\mathcal{L}_{\mathcal{R}}$, the neighboring lattice particles within the material horizon δ are given by the neighbor list $\mathcal{N}[P_i]$, defined next.

Definition 6.7 Given lattice body $\mathcal{L}_{\mathcal{R}}$ with the set of particles $\mathbb{P}$ and a material particle $P_i \in \mathbb{P}$, P_i's *neighbor list* $\mathcal{N}[P_i]$ is a function mapping the particle bond list $\mathbb{B}$ to adjacent lattice particles: $\mathcal{N}[P_i]\langle\cdot\rangle : \mathbb{B} \to \mathbb{P}$.

For a given particle $P_i \in \mathbb{P}$, if there is no adjacent particle within $\mathcal{L}_{\mathcal{R}}$ associated with bond $B_j \in \mathbb{B}$, then its absence is signaled by setting $\mathcal{N}[P_i]\langle B_j\rangle = P_\phi$, in which P_ϕ is the null particle of $\mathbb{P}$. Also, $\mathcal{N}[P_\phi]\langle B_j\rangle = P_\phi$, $\forall B_j$.

$P_j = \mathcal{N}[P_i]\langle B_j\rangle$ gives the lattice particle P_j that interacts with lattice particle P_i via bond B_j. So the neighbor list $\mathcal{N}[P_i]$ provides an ordered list of interacting lattice particles P_j, except that if a given bond B_j is on the boundary of the lattice body $\mathcal{L}_{\mathcal{R}}$, then an associated adjacent lattice particle P_j does not exist, as is indicated by its connection to the null particle P_ϕ.

Also, we may view the particle bond list $\mathbb{B}$ as the local domain associated with each particle, containing the independent variables B_i, used to represent particle-associated fields such as a bond stretch, bond force, bond plastic stretch, and bond damage parameter.

6.6 Computational lattice model for a structure

So far, we have defined the lattice body $\mathcal{L}_{\mathcal{R}}$ and its topology described by the particle bond list $\mathbb{B}$ and the neighbor list $\mathcal{N}$. To write a computer program using lattice bodies to represent a structure, we require a few more ideas, conveyed by the following definitions.

First, let us define what we mean by a "lattice model for a deformable solid structure". We start with several definitions.

Definition 6.8 A *structure* is a collection of lattice bodies $\mathcal{L}_{\mathcal{R}}$. The lattice bodies interact through *body interactions*.

Structures are composed of component lattice bodies $\mathcal{L}_{\mathcal{R}}$. We are assuming that each component of a structure is a body composed of a particular material. As each material may have its own characteristic mesoscale, each material body requires its own lattice, and has its own reference lattice spacing L.

Definition 6.9 A *material body* is an object containing a lattice body $\mathcal{L}_{\mathcal{R}}$, an associated particle list $\mathbb{P}$, a bond list $\mathbb{B}$, a neighbor list $\mathcal{N}[P_i]$ for each particle $P_i \in \mathbb{P}$, a particle mass m, and a *peridynamic force state function* $\tilde{\boldsymbol{T}}$ (defined in Chapter 8).

Each particle in a given lattice body $\mathcal{L}_{\mathcal{R}}$ has identical mass m, and identical peridynamic force state function $\underline{\tilde{\boldsymbol{T}}}$. This model thus excludes functionally-graded materials, for example, in which material properties may vary spatially within a given lattice body (although alternate choices for modeling objects could well include the class of functionally-graded materials).

Definition 6.10 A *material particle* is a particle, $P_i \in \mathcal{L}_{\mathcal{R}}$, characterized by *particle fixed attributes* and *particle alterable attributes*.

We might want to separate the fixed attributes from the alterable attributes because for parallel computing on multiple processors, the fixed attributes need not be swapped between adjacent processors in each time step, in contrast to alterable attributes. In addition, it is of value to the human modeler to clearly separate that which is fixed from that which is alterable in a given simulation.

Definition 6.11 The *particle fixed attributes* include all attributes of a material particle that do not change over the course of the simulation.

Particle fixed attributes include the particle global identifier, $P_i \in \mathbb{P}$, initial particle position $\boldsymbol{X}$, the particle material body identifier, and boundary condition codes $BCcodes = (bc_x, bc_y, bc_z)$, which provide a method to assign loads and displacement conditions to each degree of freedom. The particle mass m is assigned to the material body, and need not be assigned to each individual particle of the material body.

Definition 6.12 The *particle alterable attributes* include all attributes of a material particle that may change over the course of the simulation.

Particle alterable attributes include the current spatial position $\boldsymbol{x}(t) = (x(t), y(t), z(t))$, the current velocity $\boldsymbol{v}(t) = (v_x(t), v_y(t), v_z(t))$, perhaps rotational particle position and rotational velocity, and other particle state variables like plastic stretch parameters, damage, temperature, entropy, etc.[1]

Definition 6.13 A *peridynamic force state function* $\underline{\widetilde{\boldsymbol{T}}}$ is a specification of the peridynamic relationship between force state $\underline{\boldsymbol{T}}$ and deformation state $\underline{\boldsymbol{x}_R}$: $\underline{\boldsymbol{T}} = \underline{\widetilde{\boldsymbol{T}}}(\underline{\boldsymbol{x}_R}, \Lambda_0)$, where Λ_0 represents state variables other than particle position, like damage, plastic stretch, and temperature, that affect the forces between particles within a lattice body. (We define force states $\underline{\boldsymbol{T}}$ and deformation states $\underline{\boldsymbol{x}_R}$ in Chapter 8.)

Definition 6.14 A *body interaction* is a specification of how material particles in bodies interact, excluding internal material forces determined by the peridynamic force state function $\underline{\widetilde{\boldsymbol{T}}}$.

Body interactions are different than peridynamic force state functions $\underline{\widetilde{\boldsymbol{T}}}$. Body interactions specify all forces between particles in the structure that are not specified by the force state function $\underline{\widetilde{\boldsymbol{T}}}$. For example, contact forces between particles in different bodies (or even forces between contacting particles within the same body) are handled by body interactions.

With the structure now defined as a collection of material bodies, which are in turn defined by lattices of material particles, it is also necessary to specify how the body is loaded and how it is supported. We implement these specifications in the time integration function that updates the particle state in each time step. The boundary condition codes specified by the particle fixed attributes are used to aid in defining how loads and supports are applied to each particle.

We describe the time integration procedures in Chapter 12.

---

[1] For convenience, we may also include the global particle identifier $P_i \in \mathbb{P}$ as an alterable attribute, even though it remains fixed in a given simulation.

6.7 Modeling issues

The lattice body $\mathcal{L}_{\mathcal{R}}$ that we have described in this chapter assumes that the particle spacing L is much smaller than the least macro-dimension of the material body. Thus, any material body, even a slender beam, or, for another example, a thin shell, can be simulated as a fully 3D particle lattice, but the particle spacing L needs to be significantly smaller than the minimum transverse dimension of the beam or shell. This implies that many thousands, and perhaps millions, of material particles are required to simulate realistic structures. Our assumption is that computational resources are expanding rapidly and thus that many structures of interest can be, or may soon be, modelled as collections of three-dimensional lattice bodies. More practically, it is nonetheless often useful to model axial members as 1D lattices and problems with planar symmetry as 2D lattices, which we handle as separate modeling cases.

It is up to the modeler to decide how material bodies should interact. For example, in Chapter 14, we simulate reinforced concrete beams by modeling the concrete as a 2D or 3D material body, with each of the steel reinforcing bars as a separate 1D material body. We assume that a steel particle interacts with a concrete particle elastically but with degrading stiffness as damage to the concrete particle increases.

As another example, if a steel base plate is bonded to a concrete beam, the steel-concrete particle bond interactions can be assumed linear, with an appropriately chosen stiffness relationship.

Thus, we handle interaction between particles residing within separate material bodies using ad-hoc, problem-specific peridynamic interaction functions. One cannot think of this type of peridynamic interaction as a material property, but instead as an interfacial interaction between dissimilar material bodies. These interfacial peridynamic functions can be quite complex and nonlinear.

We discuss application of external loads and applied displacements in detail in Chapter 14. Generally, at the computational level, one applies loads only to material particles – and not to volumes, faces, edges, and vertices as is typical in a conventional CAD model. However, one can write higher-level functions that automatically simulate applied body forces, tractions, line loads, and point loads as statically equivalent loads applied to material particles within appropriate subdomains of the body.

Care is required in applying kinematic boundary conditions to material bodies. For example, applying a "pin support" may be more involved than simply fixing the location of a single particle, which would be like fixing the location of a single piece of aggregate in a real beam. Actual structures distribute kinematic supports over a finite volume, and one must similarly distribute these supports in the computational simulation. One can simulate a pin support by providing appropriate translational constraints to a set of particles residing within a spherical domain, as indicated in Fig. 6.8. One chooses the radius of the pinned support to simulate the actual support conditions.

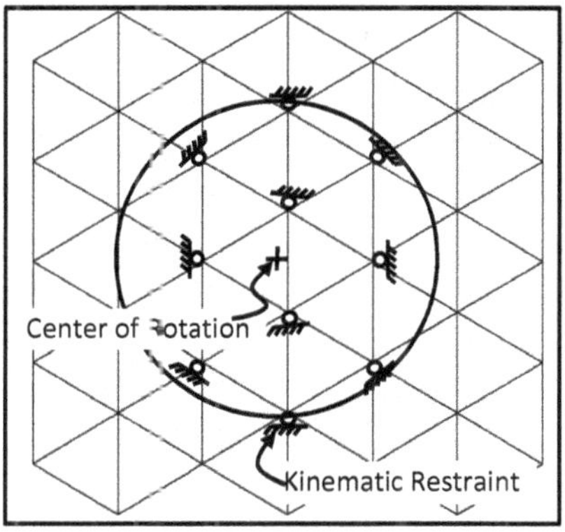

Figure 6.8. Simulation of a pinned support

6.8 Summary

In this chapter, we have presented a particle lattice model for representing structures composed of solid bodies. Given a valid CAD solid geometry and associated geometry query engine, the modeler can generate an equivalent lattice model. The lattice model will generally contain a large number of lattice particles, and these particles serve as the base domain for describing all simulation fields.

You might well ask the question: "Why not just model the material as a collection of *random* particles?" The answer is that the lattice model provides conceptual, as well as computational, simplicity. For example, given a material body with a material mass density in the reference configuration, it is easy to generate the particle lattice representation, and using this lattice, then to calculate the particle masses and peridynamic force state functions. It would be much more difficult to assign particle masses if there were no lattice structure. We can exploit the symmetry provided by the lattice.

The random particle approach has been taken by various researchers, for example (J. E. Bolander and G. S. Hong, 2002) and (Cusatis, 2001). While random particle approaches may be reasonable to model materials at the mesoscale, where material science is the primary interest, a more unified, uniform, and deterministic approach seems more efficacious for modeling macroscale structures.

The material lattice provides a length scale L for each material body that is helpful in constitutive modeling. Essentially, once we have chosen the length scale, we have specified the level of detail of the material body. For conceptual simplicity this is a good thing because it defines what smallest feature size the model has a chance of simulating. Additionally, all real materials have intrinsic length scales, and it makes no sense to model a material at a length scale smaller than the intrinsic material scale. However, materials can be modeled with lattice spacing L larger than the intrinsic material length scale – as long as the computational modeler keeps in mind that features of size smaller than L, both initial and emergent, will not be well represented.

In conclusion, the particle lattice model seems to be as simple as possible for modeling "material bodies", while still allowing the major features of both continuum mechanics of solids and fracture mechanics to emerge using the peridynamic approach.

6.9 Exercises

6.1 Show that a lattice is not a vector space over the field of real numbers $\mathbb{R}$.

6.2 Show that a lattice is a vector space over the field of integers $\mathbb{Z}$, using the conventional definitions of vector addition and scalar product.

6.3 Define the lattice basis matrix $[B^2]$ for a cubic lattice with particle spacings of L_X and L_Y aligned with the X and Y directions in $\mathbb{R}^2$.

6.4 Using the lattice basis matrix of Problem 6.3, create a lattice body $\mathcal{L}_{\mathcal{R}}$ that represents a disk of radius 5 m and centered at $\boldsymbol{X_0} = \{0\}$. Assume $L_X = 1m$, $L_Y = 1m$, $\boldsymbol{X_0} = \{0\}$, and the proper rotation matrix is $[Q] = [I]$. Plot the lattice by hand or using MatLab.

6.5 Repeat Problem 6.4, except: (a) translate the lattice by 0.5 m in the X direction; (b) first rotate the lattice counterclockwise by an angle of 15 degrees about the lattice origin, and then translate the lattice by 0.5 m in the X direction; (c) first translate the lattice by 0.5 m in the X direction and then rotate the lattice counterclockwise by an angle of 15 degrees about the lattice origin.

6.6 Determine the covolume $d(\Lambda^2)$ of the lattice of Problem 6.3 by calculating $det[B^2]$, and verify the result using simple geometry.

6.7 Using the two-dimensional hexagonal lattice basis matrix $[B^2]$ in Eq. 6.1, determine the lattice coordinates of (a) all first-nearest neighbors; and (b) all second-nearest neighbors to the particle whose lattice coordinates are $\boldsymbol{a} = \{0,0\}$. What are the reference lengths of the second-nearest neighbor bonds?

6.8 Give the three integer lattice coordinates (a_1, a_2, a_3), relative to the central particle, of each of the 18 bonds shown in Fig. 6.7. Show these coordinates in a table with 18 rows and three columns.

6.9 Using the lattice topology defined by the particle bond list $\mathbb{B}$ defined by Fig. 6.7, and the FCC lattice basis matrix $[B^3]$ in Eq. 6.1, and assuming $\boldsymbol{Q} = [I]$, $\boldsymbol{X_0} = \{0\}$ and $L = 1$, write a MatLab function that returns the coordinates $\boldsymbol{X_i}$ of all 18 first- and second-neighbors to a lattice particle located at the origin of $\mathbb{R}^3$. The output will be an 18x3 matrix with one particle per row, and the relative reference coordinates of each particle in the three columns. Use the lattice coordinates a_i from Problem 6.8 to generate coordinates $\boldsymbol{X}_i$.

6.10 According to the bond numbering particle bond list $\mathbb{B}$ defined in Fig. 6.7, show that for all particles P_0 not on the boundary of lattice body $\mathcal{L}_{\mathcal{R}}$, $\mathcal{N}[\mathcal{N}[P_0]\langle B_i\rangle]\langle B_{i+1}\rangle = P_0\,, for\ all\ i = 1,3,5,\dots,17$, where $\mathcal{N}[P_0]$ is the neighbor list of particle P_0.

6.11 Write a MatLab function, called *plotParticles*, that takes, as input, an array of points $\boldsymbol{X}$, and plots these points on the computer screen. Assume that $\boldsymbol{X}$ is of size N_P points by dimensionality N_R, where $N_R = 1, 2, or\ 3$.

6.12 Using MatLab, write a function called *disk*, that, given a 2D center location $\boldsymbol{X}_c$, a radius R, and a lattice spacing L, produces a hexagonal lattice model of a planar disk. Assume the lattice is rotated by θ degrees and translated by $\boldsymbol{X}_0$. The output of the function is an array of reference points, $\boldsymbol{X}$. Use the function *plotParticles* from Problem 6.11 to display the disk.

6.13 Using MatLab, write a function called *squareWithHole*, that, given a side dimension, a, a radius, R, and a lattice spacing, L, produces a particle model of a square plate aligned with the Cartesian coordinate axes $\boldsymbol{X}$ and centered at the origin of $\mathbb{R}^2$, with a circular hole of radius R also centered at the origin. Assume the hexagonal lattice is rotated by θ degrees and translated by $\boldsymbol{X}_0$. The output of the function is an array of reference points, $\boldsymbol{X}$. Use the function *plotParticles* from Problem 6.11 to display the lattice.

6.14 Using MatLab, write a function called *quarterSphere*, that creates the FCC lattice geometry shown in Fig. 6.6. Assume the sphere has radius R. Assume the lattice is rotated by θ_X degrees about the X-axis and translated by $\boldsymbol{X}_0$. The output of the function is an array of points, $N_P \times 4$ array $\boldsymbol{X}$, representing the lattice $\mathcal{L}_{\mathcal{R}}$. The fourth column of $\boldsymbol{X}$ defines the color (as an integer value) of each particle for plotting purposes. Modify the function *plotParticles* to display the lattice using appropriate colors for the various surfaces of the quarter sphere. Do not display the interior points.

6.15 Using the function *quarterSphere*, from Problem 6.14, assume a lattice spacing $L = 1$ and radius $R = 20$. Plot the volume of the solid geometry as the lattice is translated in the X direction by one full lattice spacing for $R = 5L$, $R = 10L$ and $R = 20L$. Also, plot the number of particles on the surface and the total number of particles as a function of one full lattice translation in the X-direction.

6.16 Repeat Problem 6.15, but this time, instead of translating the lattice in the X direction, rotate the lattice by 30 degrees about the X axis, and make the requested plots as a function of angle of rotation.

Chapter 7

Elastic Bond-Based Peridynamic Lattice Model

Chapter objectives are to:

- demonstrate the use of the lattice as a basis for elastic bond-based peridynamic modeling;
- introduce a linear-elastic bond-based peridynamic lattice model;
- develop a linear-elastic micropolar bond-based model;
- discuss the advantages and the limitations of bond-based lattice models.

We have introduced the *continuum* bond-based peridynamic model in Chapter 5. To implement the model on a computer, however, we must first *discretize* the model. In Chapter 6, we developed a lattice-based model for solids. The advantage of the lattice-based model is that it provides an already-discretized solid model that can then serve as the domain supporting deformed solids, including those with cracks.

In this chapter, assuming conditions of small bond stretch, we implement an elastic bond-based peridynamic model directly using a lattice solid model, thus avoiding any assumption of spatial continuity. The resulting model provides the pairwise peridynamic force as a function of bond stretch. As we reduce the lattice spacing, the bond-based lattice model converges to the classical Navier theory of linear elasticity. However, the lattice-based theory correctly models large rigid-body rotational deformations, while the Navier theory does not. Thus, the current theory is geometrically nonlinear.

Additionally, the current bond-based lattice theory is applicable even if the bond stretches become relatively large. This is because, aside from the lattice approximation, we are making no kinematic approximations, as is the case with the Cauchy small strain theory.

We show that the resulting elastic peridynamic model is macroscopically isotropic under homogeneous deformation, even though at the microscale the lattice is not isotropic.

Then, we develop one-dimensional, two-dimensional, and three-dimensional bond-based peridynamic elastic lattice models and we provide microelastic parameters and their relationships to the classical Young's modulus and Poisson's ratio. We find that the bond-based peridynamic lattice model cannot model three-dimensional bodies with Poisson's ratio other than one-quarter.

Next, we develop a linear elastic micropolar lattice model, which can model materials with arbitrary physically admissible Poisson's ratios.

For one-dimensional and two-dimensional lattice models, it is sufficient to include only nearest lattice neighbors in the peridynamic horizon. For three-dimensional models, in order to simulate isotropy, it is necessary also to include the second-nearest neighbors.

In addition, we discuss how to handle the conditions at the boundary of the lattice body.

7.1 Linear elastic bond-based lattice models

Using the "bond-based" continuum peridynamic model, established by Silling (Silling, 2000), as a guide, we develop in this chapter two types of linear elastic bond-based lattice models: regular and micropolar.

By *bond-based*, we mean that the mutual forces (and in the micropolar case, moments) acting between two particles are functions only of the states of the two interacting particles. This is in contrast with the *state-based* peridynamic model, to be introduced in Chapter 8, in which the mutual forces acting between two particles are assumed to be functions not only of the states of the interacting pair of particles, but also of the states of other particles in the neighborhood of the interacting particles. Researchers have used bond-based style models for decades, in the form of the discrete element model (DEM) by the geotechnical community (Radjaï, Farhang and Dubois, Frédéric, 2011). Elastic bond-based lattice models of various sorts have been presented in (Eringen, 2002), (Kroner, 1981), (Krumhansl, 1968), (Kunin, 1983), and

(Schlangen, E. and Van Mier, J.G.M., 1992). However, it appears that none of these models has focused upon providing a general and rigorous computational alternative to continuum mechanics.

We will see that the regular (non-micropolar) linear elastic bond-based peridynamic model is a one-parameter model that, for regular lattices, produces a Poisson's ratio of one-fourth for plane strain and three-dimensional problems, and produces a Poisson's ratio of one-third for plane-stress problems, and is thus insufficiently general to model materials with other Poisson's ratios. However, for many applications, a Poisson's ratio of one-fourth or one-third is perfectly adequate (concrete typically has a Poisson's ratio of 0.22, and steel has a Poisson's ratio of 0.3). For the majority of engineering problems, Poisson's ratio is simply not a critical parameter (in fact, neither the AISC Steel Construction Manual nor the ACI 318 building code even mentions Poisson's ratio – Poisson's ratio is not important for the design of most civil engineering structures).

Thus, for many structural engineering problems, the bond-based regular peridynamic model is acceptable – and is both simpler to implement and computationally more efficient than the state-based peridynamic lattice model (SPLM), described in Chapter 8.

On the other hand, the linear elastic bond-based *micropolar* model is a two-parameter model that can simulate materials with various Poisson's ratios. In addition, the micropolar model is perhaps physically more appealing than the regular model, because one would expect finite-sized particles to possess rotational as well as translational degrees of freedom. While the non-micropolar model predicts that a one-dimensional lattice body $\mathcal{L}_{\mathcal{R}}$ representing, say, a cantilever beam in a three-dimensional physical space $\mathbb{R}^3$ would be unstable in bending, the micropolar model predicts stable behavior in such a case.

One cannot always compare the micropolar model to existing classical continuum theories. The micropolar model is more comparable to the continuum micropolar theory of the Cosserat brothers (Cosserat, E. and Cosserat, F., 1909). However, the micropolar model, involving particle rotations, is more difficult to extend to large-deformation regimes, and is more involved to implement in a computer program.[a]

To give a flavor for these two (regular and micropolar) bond-based peridynamic models we present simple versions of each of them in the following two sections. To avoid unnecessary generality, and to simplify our exposition, we assume a lattice body $\mathcal{L}_{\mathcal{R}}$ based upon a close-packed face-centered cubic particle lattice Λ, as described in Chapter 6. In addition, we assume that particle interaction is only between nearest neighbors, and only if necessary, to enable isotropic material behavior, second-nearest neighbors.

7.2 Linear elastic regular lattice model

In the following subsections, we first present a one-dimensional lattice body $\mathcal{L}_{\mathcal{R}}$ defined in lattice Λ^1 for modeling straight bars, then a two-dimensional hexagonal planar lattice body $\mathcal{L}_{\mathcal{R}}$ in lattice Λ^2 for modeling plane stress and plane strain problems, and finally, a fully three-dimensional face-centered cubic close-packed lattice model based upon lattice Λ^3.

7.2.1 *One-dimensional strand*

First, let us first consider a one-dimensional lattice body $\mathcal{L}_{\mathcal{R}}$ defined with respect to lattice Λ^1 (defined by Definition 6.3) embedded within a three-dimensional physical space $\mathbb{R}^3$. With reference to Fig. 6.7 and Definition 6.6, the particle bond list is $\mathbb{B} = \{1, 2\}$. We label each particle as P_i, and the set of particles in lattice body $\mathcal{L}_{\mathcal{R}}$ is $\mathbb{P} = \{P_i \in \mathbb{Z} \mid 1 \leq P_i \leq N_P\}$. Initially, in the undeformed reference configuration, we assume that the particles in the material body form a linear strand along the X-axis (perhaps with gaps in the lattice), and each particle P_i has an initial position X_i, given by the one-dimensional form of Definition 6.3. The initial reference spacing between neighboring particles P_i and $P_j = \mathcal{N}[P_i]\langle B_j \rangle$ is constant: $L = |X_j - X_i|$. As external forces F_i, assumed oriented in the X-direction, are applied to each of the

[a] Large rotations do not behave as vectors, which causes theoretical difficulty.

particles, each particle is displaced from its initial position X_i to a new position x_i. The stretch between two adjacent particles, P_i and P_j, is defined as

$$S \equiv \frac{|x_j - x_i| - L}{L} , \tag{7.1}$$

We assume that the pairwise peridynamic force F between adjacent particles is linearly related to the stretch S between the particles:

$$F = cS , \tag{7.2}$$

or in vectorial form, (in the case where the bar represented by lattice body $\mathcal{L}_{\mathcal{R}}$ may reside in a two-dimensional or three-dimensional space $\mathbb{R}^3$)

$$\boldsymbol{F}_{ij} = cS \frac{x_j - x_i}{|x_j - x_i|} . ^{\text{b}} \tag{7.3}$$

By analogy with a linear elastic bar from Mechanics of Materials, we see that the parameter c is equal to EA, where E is the Young's modulus of the material and A is the cross-sectional area of the bar.

The stretches of the bonds attached to each particle are represented on a computer by a 2×1 matrix containing the bond stretches S_1 and S_2, along with the stipulation that the direction of each bond's deformation vector aligns with its associated current bond direction. We calculate the forces acting upon each of the particle's two bonds by the linear relationship

$$\begin{Bmatrix} F_1 \\ F_2 \end{Bmatrix} = c \begin{Bmatrix} S_1 \\ S_2 \end{Bmatrix} , \tag{7.4}$$

along with the understanding that the forces act in the direction given by Eq. 7.3.

---

[b] This is what is known as an *ordinary, mobile* material, discussed further in Chapter 8.

It is also possible to consider a strand of particles, in a two- or three-dimensional physical space $\mathbb{R}^2$ or $\mathbb{R}^3$, that need not be initially straight (as with lattice Λ^1). In this case, the direction of the peridynamic force between an adjacent pair of particles is aligned with the positions of the particles. Furthermore, the direction of the pairwise force changes as the particles move in two or three dimensions, thus rendering the relationship between force and particle position nonlinear, even if we continue to assume the relationship between the peridynamic force signed magnitude and the stretch given in Eq. 7.4 is linear. If each particle has mass, m, then we can simulate large-deformation cable dynamics using the particle dynamics explicit time integration algorithm described in Chapter 12. A strand of particles that is not straight may be used, say, to model a bent or curved reinforcing bar in a concrete structure, where the lack of internal equilibrium due to differing peridynamic force directions acting upon a particle is equilibrated by interactions with the surrounding concrete particles.

The regular (non-micropolar) bond-based model does not model bending resistance. This may in many cases be acceptable, as typically the axial stiffness of a reinforcing bar is much greater than its bending stiffness.

As an aside, there is an even simpler peridynamic model that can be useful. A *harmonic* material is one in which the *magnitude* of the peridynamic force does not change with particle position – only its *direction*. The problem of a vibrating string with initial pretension is an example of harmonic material behavior. One may model prestressing tendons in prestressed concrete beams as harmonic materials.

To clarify, what we call a *linear elastic* bond-based model may not be linear in an overall sense if the particle positions cause internal force directions between particles to change during the simulation. The model is only linear in the sense that the axial force in a bond relates linearly to the stretch of that bond.

7.2.2 *Two-dimensional planar lattice*

A two-dimensional linear elastic planar lattice body $\mathcal{L}_{\mathcal{R}}$ is useful for modeling plane stress and plane strain problems (to use classical terminology). With a hexagonal lattice Λ^2, using only bonds $\mathbb{B} = \{1, \cdots, 6\}$ shown in Fig. 6.7, a planar body with Young's modulus E and thickness t_b can be simulated by assigning c in Eq. 7.2 appropriately. We need only include the six nearest neighbors in the peridynamic horizon to simulate an isotropic homogeneous material. In consideration of radial symmetry with respect to a given particle, all six pairwise interactions must have identical stiffnesses. To develop a peridynamic lattice model that is "as close as possible" to a classical elastic model, we require, for identical homogeneous deformation fields, identical elastic energy density. The initial reference volume associated with a typical lattice particle is given by the lattice covolume $d(\Lambda^{N_R})$ times the body thickness t_b,

$$\Delta V = d(\Lambda^{N_R})t_b = \frac{\sqrt{3}}{2}L^2 t_b \,, \tag{7.5}$$

and the linear elastic strain energy, $\Delta E_{classical}$, associated with this volume in the classical model, where stress $\{\sigma\}$ and strain $\{\varepsilon\}$ are linearly related as

$$\{\sigma\}=[D]\{\varepsilon\}, \tag{7.6}$$

is found to be

$$\Delta E_{classical} = \int_{\Delta V}\left(\int_{\{\varepsilon\}} \lfloor\sigma\rfloor d\{\varepsilon\}\right) dV = \frac{1}{2}\lfloor\varepsilon\rfloor[D]\{\varepsilon\}\Delta V. \tag{7.7}$$

On the other hand, we find the elastic energy stored per particle in the peridynamic lattice to be the sum of elastic energy contributions stored in each of the six bonds:

$$\begin{aligned} \Delta E_{peridynamic} &= \sum_{i=1}^{6}\left(\int_{S=0}^{S} F_i\left(\frac{L}{2}\right) dS_i\right) \\ &= \sum_{i=1}^{6}\left(\int_{S=0}^{S} cS_i\left(\frac{L}{2}\right) dS_i\right) = c\left(\frac{L}{2}\right)\sum_{i=1}^{6}\left(\frac{1}{2}S_i^{\,2}\right) \\ &= c\left(\frac{L}{4}\right)\sum_{i=1}^{6}(S_i^{\,2}) = \lfloor S_i\rfloor\left[\frac{cL}{4}\right]\{S_i\} = \frac{cL}{4}\lfloor S_i\rfloor\{S_i\}, \end{aligned} \tag{7.8}$$

where $\left[\frac{cL}{4}\right]$ is a diagonal 6x6 matrix. Note that, in consideration of symmetry, we have associated one-half of the energy in each bond with a given particle.

For the lattice displacement field to be equivalent to a homogeneous classical strain field, we require that

$$S_i = n_{Xi}{}^2\varepsilon_{XX} + n_{Yi}{}^2\varepsilon_{YY} + n_{Xi}n_{Yi}\gamma_{XY}, \tag{7.9}$$

where n_{Xi} is the projection of bond i in the X-direction divided by bond length L, n_{Yi} is the projection of bond i in the Y-direction divided by L, and ε_{XX}, ε_{YY}, and γ_{XY} are the conventional small-strain components.[c]

Recognizing that opposing bonds on a given particle have the same stretches in a homogeneous strain field, and using the bond-numbering scheme given in Fig. 6.7, we see that

$$\begin{Bmatrix} S_1 \\ S_3 \\ S_5 \end{Bmatrix} = \begin{bmatrix} n_{X1}{}^2 & n_{Y1}{}^2 & n_{X1}n_{Y1} \\ n_{X3}{}^2 & n_{Y3}{}^2 & n_{X3}n_{Y3} \\ n_{X5}{}^2 & n_{Y5}{}^2 & n_{X5}n_{Y5} \end{bmatrix} \begin{Bmatrix} \varepsilon_{XX} \\ \varepsilon_{YY} \\ \gamma_{XY} \end{Bmatrix} \quad \text{or} \quad \{S_i\} = [N]\{\varepsilon\}\,. \tag{7.10}$$

and evaluating Eq. 7.10 numerically, assuming small strains,

$$\begin{Bmatrix} S_1 \\ S_3 \\ S_5 \end{Bmatrix} = \begin{bmatrix} 1 & 0 & 0 \\ \frac{1}{4} & \frac{3}{4} & \frac{\sqrt{3}}{4} \\ \frac{1}{4} & \frac{3}{4} & \frac{-\sqrt{3}}{4} \end{bmatrix} \begin{Bmatrix} \varepsilon_{XX} \\ \varepsilon_{YY} \\ \gamma_{XY} \end{Bmatrix} \quad \text{or} \quad \{S_i\} = [N]\{\varepsilon\}\,. \tag{7.11}$$

Finally, setting the stored strain energies of Eqs. 7.7 and 7.8 equal, and substituting in twice Eq. 7.11, (to account for all six bonds, rather than just bonds 1, 3, and 5) we obtain

$$2\left(\lfloor\varepsilon\rfloor\tfrac{cL}{4}[N]^T[N]\{\varepsilon\}\right) = \tfrac{\Delta V}{2}\lfloor\varepsilon\rfloor[D]\{\varepsilon\}. \tag{7.12}$$

---

[c] Recognizing that $S_i = \varepsilon_{X'X'}$, we can also obtain this equation from a Mohr's circle-type strain transformation.

These stored strain energies must be identical regardless of the particular value of the state of strain $\{\varepsilon\}$, and therefore we conclude that

$$\frac{cL}{2}[N]^T[N] = \frac{\Delta V}{2}[D]. \tag{7.13}$$

The left side of Eq. 7.13 is

$$\frac{cL}{2}[N]^T[N] = \frac{3cL}{16}\begin{bmatrix} 3 & 1 & 0 \\ 1 & 3 & 0 \\ 0 & 0 & 1 \end{bmatrix}, \text{ and} \tag{7.14}$$

for plane stress, the elasticity matrix $[D]$ is given by

$$\begin{Bmatrix} \sigma_{XX} \\ \sigma_{YY} \\ \tau_{XY} \end{Bmatrix} = \frac{E}{1-\nu^2}\begin{bmatrix} 1 & \nu & 0 \\ \nu & 1 & 0 \\ 0 & 0 & \frac{1-\nu}{2} \end{bmatrix}\begin{Bmatrix} \varepsilon_{XX} \\ \varepsilon_{YY} \\ \gamma_{XY} \end{Bmatrix}. \tag{7.15}$$

The only way to satisfy Eq. 7.13 is if $\nu = 1/3$ and $c = \frac{2\Delta V}{s} = \frac{\sqrt{3}}{2}ELt$.

Similarly, for plane strain,

$$\begin{Bmatrix} \sigma_{XX} \\ \sigma_{YY} \\ \tau_{XY} \end{Bmatrix} = \frac{E}{(1+\nu)(1-2\nu)}\begin{bmatrix} 1-\nu & \nu & 0 \\ \nu & 1-\nu & 0 \\ 0 & 0 & \frac{1-2\nu}{2} \end{bmatrix}\begin{Bmatrix} \varepsilon_{XX} \\ \varepsilon_{YY} \\ \gamma_{XY} \end{Bmatrix}. \tag{7.16}$$

and we find that $\nu = 1/4$ and $c = \frac{8\sqrt{3}}{15}ELt$.

With the peridynamic microelastic parameter c known, we can calculate the sum of the force vectors acting upon a given material particle, and hence the acceleration of the particle is determined using Newton's second law. Thus, the motion of a material body is determined.

Note that in this model, the force directions can change as particles move. Thus, even though the axial force in each bond is linearly related to its stretch, $F_i = cS_i$, the model produces a nonlinear relationship between applied external forces and particle positions, $\boldsymbol{x}$. This is because, according to Eq. 7.1, bond stretch S is nonlinearly related to the associated particle positions $\boldsymbol{x}$. In contrast to the classical elasticity of

Navier and Cauchy, the geometric nonlinearity arises without theoretical difficulty.

Even though we have assumed small deformations in deriving the peridynamic microelastic parameter c, we see that the axial force F_i in each bond is linearly related only to the stretch S_i in the same bond. Thus, the model is insensitive to rigid-body rotation of the body under deformation (which does not alter bond stretches), and provides accurate results if the stretch S_i and the direction of the bond force, $\frac{x_j - x_i}{|x_j - x_i|}$ are properly updated in each time step.

To summarize, we have created a regular bond-based peridynamic lattice model for planar isotropic linear elastic material behavior. However, the theory is much more general than Navier-Cauchy classical elasticity, because it accounts for large rotations. The described peridynamic material model is linear only in the sense that, for a given bond, the axial bond force relates linearly to the bond stretch, but because the bond stretch relates nonlinearly to particle positions, the overall model is nonlinear.

7.2.3 *3D face centered cubic lattice*

Following an approach very similar to that developed in the previous section for two-dimensional lattices, we now develop the microelastic constant necessary to model a homogeneous, isotropic, three-dimensional linear elastic continuum. However, in this case, we find that we cannot model an isotropic material if we include in the peridynamic force function only the twelve nearest neighbors shown in Fig. 6.7. To remedy the situation, in addition to the twelve first-nearest neighbors, we include the six second-nearest neighbors, labelled 13 through 18 in Fig. 6.7 in the particle bond list $\mathbb{B}$. In addition, in accordance with symmetry considerations, we assume that all twelve nearest neighbors have identical stiffnesses c, and all six second-nearest neighbors also have different but identical stiffnesses, c_1.

The volume of material ΔV associated with a particle in an FCC lattice Λ^3 is the covolume $d(\Lambda^3)$

$$\Delta V = d(\Lambda^3) = \frac{L^3}{\sqrt{2}}, \tag{7.17}$$

and the elasticity matrix, $[D]$, for an isotropic three-dimensional linear elastic solid is given by

$$\begin{Bmatrix} \sigma_{XX} \\ \sigma_{YY} \\ \tau_{XY} \\ \sigma_{ZZ} \\ \tau_{YZ} \\ \tau_{XZ} \end{Bmatrix} = \frac{E}{(1+\nu)(1-2\nu)} \begin{bmatrix} 1-\nu & \nu & 0 & \nu & 0 & 0 \\ \nu & 1-\nu & 0 & \nu & 0 & 0 \\ 0 & 0 & \frac{(1-2\nu)}{2} & 0 & 0 & 0 \\ \nu & \nu & 0 & 1-\nu & 0 & 0 \\ 0 & 0 & 0 & 0 & \frac{(1-2\nu)}{2} & 0 \\ 0 & 0 & 0 & 0 & 0 & \frac{(1-2\nu)}{2} \end{bmatrix} \begin{Bmatrix} \varepsilon_{XX} \\ \varepsilon_{YY} \\ \gamma_{XY} \\ \varepsilon_{ZZ} \\ \gamma_{YZ} \\ \gamma_{XZ} \end{Bmatrix}. \quad (7.18)$$

Again assuming a homogeneous strain field, the relationship, between classical strain $\boldsymbol{\varepsilon}$ and bond stretch S_i of any bond, B_i, is now

$$S_i = \lfloor n_{Xi}{}^2 \quad n_{Yi}{}^2 \quad n_{Xi}n_{Yi} \quad n_{Zi}{}^2 \quad n_{Yi}n_{Zi} \quad n_{Zi}n_{Xi} \rfloor \begin{Bmatrix} \varepsilon_{XX} \\ \varepsilon_{YY} \\ \gamma_{XY} \\ \varepsilon_{ZZ} \\ \gamma_{YZ} \\ \gamma_{XZ} \end{Bmatrix} \text{ or } \quad (7.19)$$

$$\{S_i\}[N]\{\varepsilon\},$$

where n_{Xi} is the direction cosine that bond B_i makes with the X axis, etc.

Because there are now bonds of two lengths, we must now modify Eq. 7.8 as follows:

$$\begin{aligned} \Delta E_{peridynamic} &= \textstyle\sum_{i=1}^{18} \left(\int_{S=0}^{S_i} f_i \left(\frac{L_i}{2}\right) dS_i \right) \\ &= \textstyle\sum_{i=1}^{12} \left(\int_{S=0}^{S_i} cS_i \left(\frac{L}{2}\right) dS_i \right) + \sum_{i=13}^{18} \left(\int_{S=0}^{S_i} c_1 S_i \left(\frac{\sqrt{2}L}{2}\right) dS_i \right) \\ &= \frac{1}{4} \lfloor S_i \rfloor [L_c] \{S_i\}, \end{aligned} \quad (7.20)$$

where $[L_c]$ is the 18×18 diagonal matrix whose first 12 elements along the diagonal are cL and whose last 6 elements along the diagonal are $c_1\sqrt{2}L$.

Again, equating the stored strain energy per particle between the classical and the lattice models,

$$\lfloor\varepsilon\rfloor\frac{1}{4}[N]^T[L_c][N]\{\varepsilon\}=\frac{\Delta V}{2}\lfloor\varepsilon\rfloor[D]\{\varepsilon\}, \tag{7.21}$$

where now $[N]$ is an 18×6 matrix. Recognizing that $\{\varepsilon\}$ is arbitrary, we conclude that

$$\frac{1}{4}[N]^T[L_c][N]=\frac{\Delta V}{2}[D], \tag{7.22}$$

from which the Poisson's ratio is found to be one-fourth and micro stiffnesses of the short and long bonds, are, respectively, found to be

$$c=\frac{2\sqrt{2}Es^2}{5};\ c_1=\frac{Es^2}{5}=2\sqrt{2}c\,. \tag{7.23}$$

Because Eq. 7.21 is satisfied for all homogeneous strain fields $\{\varepsilon\}$, this solution is guaranteed to produce perfectly isotropic material behavior with respect to lattice rotation for spatially homogeneous strain fields.

Although the linear elastic classical theory (Eq. 7.18) that we have employed to develop the bond-based peridynamic theory assumes small strains and Cauchy stress, the bond-based peridynamic theory is much more general in that deformations need not be small. None-the-less, the three-dimensional linear elastic bond-based theory we have developed is not sufficiently general to model materials with Poisson's ratio other than one-quarter. However, in the next section, we develop a bond-based peridynamic model that can simulate materials with Poisson's ratios other than one-quarter by also considering particle moments and rotations.

7.3 Linear elastic micropolar lattice model

So far, we have included only *forces* and particle *translations* in our bond-based models. By also including *moments* and particle *rotations* as degrees of freedom, it is possible to generalize the bond-based peridynamic model so that it will be capable of modeling isotropic linear elastic materials with adjustable Poisson's ratio. Again, to equate the bond-based peridynamic micropolar lattice model with the classical linear elastic model, the key concept is to require equivalent strain energy storage for equivalent homogeneous deformation fields.

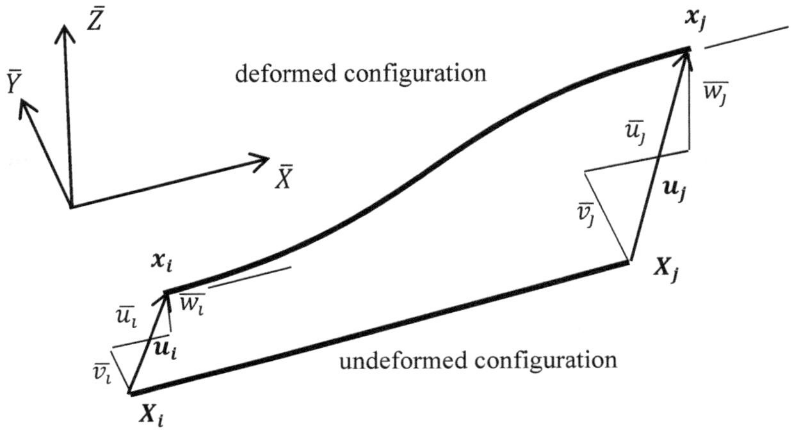

Figure 7.1. A single micropolar bond in a homogeneous deformation field.

Fig. 7.1 shows two particles, with undeformed reference locations $\boldsymbol{X}_i$ and $\boldsymbol{X}_j$, deformed locations $\boldsymbol{x}_i$ and $\boldsymbol{x}_j$, and a local coordinate system $(\bar{X}, \bar{Y}, \bar{Z})$, which is aligned with the reference locations of the two particles.[d] In the local coordinate system, we postulate that the mutual forces $\overline{\boldsymbol{F}}$ and the relative displacements $\Delta\bar{\boldsymbol{u}}$ are related as

$$\begin{Bmatrix} \bar{F}_X \\ \bar{F}_Y \\ \bar{F}_Z \end{Bmatrix} = \begin{bmatrix} c & 0 & 0 \\ 0 & d & 0 \\ 0 & 0 & d \end{bmatrix} \begin{Bmatrix} \Delta\bar{u} \\ \Delta\bar{v} \\ \Delta\bar{w} \end{Bmatrix}, \quad \text{or} \quad \{\bar{F}\} = [C]\{\Delta\bar{u}\}\,, \tag{7.24}$$

where the displacement components are $\Delta\bar{u} \equiv \bar{u}_j - \bar{u}_i$, etc. Note that, in addition to the forces acting upon the particles, there are moments, but these moments do no work, as there is no relative rotation of the particles under homogeneous deformation, as described in the next paragraph.

Consideration of translational lattice symmetry shows that for a homogeneous (spatially constant strain) deformation field, all particle

[d] The over-bar denotes "in local coordinates".

rotations must be identical. If all particle rotations are identical, then the entire material body can be rotated as a rigid body (which does not change the stored strain energy) so that all particle rotations are null, as shown in Fig. 7.1.

The stored strain energy associated with one bond is thus

$$\Delta E_{bond} = \frac{1}{2}\lfloor \Delta\bar{u} \rfloor [C] \{\Delta\bar{u}\} \,. \tag{7.25}$$

As two particles share each bond, the strain energy per particle is one-half the sum of the strain energies of all bonds attached to the particle:

$$\Delta E_{particle} = \frac{1}{2}\sum_{bonds} \frac{1}{2}\lfloor \Delta\bar{u} \rfloor [C] \{\Delta\bar{u}\} \,. \tag{7.26}$$

The displacement transforms as a vector, and the local displacement components $\{\Delta\bar{u}\}$ are related to the global displacement components Δu by a 3x3 transformation matrix as follows

$$\{\Delta\bar{u}\} = [T]\{\Delta u\} \,, \tag{7.27}$$

where $[T]$ is a matrix of direction cosines

$$[T] = \begin{bmatrix} \frac{\Delta X}{L} & -\frac{\Delta Y}{\sqrt{\Delta X^2+\Delta Y^2}} & -\frac{\Delta X \Delta Z}{L\sqrt{\Delta X^2+\Delta Y^2}} \\ \frac{\Delta Y}{L} & \frac{\Delta X}{\sqrt{\Delta X^2+\Delta Y^2}} & -\frac{\Delta Y \Delta Z}{L\sqrt{\Delta X^2+\Delta Y^2}} \\ \frac{\Delta Z}{L} & 0 & \frac{\sqrt{\Delta X^2+\Delta Y^2}}{L} \end{bmatrix}, \tag{7.28}$$

except in the special case where the bond is oriented such that the local coordinate $\bar{X}$ is parallel to the Z-direction, $\sqrt{\Delta X^2 + \Delta Y^2} = 0$, and then if the bond is directed such that $\bar{X}$ is in the positive Z-direction

$$[T] = \begin{bmatrix} 0 & 0 & 1 \\ 0 & 1 & 0 \\ -1 & 0 & 0 \end{bmatrix}, \tag{7.29}$$

and if $\bar{X}$ is directed in the negative Z direction

$$[T] = \begin{bmatrix} 0 & 0 & -1 \\ 0 & 1 & 0 \\ 1 & 0 & 0 \end{bmatrix}. \tag{7.30}$$

(This transformation strategy assures us that the $\bar{Y}$ axis will always be parallel to the XY plane.)

We can express the relative displacement vector in terms of a classical strain field as

$$\{\Delta u\} = [Q_c]\{\varepsilon\}\,. \tag{7.31}$$

or in terms of direction cosines $n_X \equiv \frac{\Delta X}{L}$, $n_Y \equiv \frac{\Delta Y}{L}$, $n_Z \equiv \frac{\Delta Z}{L}$:

$$\begin{Bmatrix} \Delta u \\ \Delta v \\ \Delta w \end{Bmatrix} = L \begin{bmatrix} n_X & 0 & {n_Y}/{2} & 0 & 0 & {n_Z}/{2} \\ 0 & n_Y & {n_X}/{2} & 0 & {n_Z}/{2} & 0 \\ 0 & 0 & 0 & n_Z & {n_Y}/{2} & {n_X}/{2} \end{bmatrix} \begin{Bmatrix} \varepsilon_X \\ \varepsilon_Y \\ \gamma_{XY} \\ \varepsilon_Z \\ \gamma_{YZ} \\ \gamma_{XZ} \end{Bmatrix}. \tag{7.32}$$

Substituting Eqs. 7.27 and 7.31 into Eq. 7.26, we obtain the following expression for the strain energy per particle

$$\Delta E_{particle} = \lfloor \varepsilon \rfloor \left[\tfrac{1}{4} \textstyle\sum_{bonds} [Q_c]^T [T]^T [C][T][Q_c] \right] \{\varepsilon\}\,. \tag{7.33}$$

This energy must be the same as the corresponding classical strain energy:

$$\lfloor \varepsilon \rfloor \left[\tfrac{1}{4} \textstyle\sum_{bonds} [Q]^T [T]^T [C][T][Q] \right] \{\varepsilon\} = \tfrac{\Delta V}{2} \lfloor \varepsilon \rfloor [D] \{\varepsilon\}, \tag{7.34}$$

from which, considering that $\{\varepsilon\}$ is arbitrary,

$$\tfrac{1}{4} \textstyle\sum_{bonds} [Q]^T [T]^T [C][T][Q] = \tfrac{\Delta V}{2} [D]\,. \tag{7.35}$$

Now, the problem is to satisfy Eq. 7.35 by choosing c and d defined in Eq. 7.24 appropriately. We find that when only the twelve shortest bonds are included in the summation on the left-hand side of Eq. 7.35, there is no solution. However, if second-nearest neighbors are also included in the summation, with the differing stiffness relation

$$\begin{Bmatrix} \bar{F}_X \\ \bar{F}_Y \\ \bar{F}_Z \end{Bmatrix} = \begin{bmatrix} c_1 & 0 & 0 \\ 0 & d_1 & 0 \\ 0 & 0 & d_1 \end{bmatrix} \begin{Bmatrix} \Delta\bar{u} \\ \Delta\bar{v} \\ \Delta\bar{w} \end{Bmatrix}, \tag{7.36}$$

then we find that there are an infinite number of solutions of the form

$$\begin{aligned} c &= -z + \frac{\sqrt{2}EL}{4(1-2\nu)(1+\nu)} \\ d &= -z + \frac{\sqrt{2}EL(1-4\nu)}{4(1-2\nu)(1+\nu)} \\ c_1 &= z + \frac{\sqrt{2}EL\nu}{4(1-2\nu)(1+\nu)} \\ d_1 &= z\,, \end{aligned} \tag{7.37}$$

where z is an arbitrary real number. For simplicity, we choose the solution where $z = 0$, so that

$$\begin{aligned} c &= \frac{\sqrt{2}EL}{4(1-2\nu)(1+\nu)} \\ d &= \frac{\sqrt{2}EL(1-4\nu)}{4(1-2\nu)(1+\nu)} \\ c_1 &= \frac{\sqrt{2}EL\nu}{4(1-2\nu)(1+\nu)} = c\nu\,. \\ d_1 &= 0 \end{aligned} \tag{7.38}$$

If we identify c and c_1 with an axial stiffnesses, $A'E'$, of the shortest and second-shortest bonds, respectively, and $d = \frac{12I'E'}{L^3}$ with the flexural stiffness of the short bonds, then, borrowing from the stiffness matrix of a 3D frame element (Gerstle, 1974), we find the relationship between generalized forces and generalized displacements in local bond coordinates as (Gerstle, K.H. 1974)

$$\begin{Bmatrix} F_{\overline{X}i} \\ F_{\overline{Y}i} \\ F_{\overline{Z}i} \\ M_{\overline{X}i} \\ M_{\overline{Y}i} \\ M_{\overline{Z}i} \\ F_{\overline{X}j} \\ F_{\overline{Y}j} \\ F_{\overline{Z}j} \\ M_{\overline{X}j} \\ M_{\overline{Y}j} \\ M_{\overline{Z}j} \end{Bmatrix} = \begin{bmatrix} \frac{c^*}{L} & & & & & & & & & & & \\ & \frac{12d^*}{L^3} & & & & & & & & & sym & \\ & & \frac{12d^*}{L^3} & & & & & & & & & \\ & & & \frac{G'J'}{L} & & & & & & & & \\ & & -\frac{6d^*}{L^2} & & \frac{4d^*}{L} & & & & & & & \\ & \frac{6d^*}{L^2} & & & & \frac{4d^*}{L} & & & & & & \\ -\frac{c^*}{L} & & & & & & \frac{c^*}{L} & & & & & \\ & -\frac{12d^*}{L^3} & & & & -\frac{6d^*}{L^2} & & \frac{12d^*}{L^3} & & & & \\ & & -\frac{12d^*}{L^3} & & \frac{6d^*}{L^2} & & & & \frac{12d^*}{L^3} & & & \\ & & & -\frac{GJ}{L} & & & & & & \frac{G'J'}{L} & & \\ & & -\frac{6d^*}{L^2} & & \frac{2d^*}{L} & & & & \frac{6d^*}{L^2} & & \frac{4d^*}{L} & \\ & \frac{6d^*}{L^2} & & & & \frac{2d^*}{L} & & -\frac{6d^*}{L^2} & & & & \frac{4d^*}{L} \end{bmatrix} \begin{Bmatrix} \bar{u}_i \\ \bar{v}_i \\ \bar{w}_i \\ \bar{\theta}_{Xi} \\ \bar{\theta}_{Yi} \\ \bar{u}_j \\ \bar{v}_j \\ \bar{w}_j \\ \bar{\theta}_{Xj} \\ \bar{\theta}_{Yj} \\ \bar{\theta}_{Zj} \end{Bmatrix}. \tag{7.39}$$

In Eq. 7.39, for the short bonds, with bond length L,

$$\begin{aligned} c^* = A'E' &= \frac{\sqrt{2}EL^2}{4(1-2\nu)(1+\nu)}, \\ d^* = I'E' &= \frac{\sqrt{2}EL^4(1-4\nu)}{48(1-2\nu)(1+\nu)} \end{aligned} \tag{7.40}$$

and for the long bonds, with bond length $\sqrt{2}L$,

$$\begin{aligned} c^* = A'E' &= \frac{\sqrt{2}EL^2\nu}{2(1-2\nu)(1+\nu)}. \\ d^* = I'E' &= 0\,. \end{aligned} \tag{7.41}$$

Note that the torsional stiffness of a bond, $G'J'$, is arbitrary in matching the classical material under homogeneous deformation, as there is no bond twist in such a case, and thus $G'J'$ can be set to any value. For nonhomogeneous deformations, however, the material behavior depends upon the chosen value for $G'J'$. We suggest using a value of $'J' = \left(\frac{E}{2(1+\upsilon)}\right)(I_Y + I_Z) = \frac{2d^*}{(2(1+\upsilon))} = \frac{d^*}{(1+\upsilon)}$, in keeping with the notion that the polar moment of inertia of a cross-sectional area is equal to the sum of the moments of inertia about the two principal axes.

It is important to understand that Eq. 7.39 is simply a linear relationship between generalized forces and generalized relative displacements. Although this relationship comes from the analogy of a frame element from solid mechanics, we should not think of physical frame elements as connecting the particles. Instead, force simply relates to deformation in a way that is analogous to that from a frame element in solid mechanics. Other types of linear force-deformation relationships are possible.

We have developed a linear peridynamic force versus bond deformation relationship for a three-dimensional FCC lattice that is consistent with the behavior of classical theory of linear isotropic elasticity. One can find similar relationships for one-dimensional lattices and for two-dimensional plane stress and plane strain hexagonal lattices. For completeness, we include these results next, without going through a detailed derivation.

For a one-dimensional strand modeling a bar with axial stiffness EA, bending stiffness, EI, and torsional stiffness, GJ, and considering only closest neighbors, the microelastic parameters in Eq. 7.39 are

$$c^* = EA; \, d^* = EI \, ; \, G'J' = GJ \, . \tag{7.42}$$

For plane stress, using a hexahedral plane lattice, and with all six nearest neighbors included, one models an isotropic linear elastic material using a micropolar lattice with axial and bending parameters

$$c^* = E'A' = \frac{ELt}{\sqrt{3}\,(1-\nu)}; \quad d^* = E'I' = \frac{E\,L^3 t(1-3\nu)}{12\sqrt{3}(1-\nu^2)} . \tag{7.43}$$

Finally, for plane strain,

$$c^* = E'A' = \frac{ELt}{\sqrt{3}\,(1+\nu)(1-2\nu)}\,;\ d^* = E'I' = \frac{E\,L^3 t(1-4\nu)}{12\sqrt{3}(1+\nu)(1-2\nu)}\,. \tag{7.44}$$

The fact that the micropolar torsional stiffness $G'J'$ of each bond is arbitrary is puzzling. $G'J'$ is arbitrary because under homogeneous deformations, in consideration of symmetry, bonds cannot twist. However, if deformations are not homogeneous, then the torsional stiffness $G'J'$ will affect the material response. Thus, the material response, in the sense of energy stored per unit volume, is affected not only by the strain, but also by the strain gradient. In the classical model, if we take the limit as the material horizon goes to zero, the micropolar moments and torques, experiencing a homogeneous deformation environment, are irrelevant. Thus, the micropolar bond-based peridynamic model is somewhat more general than the classical linear elastic model, having three independent microelastic parameters, rather than only two.

7.4 Conclusions regarding bond-based lattice models

We have implemented the non-micropolar (or regular) bond-based model in particle dynamics codes, and the model works as expected for simulating linear elastic problems. The non-micropolar bond-based peridynamic lattice model is limited to modeling materials with Poisson's ratio of one-quarter (or one-third for plane stress problems). Large deformation behavior comes for free, as the bond forces are functions only of the bond stretches, which can be easily calculated as deformation evolves and are independent of rigid-body rotation of the bond.

The micropolar model is useful for modeling problems with Poisson's ratio other than one-quarter, but is more computationally expensive and is more complicated to implement.

For a three-dimensional FCC lattice, to enable isotropic material behavior, both the regular and the micropolar model require both the twelve nearest neighbors and the six second-nearest neighbors to be included within the material horizon. Including the third-nearest neighbors is not necessary to model isotropic linear elasticity, and is in fact not recommended, as this causes the peridynamic boundary effects (further discussed shortly) to be more pronounced.

The micropolar models are capable of modeling large geometric deformations if one measures all local particle displacements, including angular displacements, with respect to a co-rotated local coordinate system associated with each bond. For two-dimensional problems, it is a simple matter to calculate bond rotations and particle rotations, as these are simply scalar quantities. However, for three-dimensional problems, it is difficult to describe bond rotations and particle rotations, because large rotations are not vectors. However, if all rotations (bond and particle rotations) remain small during the analysis, they actually behave as vectors and present no theoretical difficulty. As a practical matter, we must limit relative angles of rotation (on a given bond) to a few degrees for particle rotations to behave as vectors.

All peridynamic models, being nonlocal, feature boundary effects. Computational studies show that, as long as the material body has a minimum macro-dimension that is significantly larger than the lattice spacing, the boundary effects are small. Studies of plane stress cantilever beams with approximately 20 particles in the transverse direction and 100 particles in the axial direction show displacement variations of one or two percent, as the lattice rotates with respect to the axis of the beam. If one uses tens of thousands or more of particles, so that the lattice spacing is very small compared to the macro scale of the problem, it is our experience that the peridynamic lattice model matches classical mechanics static solutions to well within one percent, regardless of lattice orientation.

Neither the regular nor the micropolar bond-based model takes into account the volumetric deformation of the material upon inter-particle forces, and thus these models are not promising for implementing isochoric plasticity. For this reason, we move on to the more general state-based peridynamic lattice model (SPLM), described in the next chapter.

7.5 Exercises

7.1 Write a MatLab function called *cableInternalForces* to return the internal forces acting upon a one-dimensional strand of lattice particles, assumed to be positioned along the X-axis, using the following signature:

```
Function F = cableInternalForces(x, L, E, A)
    % Assume the particles are initially equally-spaced at distance L, starting with
    % particle 1 located at X = 0, particle 2 at X = L,
    % and particle Np at X = (Np − 1) × L.
    % x = [Np × 1] array containing the current particle spatial coordinates.
    % L = lattice spacing in the reference configuration
    % E = Young's modulus
    % A = area of the cable
    % F= [Np × 1] array containing the internal forces acting upon each particle.
```

7.2 Write a MatLab function called *cableSimulation* to return the spatial locations of the particles constituting a cable, assumed to be positioned along the X-axis, using the following signature. Assume that the cable is initially unstretched, and that particle 1 is restrained from motion, with a constant body force B is applied to all particles. Assuming Np = 10, L = 1, E = 1, A = 1, m = 1, and B = 1, and numSteps = 200, plot the output $xEnd$ after 200 time steps, and plot XNp for the duration of the simulation. (Hint: call *cableInternalForces* from the previous problem in each time step. Select dt as large as possible such that the solution is stable.)

```
Function [xEnd XNp] = cableSimulation (Np, L, E, A, m, B, dt, numSteps)
  % Assume the particles are initially equally-spaced at distance L, starting with
  % particle 1 located at X = 0, particle 2 at X = L,
  % and particle Np at X = (Np − 1) × L.
  % L = lattice spacing in the reference configuration
  % E = Young's modulus
  % A = area of the cable
  % m = mass, per particle
```

% B = applied body force, per particle

% xEnd = $[Np \times 1]$ array with particle spatial coordinates at time t = tEnd.

% xNp = $[numSteps \times 1]$ array with particle Np's positon x at each time step.

7.3 For a two-dimensional hexagonal lattice, with spacing L and thickness t, show that the volume occupied per lattice particle is given by $\Delta V = \frac{\sqrt{3}}{2} L^2 t$.

7.4 Assuming small strains, show that Eq. 7.9, $S_i = {n_{xi}}^2 \varepsilon_{xx} + {n_{yi}}^2 \varepsilon_{yy} + n_{xi} n_{yi} \gamma_{xy}$, is correct.

7.5 Show that for plane stress, $\nu = 1/3$ and $c = \frac{\sqrt{3}}{2} ELt$ using Eq. 7.13.

7.6 Show that for plane strain, $\nu = 1/4$ and $c = \frac{8\sqrt{3}}{15} ELt$ using Eq. 7.13.

7.7 Show how one can recast the plane strain and plane stress bond-based models described in Section 7.2.2 as state-based peridynamic models.

7.8 For a three-dimensional face-centered cubic lattice, with spacing L, show that the volume occupied per lattice particle is given by $\Delta V = \frac{L^3}{\sqrt{2}}$.

7.9 Using the regular bond-based approach outlined in Section 7.3.2, show that for a face centered cubic lattice, including only the twelve nearest neighbors within the material horizon cannot result in an isotropic elastic material.

7.10 Show that, for small strains, Eq. 7.19,

$$S_i = \lfloor {n_{xi}}^2 \quad {n_{yi}}^2 \quad n_{xi} n_{yi} \quad {n_{zi}}^2 \quad n_{yi} n_{zi} \quad n_{zi} n_{xi} \rfloor \begin{Bmatrix} \varepsilon_{xx} \\ \varepsilon_{yy} \\ \gamma_{xy} \\ \varepsilon_{zz} \\ \gamma_{yz} \\ \gamma_{zx} \end{Bmatrix},$$ is correct.

7.11 Show that for a face centered cubic lattice, $\upsilon = 1/4$, $c = \frac{2\sqrt{2}Es^2}{5}$, and $c_1 = \frac{Es^2}{5} = 2\sqrt{2}c$ using Eq. 7.22.

7.12 Section 7.3, makes the statement that "Translational symmetry considerations show that for a homogeneous (spatially constant strain) deformation field, all particle rotations must be identical." Explain why this statement is true.

7.13 Eq. 7.25 states that the strain energy stored in a single bond is given by $U_{bond} = \frac{1}{2}\lfloor\Delta\bar{u}\rfloor[C]\{\Delta\bar{u}\}$. Derive this equation.

7.14 Verify the micropolar peridynamic constants given by Eq. 7.43, $c^* = E'A' = \frac{ELt}{\sqrt{3}\,(1-\nu)}$; $d^* = E'I' = \frac{E\,L^3t(1-3\nu)}{12\sqrt{3}(1-\nu^2)}$, for a plane stress micropolar hexagonal lattice. Show that this results in isotropic behavior under homogeneous deformations.

7.15 Verify the micropolar peridynamic constants given by Eq. 7.44, $c^* = E'A' = \frac{ELt}{\sqrt{3}\,(1+\nu)(1-2\nu)}$; $d^* = E'I' = \frac{E\,L^3t(1-4\nu)}{12\sqrt{3}(1+\nu)(1-2\nu)}$, for a plane strain micropolar hexagonal lattice. Show that this results in isotropic behavior under homogeneous deformations.

Chapter 8

State-Based Peridynamic Lattice Model (SPLM)

Chapter objectives are to:

- adapt Silling's continuum state-based peridynamic model to a lattice model;
- generalize Navier's elasticity theory;
- introduce the reader to peridynamic states;
- discuss the similarities between peridynamic states and tensors;
- describe state-based peridynamic theory for particle lattices;
- clarify the relationship between state-based peridynamics and continuum mechanics;
- provide conceptual and mathematical context for the state-based peridynamic lattice model (SPLM).

In the last chapter, we saw that the bond-based peridynamic model is unable to simulate three-dimensional elastic bodies with Poisson's ratio different from one-quarter. In this chapter, we provide background and motivation for the more general state-based peridynamic theory.

We begin by reviewing Navier's theory, presented in Chapter 2, and by showing how, even without concepts of strain and stress, we can generalize the original theory to produce the Navier-Cauchy equations, which simulate elastic solids with any physically permissible Poisson's ratio. However, even with this generalization, the Navier-Cauchy theory falls short in describing materials that crack.

The state-based continuum peridynamic theory (Silling, S.A., Epton, M., Weckner, O., Xu, J., and Askari, E., 2007) corrects this shortcoming, and allows material bodies to fracture. However, this latter theory, involving as it does infinite-dimensional states and infinite numbers of particles, is impossible exactly to implement on a computer. In addition, the continuum state-based peridynamic theory is more general than needed for realistic modeling of most materials.

Therefore, we present Silling's state-based peridynamic theory, except with a fundamental modification: we assume a discrete lattice particle model, as described in Chapter 6. By changing the particle model from a continuum to a lattice, we reduce the peridynamic states to finite-dimensional objects, and at the same time reduce the quantity of material particles to a finite number. This approximation aligns with the physical reality of an actual material, which has a length scale that defines the appropriate lattice spacing.

By employing a lattice particle to model a material body, we avoid the problem of defining mass density. Instead, each lattice particle simply has a finite mass, and furthermore, for a given material body, each particle has identical mass. Also, when it comes to defining other state variables like plasticity and damage responses, these can be stored as discrete entities associated with each lattice particle, rather than as fields, continuous or discontinuous, which must somehow (say, using finite elements, Fourier series, or, well, a lattice) be defined over a real Euclidian reference space representing the body $\mathcal{R} \subset \mathbb{R}^3$.

Thus, we develop a computational state-based peridynamic lattice model (SPLM) that, while less general than Silling's state-based model, provides more direct connections between the physical world, the human mind, and numerical computer models.

8.1 Generalizing Navier's theory

We have seen that Navier's original elasticity treatise, presented in Section 2.2, resulted in only one elastic constant, ε. Can we generalize his theory to require two material constants for an isotropic linear elastic material, without having to resort to Cauchy's concepts of strain and stress?

Navier assumed that the force between any two particles was a function of the change in bond length, given by Eq. 2.9. Let us generalize this assumption slightly, and instead assume that the force between two particles is a function of bond stretch S *and also of the divergence of the displacement field*, $\boldsymbol{\nabla}\cdot\boldsymbol{u} = \frac{\partial U}{\partial X} + \frac{\partial V}{\partial Y} + \frac{\partial W}{\partial Z} = U_{i,i}$. (Note that the concept of divergence does not depend upon the concept of continuity, or upon the differentiability of the displacement field. However, in this section, we do assume that the displacement field is differentiable and continuous. We will relax this assumption later on in this chapter.) The divergence of the displacement field is equal to the change in volume per unit reference volume, or the volumetric strain. This is a somewhat arbitrary assumption; we could make others, but at least it is an isotropic assumption. Thus, the force between particles P and P' now becomes

$$F = [G(R)S + H(R)\boldsymbol{\nabla}\cdot\boldsymbol{u}]dV' \ , \tag{8.1}$$

where $H(R)$, similar to $G(R)$ in Eq. 2.8, is a spherically-symmetric function that specifies the influence of distance between P and P' upon the portion of the force governed by the divergence of the displacement field at P.

With this changed assumption, Eq. 2.16 now becomes

$$\begin{aligned}
\varepsilon \iiint_{\Omega} \{&3U_{1,1}\delta U_{1,1} + U_{1,2}\delta U_{1,2} + U_{1,2}\delta U_{2,1} + U_{2,1}\delta U_{1,2} \\
&+ U_{2,1}\delta U_{2,1} + U_{1,1}\delta U_{2,2} + U_{2,2}\delta U_{1,1} \\
&+ U_{1,1}\delta U_{1,3} + U_{1,3}\delta U_{3,1} + U_{3,1}\delta U_{1,3} \\
&+ U_{3,1}\delta U_{3,1} + U_{1,1}\delta U_{3,3} + U_{3,3}\delta U_{1,1} \\
&+ 3U_{2,2}\delta U_{2,2} + U_{2,3}\delta U_{2,3} + U_{2,3}\delta U_{3,2} \\
&+ U_{3,2}\delta U_{2,3} + U_{3,2}\delta U_{3,2} + U_{2,2}\delta U_{3,3} \\
&+ U_{3,3}\delta U_{2,2} + 3U_{3,3}\delta U_{3,3}\}d\Omega \\
+\beta \iiint_{\Omega} \{&U_{1,1}\delta U_{1,1} + U_{1,1}\delta U_{2,2} + U_{1,1}\delta U_{3,3} + U_{2,2}\delta U_{1,1} \\
&+ U_{2,2}\delta U_{2,2} + U_{2,2}\delta U_{3,3} + U_{3,3}\delta U_{1,1} + U_{3,3}\delta U_{2,2} \\
&+ U_{3,3}\delta U_{3,3}\}d\Omega \\
&- \iiint_{\Omega} B_i \delta U_i d\Omega - \iint_{\Gamma} T_i \delta U_i d\Gamma = 0.
\end{aligned} \tag{8.2}$$

where

$$\beta \equiv \left(\frac{4\pi}{3}\right) \int_0^{\infty} R^4 H(R)\, dR \ . \tag{8.3}$$

With this modification, Eq. 2.18 now becomes

$$
\begin{aligned}
0 = &-\varepsilon \iiint_{\Omega} \left\{ \begin{array}{c} (3U_{,XX}+U_{,YY}+U_{,ZZ}+2V_{,XY}+2W_{,XZ})\delta U \\ +(V_{,XX}+3V_{,YY}+V_{,ZZ}+2U_{,XY}+2W_{,YZ})\delta V \\ +(W_{,XX}+W_{,YY}+3W_{,ZZ}+2U_{,XZ}+2V_{,YZ})\delta W \end{array} \right\} d\Omega \\
&-\beta \iiint_{\Omega} \left\{ \begin{array}{c} (U_{,XX}+V_{,YX}+W_{,ZX})\delta U \\ +(U_{,XY}+V_{,YY}+W_{,ZY})\delta V \\ +(U_{,XZ}+V_{,YZ}+W_{,ZZ})\delta W \end{array} \right\} d\Omega \\
&-\iiint_{\Omega} \left\{ \begin{array}{c} B_X\delta U \\ B_Y\delta V \\ B_Z\delta W \end{array} \right\} d\Omega \\
&+\varepsilon\left[\iint_{\Gamma} (3U_{,X}+V_{,Y}+W_{,Z})n_X + (U_{,Y}+V_{,X})n_Y + (U_{,Z}+W_{,X})n_Z\right] d\Gamma\delta U \\
&+\varepsilon\left[\iint_{\Gamma} (U_{,Y}+V_{,X})n_X + (U_{,X}+3V_{,Y}+W_{,Z})n_Y + (V_{,Z}+W_{,Y})n_Z\right] d\Gamma\delta V \\
&+\varepsilon\left[\iint_{\Gamma} (U_{,Z}+W_{,X})n_X + (V_{,Z}+W_{,Y})n_Y + (U_{,X}+V_{,Y}+3W_{,Z})n_Z\right] d\Gamma\delta W \\
&+\beta\left[\iint_{\Gamma} (U_{,X}+V_{,Y}+W_{,Z})n_X\right] d\Gamma\delta U \\
&+\beta\left[\iint_{\Gamma} (U_{,X}+V_{,Y}+W_{,Z})n_Y\right] d\Gamma\delta V \\
&+\beta\left[\iint_{\Gamma} (U_{,X}+V_{,Y}+W_{,Z})n_Z\right] d\Gamma\delta Z \\
&-\iint_{\Gamma} \left\{ \begin{array}{c} T_X\delta U \\ T_Y\delta V \\ T_Z\delta W \end{array} \right\} d\Gamma ,
\end{aligned}
\tag{8.4}
$$

and Equation 2.19 becomes

$$
\begin{aligned}
-B_X &= \varepsilon(3U_{,XX}+U_{,YY}+U_{,ZZ}+2V_{,XY}+2W_{,XZ}) + \beta(U_{,XX}+V_{,YX}+W_{,ZX}), \\
-B_Y &= \varepsilon(V_{,XX}+3V_{,YY}+V_{,ZZ}+2U_{,XY}+2W_{,YZ}) + \beta(U_{,XY}+V_{,YY}+W_{,ZY}), \\
-B_Z &= \varepsilon(W_{,XX}+W_{,YY}+3W_{,ZZ}+2U_{,XZ}+2V_{,YZ}) + \beta(U_{,XZ}+V_{,YZ}+W_{,ZZ}),
\end{aligned}
\tag{8.5}
$$

Using indicial notation, an equivalent expression to Eq. 2.20 is

$$
\begin{aligned}
-B_i &= \varepsilon\left(U_{i,jj} + 2U_{j,ij}\right) + \beta\left(U_{j,ji}\right), \text{ or} \\
\varepsilon U_{i,jj} &+ (2\varepsilon + \beta)U_{j,ij} + B_i = 0 ,
\end{aligned}
\tag{8.6}
$$

which is precisely the Navier-Cauchy equation for elastostatics:

$$\mu U_{i,jj} + (\mu + \lambda) U_{j,ij} + B_i = 0 , \tag{8.7}$$

where μ and λ are the Lamé parameters, with

$$\varepsilon = \mu \text{ and } \beta = \lambda - \mu .^{a} \tag{8.8}$$

Thus we see that, without resorting to Cauchy's stress definition, it is possible to arrive at the Navier-Cauchy equations by making a correction to Navier's original assumption regarding the causes of the forces between neighboring particles.

This is a special case of the much more general correction made by Silling to extend bond-based theory to state-based theory.

Significantly, in Eq. 2.1, Navier differentiates the displacement field, thus implicitly assuming that the displacement field is continuous. Had he decided instead to make no assumption about the continuity of the displacement field, he might have arrived instead at Silling's state-based peridynamic theory, discussed next.

8.2 Reasons for state-based peridynamic theory

In Chapter 5, we saw that Silling's original continuum bond-based peridynamic model is insufficiently general to model materials with Poisson's ratios having values different than one-quarter, and the model is also inadequate to model isochoric (volume invariant) behavior such as is observed in plastic deformation. This is because the peridynamic pairwise force function $\boldsymbol{f}(\boldsymbol{\xi}, \boldsymbol{\eta})$ between particle P with reference position $\boldsymbol{X}$, and particle P' with reference position $\boldsymbol{X}'$, is assumed to depend upon the reference and deformed positions of particles P and P' only. When materials deform plastically, researchers have observed the plastic part of the deformation to be essentially isochoric. Thus, to model plasticity with the peridynamic model, it would seem that $\boldsymbol{f}$ would need

[a] The Lamé parameters are related to Poisson's Ratio ν and Young's modulus E as $\lambda = \frac{E\nu}{(1+\nu)(1-2\nu)}$ and $\mu = \frac{E}{2(1+\nu)}$ (μ is also known as the shear modulus G).

to depend upon the relative locations of all particles within the material horizon.

To remedy this situation, in (Silling, S.A., Epton, M., Weckner, O., Xu, J., and Askari, E., 2007) the original peridynamic model was extended by introducing the new concepts of "force state" $\underline{\boldsymbol{T}}$ and "deformation state" $\underline{\boldsymbol{Y}}$. In this generalization, the pairwise force function, $\boldsymbol{f}$, between two particles, P with reference position $\boldsymbol{X}$, and P' with reference position $\boldsymbol{X}'$, is not just a function of the relative reference position, $\boldsymbol{\xi}$, and relative displacement, $\boldsymbol{\eta}$, of the two particles between which $\boldsymbol{f}$ acts. Instead, the pairwise force acting between the pair of particles is now regarded as being a function of the reference and deformed positions, $\boldsymbol{X}$ and $\boldsymbol{x}$, of all particles that lie within the union of both particles' material horizons.

When two particles of different "material type" interact, the concept of "material" breaks down. Perhaps in such cases it would be better to dispense with the concept of "material" entirely, and instead talk about an "interface". In (Silling, S.A., Epton, M., Weckner, O., Xu, J., and Askari, E., 2007), spatially nonhomogeneous materials are not considered; for clarity, we also adopt this assumption in this chapter.

In the state-based peridynamic theory, as with the bond-based theory, particles P in the reference configuration are assumed to be elements of a continuous subset $\mathcal{R}$ of the three-dimensional real Cartesian space $\mathbb{R}^3$. In Chapter 6, we discussed changing this assumption and instead describing the material body as a finite set of discrete particles P that are elements of a lattice body $\mathcal{L}_{\mathcal{R}}$. Deviating from the work of (Silling, S.A., Epton, M., Weckner, O., Xu, J., and Askari, E., 2007), we assume a lattice body for the remainder of this chapter, but keep in mind that almost all of the derivations are applicable as well when assuming a continuous material space.

The state-based peridynamic theory is sufficiently general to model linear elastic materials with varying Poisson's ratio, and, also nonlinear elasticity, geometric nonlinearity, plasticity, damage, and fracture. While not addressed explicitly in this chapter, we can also accommodate rate-dependent materials using the state-based peridynamic framework.

State-based peridynamics can be generalized, and could potentially be used to model fluids and fluid-structure interaction problems, heat transfer, electromagnetic, and diffusion problems as well, as has been explored in (Gerstle, W., Silling, S., Read, D., Tewary, V., and Lehoucq, R., 2008).

State-based peridynamics is more general than continuum mechanics because it allows the model to represent discontinuous as well as continuous deformation fields. It also is able to model the dynamical behavior of materials more truly by enabling dispersion of propagating vibrational waves, which the continuum theory does not easily represent. Continuum theory provides no material length scale; the peridynamic theories correct this shortcoming. While the state based continuum peridynamic model allows discontinuous fields, it is silent on how actually to represent these fields, analytically or on a computer. The state-based peridynamic lattice model (SPLM), on the other hand, defines precisely how the physical fields are to be represented, using the lattice body $\mathcal{L}_{\mathcal{R}}$, introduced in Definition 6.4, to represent the material body.

8.3 Bond-based peridynamic states

In this chapter, we follow (Silling, S.A., Epton, M., Weckner, O., Xu, J., and Askari, E., 2007), which the reader is encouraged to study, in defining the mathematics of peridynamic states. However, we modify the theory somewhat, by changing the reference space $\mathcal{R}$ to a lattice body $\mathcal{L}_{\mathcal{R}}$, defined upon a lattice Λ^{N_R} defined in Chapter 6, rather than upon a Euclidian physical reference space $\mathbb{R}^3$. Let $\mathcal{L}_m$ be the set of mth-order tensors.

Definition 8.1 A *bond-based state of order* m associated with particle P_i is a function $\underline{\mathbf{A}}\langle\cdot\rangle : \mathbb{B} \rightarrow \mathcal{L}_m$.

This definition conveys the idea that the function $\underline{\boldsymbol{A}}$ maps any bond B_j located within bond list $\mathbb{B}$ associated with particle P_i to a tensor of order m. Thus, we clearly identify the domain and range of the function $\underline{\mathbf{A}}\langle\cdot\rangle$.

Silling et al. defined the domain of the state as "$\mathcal{H}$, a spherical neighborhood of radius δ centered at the origin in $\mathbb{R}^3$". We have changed the domain of the state instead to be the particle bond list $\mathbb{B}$, presented in Definition 6.6.

For brevity, from now on we will call a "bond-based state" a "state". The context will be sufficient to identify which type of state we are considering.

In other words, a (bond-based) state $\underline{\mathbf{A}}$ is a function whose bond list $\mathbb{B}$ is its domain and whose range is the set of all tensors of order m, $\mathcal{L}_m$. We denote the set of all states of order m as $\mathcal{A}_m$. For example, $\mathbf{T_1} = \underline{\mathbf{A}}\langle B_j\rangle$ maps the bond B_j to the vector $\mathbf{T_1} \in \mathcal{L}_1$, which is a first order tensor (a vector). To distinguish from other arguments upon which the state may depend, we always use the angle brackets $\langle\cdot\rangle$ to enclose the bond B_j on which the state operates. We use the square brackets $[\cdot]$ to identify with which particle the state is associated. In a Cartesian coordinate system $\mathbb{R}^{N_R}$, with N_R dimensions, each bond B_j of the state $\underline{\mathbf{A}} \in \mathcal{A}_m$ of order m has ${N_R}^m$ scalar components that are labelled as $\underline{\mathbf{A}}_{i_1\cdots i_j\cdots i_m}$, $i_j = 1, \cdots, N_R$.[b] Thus, if a state has N_B bonds within its bond list $\mathbb{B}$, the state requires a total of $N_B \cdot {N_R}^m$ components to be fully defined.

A bond-based state is thus a tensor field defined upon a bond list $\mathbb{B}$ domain. Vector and higher-order states are written in uppercase bold font with an underscore, for example $\underline{\mathbf{A}}$.

The following definitions require that to operate on states $\mathcal{A}_m$ they must be defined on the same particle's bond list, $\mathbb{B}$.

Definition 8.2 Let $\underline{\mathbf{A}} \in \mathcal{A}_m$ and $\underline{\mathbf{B}} \in \mathcal{A}_m$ both be associated with the particle P_i. The *sum* of the states $\underline{\mathbf{A}}$ and $\underline{\mathbf{B}}$ is the state in $\mathcal{A}_m$ defined by

$$\left(\underline{\mathbf{A}} + \underline{\mathbf{B}}\right)\langle B_j\rangle = \underline{\mathbf{A}}\langle B_j\rangle + \underline{\mathbf{B}}\langle B_j\rangle \quad \forall B_j \in \mathbb{B} \tag{8.9}$$

The *difference* of the states $\underline{\mathbf{A}}$ and $\underline{\mathbf{B}}$ is the state in $\mathcal{A}_m$ defined by $\left(\underline{\mathbf{A}} - \underline{\mathbf{B}}\right)\langle B_j\rangle = \underline{\mathbf{A}}\langle B_j\rangle - \underline{\mathbf{B}}\langle B_j\rangle$.

---

[b] The order m of a tensor in $\mathcal{L}_m$ has to do with the number of vectors being represented by the tensor; the dimensionality N_R of the space $\mathbb{R}^{N_R}$ expresses how many coordinates are required to identify a point in the space.

(Silling, S.A., Epton, M., Weckner, O., Xu, J., and Askari, E., 2007) define a *composition* of states as in the following Definition 8.3; however this definition does not apply to particle-based states, does not appear to be necessary for our purposes, and will not be used further in this book. In Definition 8.3, $\mathcal{H}$ is the set of all particles within the spherical material horizon in the reference configuration.

Definition 8.3 Let $\underline{\mathbf{A}} \in \mathcal{A}_m$ and $\underline{\mathbf{V}} \in \mathcal{A}_1$. The *composition* of the states $\underline{\mathbf{A}}$ and $\underline{\mathbf{V}}$ is a state in $\mathcal{A}_m$ defined by

$$(\underline{\mathbf{A}} \circ \underline{\mathbf{V}})\langle\boldsymbol{\xi}\rangle = \underline{\mathbf{A}}\langle\underline{\mathbf{V}}\langle\boldsymbol{\xi}\rangle\rangle \qquad \forall\boldsymbol{\xi} \in \mathcal{H}\ .^{c} \tag{8.10}$$

Fig. 5.1 defines $\boldsymbol{\xi}$ in Definition 8.3, and $\mathcal{H}$ is a spherical domain in real physical space. Note that in addition to requiring $\forall\boldsymbol{\xi} \in \mathcal{H}$, $\underline{\mathbf{V}}\langle\boldsymbol{\xi}\rangle$ would also need to be an element of $\mathcal{H}$ for this definition to make sense; otherwise $\underline{\mathbf{A}}\langle\underline{\mathbf{V}}\langle\boldsymbol{\xi}\rangle\rangle$ would be undefined.

Definition 8.4 The *point product* of two states $\underline{\mathbf{A}} \in \mathcal{A}_{m+p}$ and $\underline{\mathbf{B}} \in \mathcal{A}_p$, both associated with the same particle P_i, is a state in $\mathcal{A}_m$ defined by

$$(\underline{\mathbf{AB}})_{i_1\cdots i_m}\langle B_j\rangle = \underline{\mathbf{A}}_{i_1\cdots i_m j_1\cdots j_p}\langle B_j\rangle \underline{\mathbf{B}}_{j_1\cdots j_p}\langle B_j\rangle \qquad \forall B_j \in \mathbb{B} \tag{8.11}$$

Similarly,

$$(\underline{\mathbf{BA}})_{i_1\cdots i_m}\langle B_j\rangle = \underline{\mathbf{B}}_{j_1\cdots j_p}\langle B_j\rangle \underline{\mathbf{A}}_{j_1\cdots j_p i_1\cdots i_m}\langle B_j\rangle \qquad \forall B_j \in \mathbb{B} \tag{8.12}$$

Note that the summation (using indicial notation) occurs over all the innermost indices. If either or both of the states is a scalar state, then the point product is commutative: $\underline{\mathbf{A}\mathrm{b}} = \underline{\mathrm{b}\mathbf{A}}$. If both $\underline{\mathbf{A}}$ and $\underline{\mathbf{B}}$ are of the same order, the point product is also commutative and yields a scalar state. The point product of two states is thus another state,

[c] If $\underline{\mathbf{V}}\langle\boldsymbol{\xi}\rangle$ is not an element of $\mathcal{H}$, the operation is not defined. Indeed, if the state is evaluated at a point too close to a boundary of domain $\mathcal{R}$, $\mathcal{H}$ cannot be spherical, as some points of a spherical domain will lie outside the material body, and thus will have no reference position $\boldsymbol{X}$, rendering $\boldsymbol{\xi}$ undefined.

evaluated as the inner product of two tensors, evaluated at each particular bond B_j within the bond list $\mathbb{B}$.

We introduced the neighbor list $\mathcal{N}$ in Definition 6.7. The following definition considers particle P_j which is accessed via the neighbor list $\mathcal{N}[P_i]\langle B_j\rangle$ of particle P_i evaluated at bond B_j.

Definition 8.5 The *reference position vector state* $\underline{\mathbf{X}_\mathbf{R}} \in \mathcal{A}_1$ and the *identity scalar state* $\underline{1} \in \mathcal{A}_0$ are defined by

$$\begin{aligned} \underline{\mathbf{X}_{\boldsymbol{R}}}[P_i]\langle B_j\rangle = \boldsymbol{X}(\mathcal{N}[P_i]\langle B_j\rangle) - \boldsymbol{X}(P_i) = \boldsymbol{\xi} \qquad \forall B_j \in \mathbb{B}, \\ \underline{1}[P_i]\langle B_j\rangle = 1 \quad \forall B_j \in \mathbb{B}, \end{aligned} \tag{8.13}$$

where the vectors $\boldsymbol{\xi}$ and $\boldsymbol{\eta}$ are depicted in Fig. 5.2. If $\mathcal{N}[P_i]\langle B_j\rangle = P_\phi$, (the null particle) then because $\boldsymbol{X}(P_\phi)$ is undefined, $\underline{\mathbf{X}_{\boldsymbol{R}}}[P_i]\langle B_j\rangle$ is also undefined.

We also define the *null vector state* $\underline{\mathbf{0}} \in \mathcal{A}_1$ and the null scalar state $\underline{0} \in \mathcal{A}_0$ in the obvious way,

$$\underline{\mathbf{0}}\langle B_j\rangle = \mathbf{0}, \qquad \underline{0}\langle B_j\rangle = 0, \qquad \forall B_j \in \mathbb{B}\,. \tag{8.14}$$

Definition 8.6 The *dot product* of two states $\underline{\mathbf{A}}$ and $\underline{\mathbf{B}}$ is a tensor defined by

$$\underline{\mathbf{A}} \cdot \underline{\mathbf{B}} = \textstyle\sum_{\mathbb{B}} \underline{\mathbf{AB}}\langle B_j\rangle\,, \qquad \forall B_j \in \mathbb{B}. \tag{8.15}$$

(The elements of the sum are the point product, defined in Definition 8.4.) If $\underline{\mathbf{A}}$ and $\underline{\mathbf{B}}$ are the same order, or if either of them is a scalar state, then the dot product is commutative: $\underline{\mathbf{A}} \cdot \underline{\mathbf{B}} = \underline{\mathbf{B}} \cdot \underline{\mathbf{A}}$.

Definition 8.7 If $\underline{\mathbf{A}} \in \mathcal{A}_m$, then the *magnitude state* of $\underline{\mathbf{A}}$ is the scalar state defined by

$$|\underline{\mathbf{A}}|\langle B_j\rangle = \sqrt{(\underline{\mathbf{AA}})\langle B_j\rangle} \qquad \forall B_j \in \mathbb{B}. \tag{8.16}$$

(For any B_j, the point product under the radical sign is a non-negative scalar.)

Definition 8.8 If $\underline{\mathbf{N}}$ is a state, and if $|\underline{\mathbf{N}}| = \underline{1}$ then $\underline{\mathbf{N}}$ is a *unit state.*

Definition 8.9 The *norm* of a state $\underline{\mathbf{A}}$ is a scalar defined by

$$\|\underline{\mathbf{A}}\| = \sqrt{\underline{\mathbf{A}} \cdot \underline{\mathbf{A}}}\ .^{\mathrm{d}} \tag{8.17}$$

Definition 8.10 Define the *direction* of a state $\underline{\mathbf{A}} \in \mathcal{A}_m$ to be the state Dir $\underline{\mathbf{A}} \in \mathcal{A}_m$ given by

$$(\mathrm{Dir}\,\underline{\mathbf{A}})\langle B_j\rangle = \begin{Bmatrix} \mathbf{0} \ \ \text{if} \ |\underline{\mathbf{A}}|\langle B_j\rangle = 0 \\ \underline{\mathbf{A}}\langle B_j\rangle / |\underline{\mathbf{A}}|\langle B_j\rangle \ \ \text{otherwise} \end{Bmatrix} \quad \forall B_j \in \mathbb{B}\,. \tag{8.18}$$

Note that all of our definitions regarding peridynamic bond-based states are in terms of topological bonds, B_j, rather than in terms of reference position vectors $\boldsymbol{\xi}$, as with the states defined by (Silling, S.A., Epton, M., Weckner, O., Xu, J., and Askari, E., 2007). This set of definitions is better suited to computation on lattices because the integer particle bond list $\mathbb{B}$, together with particle neighbor lists $\mathcal{N}$, provides the necessary topological information, including boundary information, that is missing in the original state-based model, which is silent on the issue of how to handle bonds that lie outside of the body $\mathcal{R}$.

8.4 Vector states and tensors

The set of bond-based states of order m, $\mathcal{A}_m$, with vector addition defined by Definition 8.2, the scalar product using the field of real numbers, the norm provided by Definition 8.17, and the vector product defined by the dot product provided by Definition 8.6, is a real Euclidean vector space, introduced by Definition 3.3. This is despite of the fact that

---

[d] The norm is different than the magnitude, in that the norm is a scalar that is not a function of B_j.

the value of the state is in general nonlinearly related to the operand, topological bond B_j.

The notion of a bond-based vector state is similar to that of a second-order tensor, except a vector state is much more general than a tensor. While a second-order tensor provides a continuous, linear map from a vector $\boldsymbol{\xi}$ into another vector, a bond-based state is slightly different:

- A bond-based state is not a function of $\boldsymbol{\xi}$; rather it is a function of topological bond B_j.
- A bond-based state is not a continuous function of $\boldsymbol{\xi}$.
- The set of bond-based states of order m, $\mathcal{A}_m$, is ${N_R}^m \times N_B$ dimensional, where N_B is the number of bonds within the bond list $\mathbb{B}$, and N_R is the dimensionality of the physical space, while the set of second-order tensors in a $\mathbb{R}^3$ space has dimension 9.

These ideas are illustrated in Fig. 8.1.

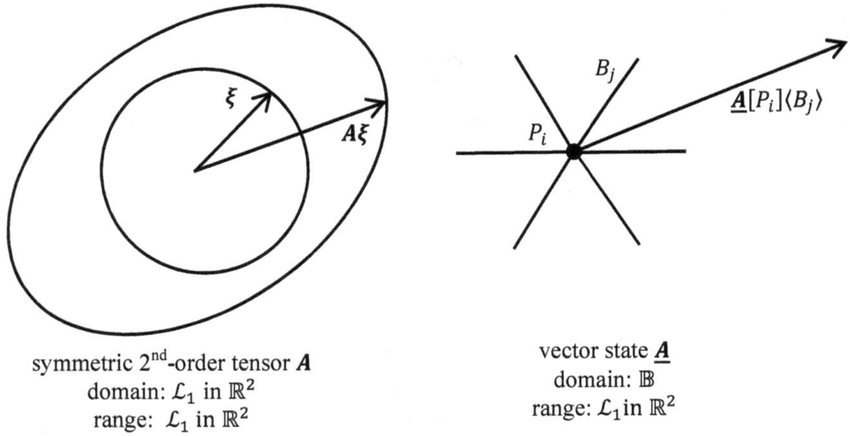

Figure 8.1. A second order tensor maps a sphere into an ellipsoid (a linear mapping); while a first-order state maps discrete numbered topological bonds B_j into vectors, as depicted schematically.

It is possible to *approximate* a vector state with a second-order tensor (called *reduction*), and to expand a second-order tensor to a vector state (called *expansion*), as follows.

Definition 8.11 Given a second order tensor $\mathbf{A}$, let $\underline{\mathcal{E}}(\mathbf{A}) \in \mathcal{A}_1$ be the vector state expanded from $\mathbf{A}$, defined by

$$\underline{\mathcal{E}}(\mathbf{A})\langle B_j\rangle = \mathbf{A}\,\underline{\mathbf{X}_R}\langle B_j\rangle = \mathbf{A}\boldsymbol{\xi} \qquad \forall B_j \in \mathbb{B}\,. \tag{8.19}$$

If $\mathcal{N}\langle B_j\rangle = P_\phi$ (the null particle) then $\underline{\mathcal{E}}(\mathbf{A})\langle B_j\rangle$ is undefined. This issue, that the material horizon $\mathcal{H}$ is not spherical close to the domain boundary, was not explicitly addressed in the original definition of *state* in (Silling, S.A., Epton, M., Weckner, O., Xu, J., and Askari, E., 2007). By changing from a reference-position-based state $\mathcal{H} \in \mathbb{R}^3$ to a bond-based state $\mathbb{B}$, the issue of undefined reference positions $\boldsymbol{X}$ for particles located outside the continuum domain $\mathcal{R}$ or outside the lattice body $\mathcal{L}_\mathcal{R}$ is explicitly addressed.

Definition 8.12 Let a scalar state $\underline{\omega} \in \mathcal{A}_0$ be given. Suppose that $\underline{\omega}$ is nonnegative on $\mathbb{B}$. Suppose further that there is a non-null subregion with $\mathbb{B}' \subset \mathbb{B}$ such that $\underline{\omega}$ is strictly positive on $\mathbb{B}'$. We call $\underline{\omega}$ an *influence function*. If $\underline{\omega}$ is an influence function and $\underline{\omega}\langle B_j\rangle$ depends only on the scalar $\left|\underline{\mathbf{X}_R}\langle B_j\rangle\right|$, then $\underline{\omega}$ is said to be spherical, and we write $\underline{\omega}\langle\xi\rangle = \underline{\omega_s}\langle|\xi|\rangle$.

Note that if $\mathcal{N}\langle B_j\rangle = P_\phi$ then $\underline{\mathbf{X}_R}\langle B_j\rangle$ is undefined and thus $\underline{\omega}\langle\xi\rangle$ is undefined. Let us adopt the convention that $\underline{\omega}\langle B_j\rangle = 0$ if $\mathcal{N}\langle B_j\rangle = P_\phi$.

Let the *dyadic product* of two vectors $\mathbf{a}$ and $\mathbf{b}$ be denoted $\mathbf{C} = \mathbf{a}\otimes\mathbf{b}$. (Thus, in rectangular coordinates, $C_{ij} = a_i b_j$.)

Definition 8.13 Let an influence function $\underline{\omega} \in \mathcal{A}_0$ be given. For any two vector states $\underline{\mathbf{A}} \in \mathcal{A}_1$ and $\underline{\mathbf{B}} \in \mathcal{A}_1$, let $\underline{\mathbf{A}} * \underline{\mathbf{B}}$ be the second order tensor defined by

$$\underline{\mathbf{A}} * \underline{\mathbf{B}} = \sum_{B_j=1}^{N_B} \underline{\omega}\langle B_j\rangle \underline{\mathbf{A}}\langle B_j\rangle \otimes \underline{\mathbf{B}}\langle B_j\rangle. \tag{8.20}$$

The tensor $\underline{\mathbf{A}} * \underline{\mathbf{B}}$ is called the tensor product of bond-based states $\underline{\mathbf{A}}$ and $\underline{\mathbf{B}}$.

Definition 8.14 Let an influence function $\underline{\omega} \in \mathcal{A}_0$ be given. Define the *shape tensor* $\mathbf{K}$ by

$$\mathbf{K} = \underline{\mathbf{X}_R} * \underline{\mathbf{X}_R} \tag{8.21}$$

where $\underline{\mathbf{X}_R} \in \mathcal{A}_1$ is introduced in Definition 8.5. Note that $\mathbf{K}$ is symmetric.

Lemma 8.1 Let an influence function $\underline{\omega} \in \mathcal{A}_0$ be given. Then $\mathbf{K}$ defined in Eq. 8.21 is positive definite.

The proof to a very similar lemma is given in (Silling, S.A., Epton, M., Weckner, O., Xu, J., and Askari, E., 2007).

Definition 8.15 Let an influence function $\underline{\omega} \in \mathcal{A}_0$ be given, and let $\underline{\mathbf{A}} \in \mathcal{A}_1$ be a vector state. Define the second order tensor by

$$\mathcal{R}\{\underline{\mathbf{A}}\} = \left(\underline{\mathbf{A}} * \underline{\mathbf{X}_R}\right) \boldsymbol{K}^{-1} \,. \tag{8.22}$$

$\mathcal{R}\{\underline{\mathbf{A}}\}$ is called the *tensor reduced from* $\underline{\mathbf{A}}$. (The existence of $\boldsymbol{K}^{-1}$ follows from Lemma 8.1.)

The provenance of the reduction of a state to a tensor defined by Eq. 8.22 needs some explanation. We provide background for this equation as follows. Assume that the tensor, $\mathcal{R}\{\underline{\mathbf{A}}\}$, reduced from the tensor state, $\underline{\mathbf{A}}$ exactly satisfies

$$\mathcal{R}\{\underline{\mathbf{A}}\} \cdot \xi(B_j) = \underline{\mathbf{A}}\langle \xi(B_j) \rangle \,, \quad \forall B_j \in \mathbb{B}, \text{(except if } \mathcal{N}\langle B_j \rangle = P_\phi) \tag{8.23}$$

which will be exactly true for all $B_j \in \mathbb{B}$ only if $\boldsymbol{\alpha}(B_j) = \underline{\mathbf{A}}\langle \xi(B_j) \rangle$ is a linear map from $\boldsymbol{\xi}$ to $\boldsymbol{\alpha}$. Next, operate dyadically on both sides of Eq. 8.23 as follows

$$(\mathcal{R}\{\underline{\mathbf{A}}\} \cdot \xi) \otimes \xi = \underline{\mathbf{A}}\langle \xi \rangle \otimes \xi \,, \tag{8.24}$$

or, due to associativity of the dyadic and inner products

$$\mathcal{R}\{\underline{\mathbf{A}}\} \cdot (\boldsymbol{\xi} \otimes \boldsymbol{\xi}) = \underline{\mathbf{A}}\langle\boldsymbol{\xi}\rangle \otimes \boldsymbol{\xi} .$$

Multiplying both sides of Eq. 8.24 by the scalar influence function, $\underline{\omega}\langle\boldsymbol{\xi}\rangle$,

$$\underline{\omega}\langle\boldsymbol{\xi}\rangle \mathcal{R}\{\underline{\mathbf{A}}\} \cdot (\boldsymbol{\xi} \otimes \boldsymbol{\xi}) = \underline{\omega}\langle\boldsymbol{\xi}\rangle \underline{\mathbf{A}}\langle\boldsymbol{\xi}\rangle \otimes \boldsymbol{\xi} . \tag{8.25}$$

$\mathcal{R}\{\underline{\mathbf{A}}\}$, being a second-order tensor, cannot be a function of $\boldsymbol{\xi}$ and it has ${N_R}^2$ components that must be satisfied at N_B locations $\boldsymbol{\xi}(B_j)$, which is clearly not in general possible. Therefore, let us try to satisfy Eq. 8.25 in an approximate sense only:

$$\textstyle\sum_{\mathbb{B}} \underline{\omega}\langle\boldsymbol{\xi}\rangle \mathcal{R}\{\underline{\mathbf{A}}\} \cdot (\boldsymbol{\xi} \otimes \boldsymbol{\xi}) = \sum_{\mathbb{B}} \underline{\omega}\langle\boldsymbol{\xi}\rangle \underline{\mathbf{A}}\langle\boldsymbol{\xi}\rangle \otimes \boldsymbol{\xi} , \tag{8.26}$$

(including only the terms for which $\boldsymbol{\xi}$ exists) and since $\mathcal{R}\{\underline{\mathbf{A}}\}$ is not a function of bonds $b_j \in \mathbb{B}$, it can be removed for the summation:

$$\mathcal{R}\{\underline{\mathbf{A}}\} \cdot \textstyle\sum_{\mathbb{B}} \underline{\omega}\langle\boldsymbol{\xi}\rangle (\boldsymbol{\xi} \otimes \boldsymbol{\xi}) = \sum_{\mathbb{B}} \underline{\omega}\langle\boldsymbol{\xi}\rangle \underline{\mathbf{A}}\langle\boldsymbol{\xi}\rangle \otimes \boldsymbol{\xi} . \tag{8.27}$$

Finally,

$$\mathcal{R}\{\underline{\mathbf{A}}\} = \left[\textstyle\sum_{\mathbb{B}} \underline{\omega}\langle\boldsymbol{\xi}\rangle \underline{\mathbf{A}}\langle\boldsymbol{\xi}\rangle \otimes \boldsymbol{\xi}\right] \cdot \left[\sum_{\mathbb{B}} \underline{\omega}\langle\boldsymbol{\xi}\rangle (\boldsymbol{\xi} \otimes \boldsymbol{\xi})\right]^{-1}, \tag{8.28}$$

which is the same as Eq. 8.22.

Note that in Eq. 8.28 all terms in which $\boldsymbol{\xi}$ is undefined are left out of the summations because for these terms $\underline{\omega}\langle\boldsymbol{\xi}\rangle = 0$.

The fact that, for any second order tensor $\mathbf{A}$, $\mathcal{R}\{\underline{\mathcal{E}}(\mathbf{A})\} = \mathbf{A}$, is proven in (Silling, S.A., Epton, M., Weckner, O., Xu, J., and Askari, E., 2007). However, $\underline{\mathcal{E}}\{\mathcal{R}(\underline{\mathbf{A}})\} = \underline{\mathbf{A}}$ is not in general true, so the expansion and reduction functions are only approximately inverses of each other.

In the reduction of a state to a tensor, information is lost; the reduction is only an approximation, and thus one should avoid the reduction if possible.

8.5 State fields

A state *field* is a state-valued function of material particle P_i (in the material space) and time t. The location and time at which particular values (states) of a state field are evaluated are shown in square brackets. For example $\underline{\mathbf{A}}[P_i, t]$ is the value of a state field evaluated at a particular particle and time.

We can also define functions of states. For example, consider the position- and time-dependent vector state-valued function $\underline{\boldsymbol{\Psi}}[P_i, t](\underline{\mathbf{A}})$. We indicate the *value* $\mathbf{v}$ of such a function of a state at a particular particle P_i and time t, and evaluated at reference bond B_j as

$$\mathbf{v} = \underline{\boldsymbol{\Psi}}[P_i, t](\underline{\mathbf{A}})\langle B_j \rangle \,. \tag{8.29}$$

This may seem a bit confusing at first, but practice with physical and computational applications will make the notation more intuitive.

8.6 State based peridynamic theory

In the (bond-based) state based theory, Newton's Second Law expressed in peridynamic terms, Eq. 5.5, is replaced with

$$m\ddot{\boldsymbol{x}}(P_i, t) = \textstyle\sum_{\mathbb{B}} \left\{ \underline{\mathbf{T}}[P_i, t]\langle B_j \rangle - \underline{\mathbf{T}}[P', t]\langle B_j{}' \rangle \right\} + \boldsymbol{b}(P_i, t), \tag{8.30}$$

where m is the mass of particle $P_i \in \mathcal{L}_{\mathcal{R}}$, $\mathbb{B}$ is the peridynamic neighborhood of particle P_i, $\boldsymbol{b}(P_i, t)$ is the external force applied to particle P_i at time t, and where $\underline{\mathbf{T}}[P_i, t]\langle B_j \rangle$ is the *bond-based force vector state field* evaluated at particle P_i and at time t with respect to bond B_j. Note that $P' = \mathcal{N}[P_i]\langle B_j \rangle$, and, using the bond-numbering scheme shown in Fig. 6.7, $B_j' = B_j + 1$ if B_j is odd and $B_j' = B_j - 1$ if B_j is even. Thus B_j and $B_j{}'$ are complementary (opposing) bonds. If $\mathcal{N}[P_i]\langle B_j \rangle = P_\phi$, then both $\underline{\mathbf{T}}[P_i, t]\langle B_j \rangle = 0$ and $\underline{\mathbf{T}}[P', t]\langle B_j{}' \rangle = 0$.

How the force state $\underline{\mathbf{T}}$ is determined is the subject of Section 8.8 and Chapters 9, 10, and 11.

To make Eq. 8.30 more concise,

$$m\ddot{x}(P) = \sum_{\mathbb{B}}\{\underline{\mathbf{T}} - \underline{\mathbf{T}}'\} + \boldsymbol{b}(P_i) \ , \tag{8.31}$$

where

$$\underline{\mathbf{T}} = \underline{\mathbf{T}}[P_i]\langle B_j\rangle \text{ and } \underline{\mathbf{T}}' = \underline{\mathbf{T}}[\mathcal{N}[P_i]\langle B_j\rangle]\langle B_j'\rangle. \tag{8.32}$$

One can show that the terms $\underline{\mathbf{T}} - \underline{\mathbf{T}}'$ in the summation in Eq. 8.31 automatically satisfy the requirement that internal forces sum to zero on lattice body $\mathcal{L}_{\mathcal{R}}$ so that Newton's second law applied to $\mathcal{L}_{\mathcal{R}}$ as a whole is satisfied:

$$\sum_{\mathbb{L}} m\ddot{x}(P_i,t) = \sum_{\mathcal{L}_{\mathcal{R}}} \boldsymbol{b}(P_i,t) \quad \forall t \geq 0 \ . \tag{8.33}$$

However, satisfaction of conservation of angular momentum on the lattice body $\mathcal{L}_{\mathcal{R}}$

$$\sum_{\mathbb{L}} \boldsymbol{x}(P_i,t) \times \left(m\ddot{\boldsymbol{x}}(P_i,t)\right) = \sum_{\mathcal{L}_{\mathcal{R}}} \boldsymbol{x}(P_i,t) \times \boldsymbol{b}(P_i,t) \ , \tag{8.34}$$

requires restrictions on the choice of $\underline{\mathbf{T}}[P_i,t]$, as discussed in the next section.

8.7 Deformation state and state-based constitutive models

Definition 8.17 Let $\underline{\boldsymbol{x}_R}$ be the *deformation vector state field* defined by

$$\underline{\boldsymbol{x}_R}[P_i]\langle B_j\rangle = \boldsymbol{x}(\mathcal{N}[P_i]\langle B_j\rangle) - \boldsymbol{x}(P_i) = \boldsymbol{\xi} + \boldsymbol{\eta} \quad \forall B_j \in \mathbb{B}. \tag{8.35}$$

Thus $\underline{\boldsymbol{x}_R}[P_i]\langle B_j\rangle = \boldsymbol{\xi} + \boldsymbol{\eta}$ is the image of the bond $\boldsymbol{\xi}$ under deformation, as shown in Fig. 8.2. If $\mathcal{N}[P_i]\langle B_j\rangle = P_\phi$ then $\boldsymbol{x}(\mathcal{N}[P_i]\langle B_j\rangle)$ is undefined and so $\underline{\boldsymbol{x}_R}[P_i]\langle B_j\rangle$ is undefined.

Also, define the reference position scalar magnitude state field $\underline{\mathrm{X}_R}$ and the deformation scalar magnitude state field $\underline{x_R}$ by

$$\underline{\mathrm{X}_R} = \left|\underline{\mathbf{X}_R}\right|, \quad \underline{x_R} = \left|\underline{\boldsymbol{x}_R}\right| . \tag{8.36}$$

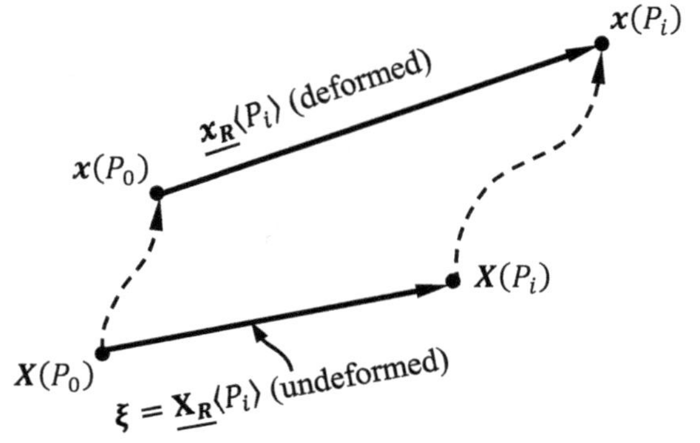

Figure 8.2. Reference undeformed and deformed vector states.

We illustrate these vector states in Fig. 8.2. We assume that

$$\underline{x_R}[P_i]\langle B_j\rangle \neq 0 \qquad \forall B_j \in \mathbb{B}, \quad \forall P_i \in \mathbb{P}. \tag{8.37}$$

which means that a particle P_i and its neighbor $P_j = \mathbb{N}[P_i]\langle B_j\rangle$ can never occupy the same point in space as the deformation progresses.[e]

A constitutive model in the state-based peridynamic theory is a relation that provides values for the force vector state field in terms of the deformation vector state field and possibly other variables as well. We write the general form of the constitutive model as

$$\underline{\mathbf{T}} = \underline{\tilde{\mathbf{T}}}\left(\underline{\mathbf{X}_R}, \underline{\boldsymbol{x}_R}, \Lambda_0\right), \tag{8.38}$$

[e] However, two particles that are *not* connected by a bond might well occupy the same location in space. We prevent particle overlap by using "short-range forces", which prevent interpenetration of particles. Such short-range forces are "interfacial", rather than being within the scope of a *material* model.

where $\underline{\tilde{\mathbf{T}}} : \mathcal{A}_1 \times \mathcal{A}_1 \times \mathcal{A}_{\Lambda_0} \to \mathcal{A}_1$ is a vector state-valued function of a vector state. Λ_0 represents all variables other than the current and reference deformation states that $\underline{\mathbf{T}}$ may depend upon for some particular material; $\mathcal{A}_{\Lambda_0}$ represents the set of all such variables.

Definition 8.18 If, for a given material, $\underline{\mathbf{T}}$ depends only upon $\underline{\mathbf{X}_R}$ and $\underline{\boldsymbol{x}_R}$, then the material is called *simple*, and we write

$$\underline{\mathbf{T}} = \underline{\tilde{\mathbf{T}}}\left(\underline{\mathbf{X}_R}\,,\ \underline{\boldsymbol{x}_R}\right), \tag{8.39}$$

for some function $\underline{\tilde{\mathbf{T}}} : \mathcal{A}_1 \times \mathcal{A}_1 \to \mathcal{A}_1$.

Thus, elastic materials are simple, but plastic materials and materials with damage, which depend upon variables describing the plastic and damage variables associated with the deformation, are not simple.

Proposition 8.1 Suppose a constitutive model, of the form $\underline{\mathbf{T}} = \underline{\hat{\mathbf{T}}}\left(\underline{\mathbf{X}_R}\,,\ \underline{\boldsymbol{x}_R}\,,\ \Lambda_0\right)$ is such that

$$\textstyle\sum_{\mathbb{B}} \underline{\boldsymbol{x}_R}\langle B_j\rangle \times \underline{\mathbf{T}}\langle B_j\rangle = \mathbf{0} \qquad \forall B_j \in \mathbb{B}\,. \tag{8.40}$$

Then the balance of angular momentum (Eq. 5.8) holds for any deformation of a body $\mathcal{L}_{\mathcal{R}}$ with this constitutive model.

The proof to a similar statement is given in (Silling, S.A., Epton, M., Weckner, O., Xu, J., and Askari, E., 2007).

Definition 8.19 Let the deformed direction vector state $\underline{\mathbf{M}}$ be the unit state valued function defined by

$$\underline{\mathbf{M}}\left(\underline{\boldsymbol{x}_R}\right) = \operatorname{Dir} \underline{\boldsymbol{x}_R} \quad \forall\, \underline{\boldsymbol{x}_R} \in \mathcal{A}_1\,, \tag{8.41}$$

thus $\underline{\mathbf{M}}\left(\underline{\boldsymbol{x}_R}[P_i]\right)\langle B_j\rangle$ is a unit vector that is directed from the deformed position $\boldsymbol{x}_i$ of particle P_i toward the deformed position $\boldsymbol{x}_j$ of particle $P_j = \mathcal{N}[P_i]\langle B_j\rangle$.

Definition 8.20 If a material has the property that for any deformation state there exists a scalar state $\underline{t}$ such that

$$\underline{\mathbf{T}} = \underline{t}\,\underline{\mathbf{M}}\,, \tag{8.42}$$

then the material is called *ordinary*, and $\underline{t}$ is called the *scalar force state field*. Otherwise, we call the material *non-ordinary*. We show ordinary and non-ordinary material models in Fig. 8.3.

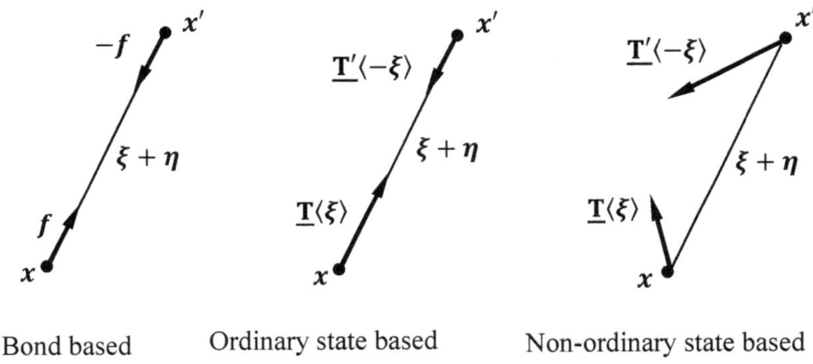

Figure 8.3. Bond based, ordinary, and non-ordinary peridynamic material models. All three models guarantee conservation of linear momentum, and the first two guarantee conservation of angular momentum.

In an ordinary material, Eq. 8.42 implies that

$$\left(\mathrm{Dir}\,\underline{\mathbf{T}}\right)\langle B_j\rangle = \left(\mathrm{Dir}\,\underline{\boldsymbol{x}_R}\right)\langle B_j\rangle \quad \text{whenever} \quad \underline{t}\langle B_j\rangle \neq 0\,. \tag{8.43}$$

Proposition 8.2 Let $\mathcal{L}_{\mathcal{R}}$ be a lattice body composed of an ordinary material and subjected to a body force field $\boldsymbol{b}$. Let $\underline{\mathbf{T}}$ be the force vector state field defined upon $\mathcal{L}_{\mathcal{R}}$. Then Eq. 8.43 holds, and the body's angular momentum is conserved.

We leave the proof to the reader as Problem 8.13.

We can show that the bond-based theory is a special case of the state-based theory.

An ordinary material has a vector force state that we can express as

$$\underline{\mathbf{T}} = \underline{t}\,\underline{\mathbf{M}}\,, \text{ with } \underline{t} = \hat{\underline{t}}\left(\underline{\boldsymbol{x}_R}\right) \tag{8.44}$$

For some function $\hat{\underline{t}} : \mathcal{A}_1 \to \mathcal{A}_0$.

Definition 8.21 If an ordinary material has the property that there exists a scalar state-valued function $\hat{\underline{t}} : \mathcal{A}_1 \to \mathcal{A}_0$ such that

$$\hat{\underline{t}}\left(\underline{\boldsymbol{x}_R}\,,\underline{\mathbf{X}_R}\right) = \underline{\tau}\left(\left|\underline{\boldsymbol{x}_R}\right|,\left|\underline{\mathbf{X}_R}\right|\right) \qquad \forall\, \underline{\boldsymbol{x}} \in \mathcal{A}_1\,, \forall\, \underline{\boldsymbol{X}} \in \mathcal{A}_1 \tag{8.45}$$

then the material is called *mobile* and $\underline{\tau}$ is called the mobile force function.

Mobility does not mean that the material is bond-based. It instead means that the magnitudes of the bond forces in $\underline{\mathbf{T}}$ depend upon the bond stretches but not upon their rotations. In a mobile material, the force in a bond may depend upon the stretches of all other bonds. A mobile material is by definition ordinary. However, an ordinary material is not necessarily mobile. Ordinariness only specifies the direction of the bond forces, while mobility specifies what these bond forces can depend upon.

8.8 Adaptation of classical material models

The deformation gradient $\boldsymbol{F}$ was introduced in Eq. 3.24 as $d\boldsymbol{x} = \boldsymbol{F}d\boldsymbol{X}$. According to Definition 8.15, for a specified influence function $\underline{\omega}\langle\boldsymbol{\xi}\rangle$, we can obtain the corresponding *approximate* deformation gradient from the deformation vector state, $\underline{\boldsymbol{x}_R}$ as

$$\bar{\mathbf{F}}\left(\underline{\boldsymbol{x}_R}\right) = \mathcal{R}\left\{\underline{\boldsymbol{x}_R}\right\} = \left(\underline{\boldsymbol{x}_R} * \underline{\mathbf{X}_R}\right)\boldsymbol{K}^{-1}. \tag{8.46}$$

With (an approximation of) the deformation gradient $\bar{\mathbf{F}}$ known, one can obtain the first Piola stress tensor $\boldsymbol{\sigma}$ from a hyperelastic (or similar) continuum model. With $\boldsymbol{\sigma}$ now known, one can recover the (approximate) force state using the equation

$$\underline{\mathbf{T}} = \underline{\omega}\boldsymbol{\sigma}\boldsymbol{K}^{-1}\boldsymbol{\xi}\ , \text{ or equivalently, } \ \underline{\mathbf{T}} = \underline{\omega\mathcal{E}}(\boldsymbol{\sigma}\boldsymbol{K}^{-1}). \tag{8.47}$$

The details of the derivation of Eq. 8.47 are given in (Silling, S.A., Epton, M., Weckner, O., Xu, J., and Askari, E., 2007).

Thus, one can approximate any hyperelastic continuum model using a state-based peridynamic model. However, this approximation is good only so long as the deformation gradient $\bar{\mathbf{F}}$ does not vary significantly within the material horizon, $\mathcal{N}[P_i]$. There are also unresolved questions about the relationship between continuum mechanics and peridynamics at the boundaries of the lattice body $\mathcal{L}_{\mathcal{R}}$. At the boundaries, the two approaches are difficult to compare.

As we shall see in the Chapters 9 through 11, we realize the real power of the state-based peridynamic lattice model not by simply adapting the SPLM to legacy continuum constitutive models, but instead by developing constitutive models suited directly to the peridynamic state-based peridynamic lattice model.

8.9 Summary

This chapter has presented a stated-based peridynamic model for the deformation of material bodies described by lattice bodies $\mathcal{L}_{\mathcal{R}}$. The theory presented here is somewhat different from the continuum state-based peridynamic theory introduced in (Silling, S.A., Epton, M., Weckner, O., Xu, J., and Askari, E., 2007).

The essential difference is that in this chapter we have distinguished between the material description and the initial reference description of the material body. Furthermore, we define the material description as a lattice; thus, we provide more regularity than one would find with a random particle model or with a discrete element model.

While the peridynamic theories published by Silling and his coworkers are continuum models, the lattice version presented in this chapter is a discrete model. Thus, we can more readily implement the current version on a computer, with no further spatial discretization decisions necessary.

An analyst steeped in the finite element approach to computational analysis might be wondering why no mesh convergence study is mandated. The answer is that the lattice spacing L chosen is a material property, and to refine the lattice further would not bring one closer to the physical solution to the problem, although it might bring one closer to the continuum mechanics solution (which itself ceases to be applicable at scales smaller than the mesoscale).

The continuum mechanics solution, however, is not what we are seeking. Instead, we are seeking a solution that is more general and more realistic than the continuum mechanics solution. We are looking for a solution that will be sufficiently general to capture a wider range of behaviors than the continuum approach can represent.

In the next chapter, we present an elastic constitutive model for the state-based peridynamic lattice model (SPLM).

8.10 Exercises

8.1 In Eq. 8.1, what are the units of $G(R)$ and of $H(R)$?

8.2 Derive Eq. 8.2. (Hint: follow the parallel derivation in Section 2.3.)

8.3 Derive Eq. 8.3.

8.4 Derive the equation for surface tractions, equivalent to Eq. 2.23, using the generalized Navier theory of Section 8.1, in terms of ε and β.

8.5 Solve the Navier-Cauchy equation, Eq. 8.7, assuming an elastic shaft of length L, and diameter D, whose major axis is aligned with the X_1 axis. The shaft is restrained at its left end, located at $X_1 = 0$ and is extended by an amount $\Delta \ll L$ in the X_1 direction at its right end, located at $X_1 = L$. Assume the cylindrical surface of the shaft is constrained from moving in the X_2 and X_3 directions. Find the displacement field $U_i(X_1, X_2, X_3)$.

8.6 For the lattice particle shown in Fig. 6.7, find the shape tensor $\mathbf{K}$ assuming an influence function $\underline{\omega} = \underline{1}$. Assume the radius of the material horizon is $\delta = 1.1L$, so that 12 bonds are included in the neighbor list $\mathbb{B}$.

8.7 What is the dimensionality of a bond-based state of order two in $\mathbb{R}^2$ Euclidian physical space, defined upon a hexagonal lattice Λ^2 with a peridynamic horizon $\delta = 1.9L$, where L is the lattice spacing?

8.8 Evaluate the state $\underline{\mathbf{A}} = \underline{\mathbf{X}_R\mathbf{X}_R} + \underline{1}L^2$ for a planar hexagonal lattice with lattice spacing L. Include only the six nearest neighbors in the bond list $\mathbb{B}$. (Hint: represent the state $\underline{\mathbf{A}}$ using a 6x1 matrix, where each row of the matrix represents a bond B_j, and each column represents a scalar. Use MatLab.)

8.9 Evaluate the norm $\left\|\underline{\mathbf{X}_R}\right\|$ of the state $\underline{\mathbf{X}_R}$ for the planar hexagonal lattice shown in Fig. 6.7 with lattice spacing L. Include only the six nearest neighbors in the bond list $\mathbb{B}$. (Hint: represent the state $\underline{\mathbf{X}_R}$ using a 6x2 matrix, where each row of the matrix represents a bond B_j, and each column represents a vector component. Use MatLab.)

8.10 For the three-dimensional lattice particle shown in Fig. 6.7, find the shape tensor $\mathbf{K}$ assuming an influence function $\underline{\omega} = \underline{1}$. Assume $\delta = 1.5L$, so that 18 particles are included in $\mathcal{X}[P_0]$.

8.11 Define a vector state $\underline{\mathbf{A}} = 3\underline{\mathbf{X}_R} + \underline{L\hat{\mathbf{e}}_1}$. Reduce this vector state to a second-order tensor $\boldsymbol{B} = \mathcal{R}\{\underline{\mathbf{A}}\}$ using Definition 8.15. Assume a hexagonal planar lattice shown in Fig. 6.7, including only the six closest bonds, with influence function $\underline{\omega} = \underline{1}$.Use MatLab.

8.12 Expand the tensor $\boldsymbol{B}$ found in Problem 8.11 back to a vector state $\underline{\mathbf{A}} = \underline{\mathcal{E}}(\mathbf{B})$. Does this vector state turn out to be $3\underline{\mathbf{X}_R} + \underline{L\hat{\mathbf{e}}_1}$? Use MatLab.

8.13 Prove Proposition 8.2.

8.14 The deformation field of a two-dimensional plate is $x_1 = .002X_1, x_2 = .001X_1X_2$. Determine the deformation gradient, $\boldsymbol{F}$. Expand the deformation gradient $\boldsymbol{F}$ to the peridynamic state $\underline{\boldsymbol{x_R}}$ defined upon the hexagonal lattice shown in Fig. 6.7, of spacing $L = 1$. Represent the state as a 6×2 matrix, where each row of the matrix represents a bond, and each column represents a deformation component.

8.15 A two-dimensional peridynamic state $\underline{\boldsymbol{x_R}} = \begin{bmatrix} .001 & 0 \\ -.001 & 0 \\ .0005 & .0003 \\ -.0005 & -.0003 \\ -.0006 & .0004 \\ .0006 & -.0004 \end{bmatrix} L$ is defined upon the hexagonal lattice shown in Fig. 6.7, of spacing $L = 1$ unit of length. Use Eq. 8.46 to convert to an approximately equivalent deformation gradient $\bar{\mathbf{F}}\left(\underline{\boldsymbol{x_R}}\right)$. For each of the six bonds, what is the percent error in deformation magnitude, defined as $E = \frac{\left|\bar{\mathbf{F}}\left(\underline{\boldsymbol{x_R}}\right)\underline{\mathbf{X_R}}\langle B_i\rangle\right| - \left|\underline{\boldsymbol{x_R}}\langle B_i\rangle\right|}{\left|\underline{\boldsymbol{x_R}}\langle B_i\rangle\right|}$? Finally, expand the approximate deformation gradient $\bar{\mathbf{F}}\left(\underline{\boldsymbol{x_R}}\right)$ back to the approximate deformation state $\overline{\underline{\boldsymbol{x_R}}}$. What is the percent error in each of the components of $\overline{\underline{\boldsymbol{x_R}}}$ with respect to each of the components of $\underline{\boldsymbol{x_R}}$?

8.16 Develop a two-dimensional peridynamic lattice model for linear isotropic Fourier heat flux: $\boldsymbol{q} = -k\boldsymbol{\nabla}T$, where $\boldsymbol{q}$ is the local heat flux, in units of $J/(s \cdot m^2)$, k is the material's conductivity, in units of $J/(s \cdot m \cdot {}^oK)$, and $\boldsymbol{\nabla}T$ is the spatial temperature gradient, in units of ${}^oK/m$. Will a bond-based model be sufficient, or does the model need to be state-based?

Chapter 9

Elastic SPLM

Chapter objectives are to:

- describe the theoretical basis of the elastic state-based peridynamic lattice model (SPLM);
- develop axial, planar, and three-dimensional elastic models;
- show how the materially-linear model is capable of modeling geometrically nonlinear behavior;
- discuss model behavior at domain boundaries;
- indicate how the SPLM might be computationally implemented.

In this chapter, we develop an elastic state-based peridynamic lattice model (SPLM). We defined the lattice Λ^{N_R} in Definition 6.3, where N_R is the dimensionality of the space spanned by the lattice. In the reference configuration, for one-dimensional problems we assume a straight row of equally spaced particles Λ^1, a close-packed hexagonal lattice Λ^2 for two-dimensional problems, and a face-centered cubic close-packed lattice Λ^3 for three-dimensional problems. We show that the SPLM is capable of simulating a linear elastic classical continuum, and furthermore that it is capable of simulating geometrically nonlinear large-deformation behavior. The SPLM presented in this chapter is "regular" as opposed to micropolar, in that we do not consider particle rotations as degrees of freedom; only particle translations are included as degrees of freedom. Furthermore, we assume that the material is *simple*, *ordinary*, and *mobile*, as defined by Definitions 8.18, 8.20 and 8.21, respectively.

The *linear* elastic SPLM assumes a linear relationship between axial force associated with bond B_j and all of the bond stretches S_j associated with the bond list $\mathbb{B}$. Under the assumption of small deformations (relative to lattice spacing L), bond stretches S_j may be assumed to be linearly related to particle spatial locations $\boldsymbol{x}$. However, because under large deformations bond stretches are nonlinearly related to particle spatial positions, the "linear elastic" model is nonlinear in the overall sense that statically applied forces are not linearly related to particle displacements, as in Navier-Cauchy linear elasticity.

One can easily implement nonlinear elastic relations between bond force and bond stretch, but we do not attempt to develop such relations in this book. Rubber-like materials undergoing large deformations would require such a nonlinear elastic constitutive model. There appears to be no particular theoretical or practical difficulty in developing nonlinear force-stretch relations if and when they might be needed. The approach to developing such nonlinear models is essentially an empirical data regression to assign constitutive model parameters in such a way that the peridynamic simulation of a laboratory test matches laboratory results.

One calls this approach a *phenomenological model*: a model that replicates the behavior of laboratory test phenomena without regard to their fundamental origins.

For one-dimensional microelastic peridynamic constitutive models, as with the bond-based model developed in Chapter 7, we find that it is sufficient to consider forces and stretches between the two nearest particle neighbors only, so with reference to Definition 6.6 and Fig. 6.7, the particle bond list is $\mathbb{B} = \{1, 2\}$. For two-dimensional problems, the deformation vector state $\underline{\boldsymbol{x}_R}$ (defined in Definition 8.17) is adequately described by including the spatial positions of the six nearest neighbors in the peridynamic constitutive model, so the particle bond list is $\mathbb{B} = \{1, \cdots, 6\}$. For three-dimensional models, to enable isotropic material behavior, as with the bond-based model described in the last chapter, we find that it is necessary to include the twelve nearest neighbors as well as the six second-nearest neighbors depicted in Fig. 6.7. Thus $\mathbb{B} = \{1, \cdots, 18\}$.

The force state is then determined through the microelastic constitutive function $\underline{\widetilde{\mathbf{T}}}$ given by Eq. 8.38, $\underline{\mathbf{T}} = \underline{\widetilde{\mathbf{T}}}\left(\underline{\mathbf{X}_R}, \underline{\boldsymbol{x}_R}, \Lambda_0\right)$, of the deformation state.[a] Because for a given lattice Λ^{N_R} the reference state $\underline{\mathbf{X}_R}$ does not change from particle to particle nor with respect to time, and Λ_0 is null because we here assume an elastic model and the force state does

---

[a] See Definitions 8.5, 8.17, and 8.18 for the definitions of $\underline{\mathbf{X}_R}$, $\underline{\boldsymbol{x}_R}$, Λ, and $\underline{\widetilde{\mathbf{T}}}$.

not depend upon any other state variables, the elastic state-based peridynamic constitutive model simplifies to $\underline{\mathbf{T}} = \underline{\tilde{\mathbf{T}}}\left(\underline{\boldsymbol{x}_R}\right)$.

The state-based peridynamic theory presented in Chapter 8 describes the initial and current spatial deformation states $\underline{\mathbf{X}_R}$ and $\underline{\boldsymbol{x}_R}$ and the force state $\underline{\boldsymbol{T}}$. Because the SPLM we develop here is ordinary and mobile, and additionally each particle has only eighteen bonds, these SPLM states, rather than being infinite-dimensional, are eighteen-dimensional. Given the particle positions $\boldsymbol{x}$, the neighbor list $\mathcal{N}[P_i]\langle B_j\rangle$ for each particle P_i, and the specification that this is a mobile material, it is sufficient to represent the force state $\underline{\boldsymbol{T}}$ by the *signed magnitudes* of the eighteen bond forces. Similarly, we adequately represent the deformation state $\underline{\boldsymbol{x}_R}$ by just the stretches of the eighteen bonds in the bond list $\mathbb{B}$.[b] It is not necessary to store directional information with the state, because this information can be calculated in each time step from the changing particle spatial positions, $\boldsymbol{x}$. Thus, rather than employing force states $\underline{\boldsymbol{T}}$ and deformation states $\underline{\boldsymbol{x}_R}$, we are justified in instead using an 18×1 lattice-based stretch matrix $\{S\}$ and the corresponding 18×1 lattice-based force matrix $\{T\}$, with the directions of the eighteen bond forces in the force state $\underline{\boldsymbol{T}}$ calculated from current particle positions $\boldsymbol{x}$ for large deformation analysis, or from the reference particle positions $\boldsymbol{X}$ for small-deformation analysis.[c]

In summary, we start by defining the lattice-based stretch matrix $\{S\}$ and the lattice-based force matrix $\{T\}$. $\{S\}$ and $\{T\}$, together with an explanation of their meanings as well as the lattice referential and current configurations, provide the necessary information to recreate the deformation state and the force state defined in the state-based peridynamic continuum model described in Chapter 8. A linear relation is then assumed between the force matrix and the stretch matrix:

---

[b] The force state $\underline{\boldsymbol{T}}$ and the deformation state $\underline{\boldsymbol{x}_R}$ of a 3D lattice particle with eighteen bonds are 18x3=54 dimensional, if three force and three deformation components are included per bond. However, since we know the current positions $\boldsymbol{x}$ of the two particles connected by each bond, and this is a mobile material, it is redundant to store the components; rather the magnitudes of each bond force and stretch vector are sufficient.

[c] The additional computational effort required for large deformation analysis is small, so we recommend using only large deformation analysis. This simplifies our conceptual understanding of the SPLM.

$\{T\} = [K]\{S\}$. There are many forms of the peridynamic microelastic stiffness matrix $[K]$ that will enable the state based peridynamic lattice model to simulate a particular classical elastic continuum. We choose a form of $[K]$ that determines the force in each bond by using one parameter a, which takes into account the stretch of that bond, as well as another parameter b, which takes into account the effect of the aggregate stretch of all bonds connected to the particle. In this sense, the SPLM is a two-parameter model, similar to the classical Navier-Cauchy isotropic elasticity model, and in particular, like the generalized Navier theory presented in Section 8.1.

9.1 SPLM stretch and SPLM force

For a three-dimensional FCC close-packed lattice, as shown in Fig. 6.7, it is necessary to include nearest neighbors as well as second-nearest neighbors. We define the *stretch matrix* associated with particle P_i as

$$\{S\}_i \equiv \begin{Bmatrix} S_1 \\ \vdots \\ S_j \\ \vdots \\ S_{18} \end{Bmatrix}_i , \tag{9.1}$$

where each of the components, S_j, of the stretch state, represents the stretch of an individual bond B_j, and is defined as

$$S_j \equiv \frac{\left|\underline{\boldsymbol{x}_R}[P_i,t]\langle B_j\rangle\right| - \left|\underline{\mathbf{X}_R}[P_i,t]\langle B_j\rangle\right|}{\left|\underline{\mathbf{X}_R}[P_i,t]\langle B_j\rangle\right|} = \frac{L^* - L_0}{L_0} . \tag{9.2}$$

Here, L_0 is the length of the bond between particle i and particle j in the reference configuration,

$$L_0 \equiv \sqrt{(X_j - X_i)^2 + (Y_j - Y_i)^2 + (Z_j - Z_i)^2} , \tag{9.3}$$

and L^* is the length of the same, but deformed bond,

$$L^* \equiv \sqrt{(x_j - x_i)^2 + (y_j - y_i)^2 + (z_j - z_i)^2}\,.^{\text{d}} \tag{9.4}$$

The small-deformation, or reference, bond direction for bond B_j is the vector of direction cosines, defined as

$$\{n_S\} \equiv \begin{Bmatrix} \frac{(X_j - X_i)}{L_0} \\ \frac{(Y_j - Y_i)}{L_0} \\ \frac{(Z_j - Z_i)}{L_0} \end{Bmatrix}, \tag{9.5}$$

or $\{n_S\} = \left(\text{Dir}\,\underline{\boldsymbol{X_R}}\right)\langle B_j\rangle$ according to Definitions 8.5 and 8.10,

and the large-deformation, or spatial, bond direction unit vector is defined as

$$\{n_L\} \equiv \begin{Bmatrix} \frac{(x_j - x_i)}{L^*} \\ \frac{(y_j - y_i)}{L^*} \\ \frac{(z_j - z)}{L^*} \end{Bmatrix}, \text{ or } \{n_L\} = \left(\text{Dir}\,\underline{\boldsymbol{x_R}}\right)\langle B_j\rangle. \tag{9.6}$$

So the deformation state of particle i is completely defined by $\{S\}_i$ and either $\{n_S\}$ or $\{n_L\}$ for all bonds B_j (Dir $\underline{\boldsymbol{X_R}}$ or Dir $\underline{\boldsymbol{x_R}}$ in terms of states) and the force state is completely defined by $\{T\}_i$ and either $\{n_S\}$ or $\{n_L\}$ for all bonds, B_j, otherwise described as Dir $\underline{\boldsymbol{X_R}}$ or Dir $\underline{\boldsymbol{x_R}}$.

Note that the stretch matrix $\{S\}$ is affected neither by bond rotations in general, nor by rigid-body rotations in specific, but that the large-deformation bond direction state Dir $\underline{\boldsymbol{x_R}}$ changes as the body undergoes rigid-body rotation.

The *stretch matrix*, thus defined, is different than the *deformation state* described in continuum state-based peridynamic theory, in that each of the eighteen components S_j of the stretch matrix in Eq. 9.1 is a scalar

---

[d] In this chapter, to simplify the notation, we replace X_1, X_2, and X_3 with X, Y, and Z.

function of the signed magnitude of the deformation $\underline{\boldsymbol{x}_R}\langle B_j\rangle$ of the associated bond B_j, but is not a function of the direction of the deformation $\underline{\boldsymbol{x}_R}\langle B_j\rangle$. In the present lattice-based theory, we assume that the bond force *direction* is in the direction of the bond in its reference configuration for geometrically *linear* behavior, and in its *spatial* configuration for geometrically *nonlinear* behavior. With an explicit time integration method, there is little advantage in linearizing the problem, and thus geometrically nonlinear behavior, being more physically descriptive, is in most cases preferred.

The force matrix $\{T\}$ associated with particle P_i is defined as *one-half* of the vector of peridynamic (signed) force magnitudes $\{F_j\}_i$ acting in each bond direction (either $\{n_S\}$ or $\{n_L\}$), directed away from particle P_i:

$$\{T\}_i \equiv \frac{1}{2}\begin{Bmatrix} F_1 \\ \vdots \\ F_j \\ \vdots \\ F_{18} \end{Bmatrix}_i . \tag{9.7}$$

The stretch matrix $\{S\}_i$, together with the spatial bond directions $\{n_L\}$ (defined by current particle locations), provides the same information as the deformation state $\underline{\boldsymbol{x}_R}$ in the state-based peridynamic theory described in Chapter 8. Similarly, the force matrix, $\{F\}_i$, together with the directions $\{n_S\}_i$ or $\{n_L\}_i$, provides identical information as the force state, $\underline{\boldsymbol{T}}$. We choose to use *matrices*, rather than *states*, in this chapter because matrices are easier to represent on a computer than states and we can easily compute the bond directions directly from particle positions.[e] However, the concept of "peridynamic state" is still useful mathematically and conceptually; therefore, we do not want entirely to drop the idea of "state" from our discussion.

---

[e] For the geometrically linear model, one can compute the bond direction vector just once, before entering the time-integration loop; on the other hand, for the geometrically nonlinear model, one must recompute the bond direction vector in each time step. However, these calculations are minor compared with other force calculations performed, especially when damage and plasticity are included.

In accordance with Eq. 8.30, we define the magnitude $F_i\langle B_j\rangle$ of the pairwise force $\boldsymbol{F}_i\langle B_j\rangle$ acting upon particle P_i due to interaction with its neighbor $\mathcal{N}[P_i]\langle B_j\rangle$ as

$$F_i\langle B_j\rangle \equiv \left|\underline{\mathbf{T}}[P_i]\langle B_j\rangle\right| - \left|\underline{\mathbf{T}}[\mathcal{N}[P_i]\langle B_j\rangle]\langle B_j{}'\rangle\right|, \tag{9.8}$$

where $B_j{}'$ is the complementary (opposing) bond to B_j. Thus according to the bond numbering scheme of Fig 6.7, B_j and $B_j{}'$ are consecutive integers, and if B_j is an odd number, then its complement is $B_j{}' = B_j + 1$, while if B_j is an even number, then $B_j{}' = B_j - 1$. Because the material is mobile, we assume the pairwise force $F_i\langle B_j\rangle$ acts in a direction collinear with the current locations of the two particles, with tension being positive and compression being negative. Eq. 9.8, together with symmetry considerations, show that in a spatially homogeneous force state field, the magnitude force state $\left|\underline{\mathbf{T}}[P_i]\langle B_j\rangle\right|$ is one-half of the force $F_i\langle B_j\rangle$ acting upon particle P_i through bond B_j.

If the deformation field is spatially non-homogeneous, then the two methods for describing deformation, the lattice-based stretch matrix $\{S\}$ and the continuum-based deformation gradient $\boldsymbol{F}$ (with associated strain $\boldsymbol{\varepsilon}$) are not directly comparable. Similarly, if the stress field varies spatially, then the force matrix $\{T\}$ and the stress tensor $\boldsymbol{\sigma}$ are not comparable. However, for spatially homogeneous strain and stress fields, and associated homogenous force states and stretch states, it is possible to relate these alternate solid models, as is accomplished in the following sections.

9.2 Relationship between SPLM stretch and strain

Given a global XYZ Cartesian coordinate system, representing physical space, let us express the strain and stress tensors at a point in the material continuum of classical mechanics using the following (6×1) matrices:

$$\{\boldsymbol{\varepsilon}\} = \begin{Bmatrix} \varepsilon_{XX} \\ \varepsilon_{YY} \\ \gamma_{XY} \\ \varepsilon_{ZZ} \\ \gamma_{YZ} \\ \gamma_{XZ} \end{Bmatrix}; \quad \text{similarly,} \quad \{\boldsymbol{\sigma}\} = \begin{Bmatrix} \sigma_{XX} \\ \sigma_{YY} \\ \tau_{XY} \\ \sigma_{ZZ} \\ \tau_{YZ} \\ \tau_{XZ} \end{Bmatrix}. \tag{9.9}$$

We shall call the representations in Eq. 9.9 the *strain matrix* and *stress matrix*, respectively. (The order of the components in Eq. 9.9, while non-standard, simplifies the description of one- and two-dimensional plane stress and plane strain models.)

If we have a spatially constant strain field, $\boldsymbol{\varepsilon}$, and assuming that all strain components are "small", then the stretch of any bond B_i with reference orientation given by the direction cosines with respect to the XYZ global axes, $\{n_S\}$, is given by

$$S = n_X{}^2\varepsilon_{XX} + n_Y{}^2\varepsilon_{YY} + n_Z{}^2\varepsilon_{ZZ} + n_X n_Y \gamma_{XY} + n_Y n_Z \gamma_{YZ} + n_Z n_X \gamma_{XZ}\,. \tag{9.10}$$

We define $[N]$ as the matrix that maps the classical (small) strain matrix to the SPLM stretch matrix:

$$\begin{Bmatrix} S_1 \\ \vdots \\ S_j \\ \vdots \\ S_{18} \end{Bmatrix} = \begin{bmatrix} n_{X1}{}^2 & n_{Y1}{}^2 & n_{X1}n_{Y1} & n_{Z1}{}^2 & n_{Y1}n_{Z1} & n_{X1}n_{Z1} \\ & & & \vdots & & \\ n_{Xj}{}^2 & n_{Yj}{}^2 & n_{Xj}n_{Yj} & n_{Zj}{}^2 & n_{Yj}n_{Zj} & n_{Xj}n_{Zj} \\ & & & \vdots & & \\ n_{X18}{}^2 & n_{Y18}{}^2 & n_{X18}n_{Y18} & n_{Z18}{}^2 & n_{Y18}n_{Z18} & n_{X18}n_{Z18} \end{bmatrix} \begin{Bmatrix} \varepsilon_{xx} \\ \varepsilon_{yy} \\ \gamma_{xy} \\ \varepsilon_{zz} \\ \gamma_{yz} \\ \gamma_{zx} \end{Bmatrix}, \tag{9.11}$$

or

$$\{S\} = [N]\{\varepsilon\}. \tag{9.12}$$

Here, n_{Xj} is the direction cosine between bond number B_j and the X axis, and so on.

For a 3D particle in an FCC close-packed lattice, with eighteen links positioned as shown in Fig. 6.7, evaluating Eq. 9.12, we find that

$$[N] = \begin{bmatrix} 1 & 0 & 0 & 0 & 0 & 0 \\ 1 & 0 & 0 & 0 & 0 & 0 \\ 1/4 & 3/4 & \sqrt{3}/4 & 0 & 0 & 0 \\ 1/4 & 3/4 & \sqrt{3}/4 & 0 & 0 & 0 \\ 1/4 & 3/4 & -\sqrt{3}/4 & 0 & 0 & 0 \\ 1/4 & 3/4 & -\sqrt{3}/4 & 0 & 0 & 0 \\ 1/4 & 1/12 & \sqrt{3}/12 & 2/3 & \sqrt{2}/6 & \sqrt{6}/6 \\ 1/4 & 1/12 & \sqrt{3}/12 & 2/3 & \sqrt{2}/6 & \sqrt{6}/6 \\ 1/4 & 1/12 & -\sqrt{3}/12 & 2/3 & \sqrt{2}/6 & -\sqrt{6}/6 \\ 1/4 & 1/12 & -\sqrt{3}/12 & 2/3 & \sqrt{2}/6 & -\sqrt{6}/6 \\ 0 & 1/3 & 0 & 2/3 & -\sqrt{2}/3 & 0 \\ 0 & 1/3 & 0 & 2/3 & -\sqrt{2}/3 & 0 \\ 0 & 2/3 & 0 & 1/3 & \sqrt{2}/3 & 0 \\ 0 & 2/3 & 0 & 1/3 & \sqrt{2}/3 & 0 \\ 1/2 & 1/6 & \sqrt{3}/6 & 1/3 & -\sqrt{2}/6 & -\sqrt{6}/6 \\ 1/2 & 1/6 & \sqrt{3}/6 & 1/3 & -\sqrt{2}/6 & -\sqrt{6}/6 \\ 1/2 & 1/6 & -\sqrt{3}/6 & 1/3 & -\sqrt{2}/6 & \sqrt{6}/6 \\ 1/2 & 1/6 & -\sqrt{3}/6 & 1/3 & -\sqrt{2}/6 & \sqrt{6}/6 \end{bmatrix}. \quad (9.13)$$

Note that starting with the first row in Eq. 9.13, every other row is repeated. This shows that, under the assumption of spatially homogeneous deformation, opposing bonds have identical stretch. If we strike out the even numbered rows, and ignore the last six lines of Eq. 9.13, we obtain the invertible linear relationship

$$\begin{Bmatrix} \bar{S}_1 \\ \bar{S}_3 \\ \bar{S}_5 \\ \bar{S}_7 \\ \bar{S}_9 \\ \bar{S}_{11} \end{Bmatrix} = \begin{bmatrix} 1 & 0 & 0 & 0 & 0 & 0 \\ 1/4 & 3/4 & \sqrt{3}/4 & 0 & 0 & 0 \\ 1/4 & 3/4 & -\sqrt{3}/4 & 0 & 0 & 0 \\ 1/4 & 1/12 & \sqrt{3}/12 & 2/3 & \sqrt{2}/6 & \sqrt{6}/6 \\ 1/4 & 1/12 & -\sqrt{3}/12 & 2/3 & \sqrt{2}/6 & -\sqrt{6}/6 \\ 0 & 1/3 & 0 & 2/3 & -\sqrt{2}/3 & 0 \end{bmatrix} \begin{Bmatrix} \varepsilon_{xx} \\ \varepsilon_{yy} \\ \gamma_{xy} \\ \varepsilon_{zz} \\ \gamma_{yz} \\ \gamma_{zx} \end{Bmatrix}, \text{ or} \quad (9.14)$$

$$\{\bar{S}\} = [\bar{N}]\{\varepsilon\}$$

where $\bar{S}_i$ are the average, and equal, stretches in opposing shortest bonds under the assumption of small deformation. By inverting $[\bar{N}]$ in Eq. 9.14, we see that, assuming spatially homogeneous strain conditions, if the average stretches of all collinear nearest neighbors are known, then the equivalent strain can be found.[f]

However, in general, given an arbitrary stretch matrix $\{S\}$, it is not possible to satisfy Eq. 9.12 except in the special case of homogeneous deformation. Thus, the SPLM is in a sense more general than the classical theory of elasticity, because it allows discrete jumps in stretch; however, only at discrete lattice points.

To account for plastic deformation, we decompose the stretch matrix into two parts: an elastic component $\{S^e\}$ and a plastic component, $\{S^p\}$:

$$\{S\} = \{S^e\} + \{S^p\}\,. \tag{9.15}$$

The plastic stretch of each bond is stored as a particle attribute, and we compute the total stretch from spatial particle positions, using Eq. 9.2. Thus, we compute the elastic stretch as

$$\{S^e\} = \{S\} - \{S^p\}\,. \tag{9.16}$$

We are concerned only with the elastic behavior in this chapter; we will discuss plastic behavior further in Chapter 10.

9.3 Relationship between SPLM force and stress

At a given lattice point, assuming a spatially homogeneous stress field $\boldsymbol{\sigma}$, a relationship is sought between the SPLM force matrix, $\{T\}$, and the classical stress matrix, $\{\sigma\}$. To find this relationship, we assert that, for a given volume of material, the internal virtual work under kinematically equivalent virtual deformations of both the classical model and the SPLM model must be equal:

---

[f] If the deformation field is not homogeneous, strain matrix $\{\varepsilon\}$ is not unique, and there is no exact relationship between stretch matrix $\{S\}$ and $\{\varepsilon\}$.

$$\delta W_{Classical} = \delta W_{SPLM}. \tag{9.17}$$

For a specified volume of material ΔV the internal virtual work due to an infinitesimal virtual strain $\{\delta\varepsilon\}$ is

$$\delta W_{Classical} = \lfloor\sigma\rfloor\{\delta\varepsilon\}\Delta V. \tag{9.18}$$

(Note that the stress $\lfloor\sigma\rfloor$ remains constant during the infinitesimal virtual strain $\{\delta\varepsilon\}$ and thus no factor of two in the denominator is required, as would be the case with the total stored strain energy.)

On the other hand, the virtual work associated with a single particle in the SPLM model undergoing infinitesimal stretch is

$$\delta W_{SPLM} = \sum F_i \delta S_i\left(\frac{L_i}{2}\right) = \lfloor F\rfloor\frac{[L_i]}{2}\{\delta S\}\ . \tag{9.19}$$

where $[L_i]$ is the diagonal matrix of reference bond lengths:

$$[L_i] = \begin{bmatrix} L_1 & 0 & \cdots & 0 \\ 0 & L_2 & & \vdots \\ \vdots & & \ddots & 0 \\ 0 & \cdots & 0 & L_k \end{bmatrix}. \tag{9.20}$$

Because two particles share each bond, only one-half of the internal virtual work stored in each bond is attributed to each of the two particles connected by the bond, hence the factor of two in the denominator of Eq. 9.19. (The bond force remains constant under the infinitesimal virtual stretch.)

Combining Eqs. 9.17, 9.18, and 9.19, we obtain

$$\lfloor\sigma\rfloor\{\delta\varepsilon\}\Delta V = \frac{1}{2}\lfloor F\rfloor[L_i]\{\delta S\}. \tag{9.21}$$

Expressing $\{\delta S\}$ as $[N]\{\delta\varepsilon\}$, and inserting this into Eq. 9.21, we get

$$\lfloor\sigma\rfloor\{\delta\varepsilon\}\Delta V = \frac{1}{2}\lfloor F\rfloor[L_i][N]\{\delta\varepsilon\}. \tag{9.22}$$

If we recognize that Eq. 9.22 must hold for each and every arbitrary virtual displacement $\{\delta\varepsilon\}$, then we conclude that

$$\lfloor\sigma\rfloor\Delta V = \frac{1}{2}\lfloor F\rfloor[L_i][N], \tag{9.23}$$

and solving for $\lfloor\sigma\rfloor$ and transposing both sides of the equation, we obtain

$$\{\sigma\} = \frac{1}{2\Delta V}[N]^T[L_i]\{F\} = [M]\{F\}, \tag{9.24}$$

with

$$[M] \equiv \frac{1}{2\Delta V}[N]^T[L_i]. \tag{9.25}$$

For a three-dimensional close-packed FCC lattice, the covolume $d(\Lambda^{N_R})$ gives the volume per particle:

$$\Delta V = d(\Lambda^3) = \frac{L^3}{\sqrt{2}}. \tag{9.26}$$

where L is the lattice spacing. Evaluating Eq. 9.25 for the three-dimensional FCC lattice, we obtain

$$[M] = \frac{1}{L^2}\begin{bmatrix} \frac{\sqrt{2}}{2} & \frac{\sqrt{2}}{2} & \frac{\sqrt{2}}{8} & \frac{\sqrt{2}}{8} & \frac{\sqrt{2}}{8} & \frac{\sqrt{2}}{8} & \frac{\sqrt{2}}{8} & \frac{\sqrt{2}}{8} & \frac{\sqrt{2}}{8} & \cdots \\ 0 & 0 & \frac{3\sqrt{2}}{8} & \frac{3\sqrt{2}}{8} & \frac{3\sqrt{2}}{8} & \frac{3\sqrt{2}}{8} & \frac{\sqrt{2}}{24} & \frac{\sqrt{2}}{24} & \frac{\sqrt{2}}{24} & \cdots \\ 0 & 0 & \frac{\sqrt{6}}{8} & \frac{\sqrt{6}}{8} & -\frac{\sqrt{6}}{8} & -\frac{\sqrt{6}}{8} & \frac{\sqrt{6}}{24} & \frac{\sqrt{6}}{24} & -\frac{\sqrt{6}}{24} & \cdots \\ 0 & 0 & 0 & 0 & 0 & 0 & \frac{\sqrt{2}}{3} & \frac{\sqrt{2}}{3} & \frac{\sqrt{2}}{3} & \cdots \\ 0 & 0 & 0 & 0 & 0 & 0 & \frac{1}{6} & \frac{1}{6} & \frac{1}{6} & \cdots \\ 0 & 0 & 0 & 0 & 0 & 0 & \frac{\sqrt{3}}{6} & \frac{\sqrt{3}}{6} & -\frac{\sqrt{3}}{6} & \cdots \end{bmatrix} \tag{9.27}$$

$$
\left.\begin{matrix}
\cdots & \frac{\sqrt{2}}{8} & 0 & 0 & 0 & 0 & \frac{1}{2} & \frac{1}{2} & \frac{1}{2} & \frac{1}{2} \\
\cdots & \frac{\sqrt{2}}{24} & \frac{\sqrt{2}}{6} & \frac{\sqrt{2}}{6} & \frac{2}{3} & \frac{2}{3} & \frac{1}{6} & \frac{1}{6} & \frac{1}{6} & \frac{1}{6} \\
\cdots & -\frac{\sqrt{6}}{24} & 0 & 0 & 0 & 0 & \frac{\sqrt{3}}{6} & \frac{\sqrt{3}}{6} & -\frac{\sqrt{3}}{6} & -\frac{\sqrt{3}}{6} \\
\cdots & \frac{\sqrt{2}}{3} & \frac{\sqrt{2}}{3} & \frac{\sqrt{2}}{3} & \frac{1}{3} & \frac{1}{3} & \frac{1}{3} & \frac{1}{3} & \frac{1}{3} & \frac{1}{3} \\
\cdots & \frac{1}{6} & -\frac{1}{3} & -\frac{1}{3} & \frac{\sqrt{2}}{3} & \frac{\sqrt{2}}{3} & -\frac{\sqrt{2}}{6} & -\frac{\sqrt{2}}{6} & -\frac{\sqrt{2}}{6} & -\frac{\sqrt{2}}{6} \\
\cdots & -\frac{\sqrt{3}}{6} & 0 & 0 & 0 & 0 & -\frac{\sqrt{6}}{6} & -\frac{\sqrt{6}}{6} & \frac{\sqrt{6}}{6} & \frac{\sqrt{6}}{6}
\end{matrix}\right] .
$$

Thus, we see that the stress matrix, $\{\sigma\}$, is linearly related to the force matrix, $\{F\}$, as $\{\sigma\} = [M]\{F\}$, under the conditions of the derivation, which is to say, under small deformations, in a spatially homogeneous medium, away from the boundary, and assuming spatially homogeneous stress and strain conditions.

The SPLM force matrix, having eighteen components, is more general than the classical Cauchy stress, which has only six independent components. On the other hand, the Cauchy stress is defined at all points in a continuum, while the SPLM force matrix is defined only at lattice points. Thus, the SPLM is in one sense more general (allowing discontinuous SPLM force state fields) and in another sense less general (defining SPLM force state only at lattice points) than the continuum elasticity model. On the other hand, the continuum model is inapplicable and predicts spurious detail at size scales smaller than the material size scale (which may be defined by the lattice spacing L), and thus assigns an improper level of detail to the stress and strain distributions at that scale.

9.4 Linear elastic constitutive relation

Let us assume spatially homogeneous small-strain conditions. For a given lattice particle, we assume the force matrix $\{F\}$ is a linear function the elastic component of the stretch matrix $\{S^e\}$:

$$\{F\} = [K]\{S^e\} \tag{9.28}$$

We call $[K]$ the *micro-elastic SPLM stiffness matrix.*

From Navier-Cauchy elasticity theory, the symmetric constitutive matrix $[D]$ relates stress to (elastic) strain,

$$\{\sigma\} = [D]\{\varepsilon^e\} . \tag{9.29}$$

Premultiplying both sides of Eq. 9.28 by $[M]$, we obtain

$$[M]\{F\} = [M][K]\{S^e\} , \tag{9.30}$$

and substituting Eq. 9.12 into Eq. 9.30 produces

$$[M]\{F\} = [M][K][N]\{\varepsilon^e\} . \tag{9.31}$$

If we now substitute Eq. 9.24 into Eq. 9.31, we get

$$\{\sigma\} = [M][K][N]\{\varepsilon^e\} . \tag{9.32}$$

Finally, comparing Eqs. 9.29 and 9.32, and recognizing that the two equations must produce identical stresses for arbitrary strains, we conclude that

$$[D] = [M][K][N] . \tag{9.33}$$

Given $[D]$, because $[M]$ and $[N]$ are not square matrices, $[K]$ is not uniquely determined. However, we can assume the form[g] of $[K]$, and then solve for the elements of $[K]$ in terms of the two independent classical elastic parameters, Young's modulus E and Poisson's ratio ν.

We next develop a micro-elastic peridynamic stiffness matrix $[K]$ for the three-dimensional SPLM (many other stiffness matrices $[K]$ are possible). We assume that a three-dimensional lattice particle interacts with its twelve nearest neighbors as well as with its six second-nearest neighbors. (We can show that using only the twelve nearest neighbors is insufficient to allow isotropic material behavior.) For the twelve nearest

---

[g] Various forms of $[K]$ can be assumed, resulting in the same macro-behavior, but differing distributions of bond forces for the same deformation state.

neighboring bonds, of reference length L, we assume that force matrix component F_j is equal to the elastic stretch state S_j multiplied by the constant a plus $12b$ times the average elastic stretch of all twelve nearest neighbors, S_{AVG} :

$$F_j = aS_j + 12bS_{AVG} \,. \tag{9.34}$$

where

$$S_{AVG} = \frac{1}{12}\sum_{j=1}^{12} S_j \,. \tag{9.35}$$

For the six second-nearest neighboring bonds, of reference length $\sqrt{2}L$, we assume that force matrix component F_j in bond B_j is simply equal to its own stretch, S_j, multiplied by the constant c:

$$F_j = cS_j \,. \tag{9.36}$$

From Eqs. 9.34, 9.35, and 9.36 we produce a symmetric $[K]$ matrix with 18 rows and 18 columns:

$$[K] = \begin{bmatrix} \underset{(12\times 12)}{\begin{bmatrix} a+b & b & \cdots & b \\ b & a+b & \cdots & b \\ \vdots & \vdots & \ddots & \vdots \\ b & b & \cdots & a+b \end{bmatrix}} & \underset{(12\times 6)}{[0]} \\ \underset{(6\times 12)}{[0]} & \underset{(6\times 6)}{\begin{bmatrix} c & 0 & \cdots & 0 \\ 0 & c & & \vdots \\ \vdots & & \ddots & 0 \\ 0 & \cdots & 0 & c \end{bmatrix}} \end{bmatrix} . \tag{9.37}$$

In three-dimensions, the classical constitutive matrix $[D]$, in terms of E and ν, is

$$[D] = \frac{E}{(1+v)(1-2v)}\begin{bmatrix} (1-v) & v & 0 & v & 0 & 0 \\ v & (1-v) & 0 & v & 0 & 0 \\ 0 & 0 & \frac{(1-2v)}{2} & 0 & 0 & 0 \\ v & v & 0 & (1-v) & 0 & 0 \\ 0 & 0 & 0 & 0 & \frac{(1-2v)}{2} & 0 \\ 0 & 0 & 0 & 0 & 0 & \frac{(1-2v)}{2} \end{bmatrix}. \quad (9.38)$$

Using Eq. 9.33, $[D] = [M][K][N]$, together with Eqs. 9.13, 9.26, 9.37 and 9.38, we obtain:

$$\frac{E}{(1+v)(1-2v)}\begin{bmatrix} (1-v) & v & 0 & v & 0 & 0 \\ v & (1-v) & 0 & v & 0 & 0 \\ 0 & 0 & \frac{(1-2v)}{2} & 0 & 0 & 0 \\ v & v & 0 & (1-v) & 0 & 0 \\ 0 & 0 & 0 & 0 & \frac{(1-2v)}{2} & 0 \\ 0 & 0 & 0 & 0 & 0 & \frac{(1-2v)}{2} \end{bmatrix} \quad (9.39)$$

$$= \frac{\sqrt{2}}{L^2}\begin{bmatrix} \left(\frac{5a}{4}+8b+\frac{c}{\sqrt{2}}\right) & \left(\frac{5a}{12}+8b+\frac{\sqrt{2}c}{6}\right) & 0 & \left(\frac{a}{3}+8b+\frac{\sqrt{2}c}{3}\right) & \left(\frac{\sqrt{2}a}{12}-\frac{c}{3}\right) & 0 \\ \left(\frac{5a}{12}+8b+\frac{\sqrt{2}c}{6}\right) & \left(\frac{5a}{4}+8b+\frac{c}{\sqrt{2}}\right) & 0 & \left(\frac{a}{3}+8b+\frac{\sqrt{2}c}{3}\right) & \left(-\frac{\sqrt{2}a}{12}+\frac{c}{3}\right) & 0 \\ 0 & 0 & \left(\frac{5a}{12}+\frac{\sqrt{2}c}{6}\right) & 0 & 0 & \left(\frac{\sqrt{2}a}{12}-\frac{c}{3}\right) \\ \left(\frac{a}{3}+8b+\frac{\sqrt{2}c}{3}\right) & \left(\frac{a}{3}+8b+\frac{\sqrt{2}c}{3}\right) & 0 & \left(\frac{4a}{3}+8b+\frac{\sqrt{2}c}{3}\right) & 0 & 0 \\ \left(\frac{\sqrt{2}a}{12}-\frac{c}{3}\right) & \left(-\frac{\sqrt{2}a}{12}+\frac{c}{3}\right) & 0 & 0 & \left(\frac{a}{3}+\frac{\sqrt{2}c}{3}\right) & 0 \\ 0 & 0 & \left(\frac{\sqrt{2}a}{12}-\frac{c}{3}\right) & 0 & 0 & \left(\frac{a}{3}+\frac{\sqrt{2}c}{3}\right) \end{bmatrix}.$$

Setting corresponding elements in Eq. 9.39 equal to each other (i.e. $D_{11} = (MKN)_{11}$, $D_{12} = (MKN)_{12}$ and $D_{66} = (MKN)_{66}$) produces three independent linear equations with three unknowns, a, b, and c. Solving for a, b, and c:

$$a = \frac{EL^2}{\sqrt{2}(v+1)}, \quad (9.40)$$

$$b = \frac{\sqrt{2}EL^2(1-4v)}{32(2v-1)(v+1)}, \tag{9.41}$$

$$c = \frac{EL^2}{4(v+1)} = \frac{1}{2\sqrt{2}}a\,. \tag{9.42}$$

Note that the parameter c in Eq. 9.42 is dependent upon parameter a. Thus, there are only two independent microelastic peridynamic material parameters, a and b, that need be determined if Young's modulus E and Poisson's ratio v are given.

Under spatially homogeneous small deformation fields, the micro elastic model produces precisely the same macro-elastic response as the Navier-Cauchy model for elasticity. The bond force response magnitudes, being dependent only upon bond stretch magnitudes (and not bond directions), are independent of rotations and translations of the deformed lattice.

For spatially nonhomogeneous deformation fields, the SPLM is nonlocal, and gives different results than the classical elasticity model, which is local. We can speculate about which of the two models is superior for representing real materials; the answer is arguable.

9.5 Boundary particles

Consider a three-dimensional lattice body $\mathcal{L}_{\mathcal{R}}$ in Λ^3. Such a body is bounded, and thus has boundary particles. The state-based peridynamic theory (Silling, S.A., Epton, M., Weckner, O., Xu, J., and Askari, E., 2007) says nothing about how to handle boundary particles. In this section, we propose a method to handle boundary particles.

If a particle P_i is on the boundary of $\mathcal{L}_{\mathcal{R}}$, then at least one of its bonds $B_j \in \mathbb{B}$ will be the null particle $\mathcal{N}[P_i]\langle B_j \rangle = P_\phi$; (see Definition 6.4). In this case, we assume the stretch S_j of the bond on the boundary is the same as the stretch of its complementary bond, except that if the complementary bond does not exist, then we assume that the stretches of both bonds are zero. These assumptions arise by assuming that the deformation field at the boundary is as uniform as possible; we could also make other assumptions.

Also, if a bond B_j of particle P_i is connected to the null particle $\mathcal{N}[P_i]\langle B_j\rangle = P_\phi$, we will assume that there is no force F_j acting in association with that bond; otherwise Newton's 3rd law would be violated.

Thus, particles on the boundary of $\mathcal{L}_\mathcal{R}$ have somewhat different constitutive behavior than particles on the interior.

These assumptions are rather ad hoc; we could make others. These assumptions seem to produce physically reasonable behavior at lattice domain boundaries. If the number of boundary particles is small relative to the number of interior particles, the effect of the boundary assumptions upon overall elastic behavior will be minor.

9.6 Uniaxial case

So far, we have assumed a three-dimensional geometric domain. However, it is efficacious to develop "uniaxial", "plane stress" and "plane strain" SPLM models as well. In this section, we develop the uniaxial model. This model is useful for the modeling of reinforcing bars, for example.

In the uniaxial case, we assume that only the forces in bonds B_1 and B_2, shown in Fig. 6.7, those aligned with the strand, participate in driving particle motion. Due to symmetry, out-of axis bond forces, coming in equal and opposite pairs, cannot have an effect upon in-axis motion. However, in the state-based model (in contrast to the bond-based peridynamic model of Chapter 5), out-of-axis bond *stretches* can have an effect upon in-axis bond forces.

In a "uniaxial" problem, one can make various assumptions about the out-of-axis constraint condition. In the classical "uniaxial stress" theory, for example, we assume that the out-of-axis *stress components* are null; in another case, usually not explicitly considered in textbooks concerning the mechanics of deformable bodies, one assumes that the out-of-axis *strain components* are null (geotechnical engineers call this confined compression).

Let us assume that we want to model a uniaxial member of reference cross-sectional area A_{BAR}. Let us consider a case of uniaxial stress (with null out-of-axis stresses). Furthermore, we assume spatially homogeneous strain and stress fields along the length of the bar.

Consideration of static equivalency between the classical model and the SPLM model in the X-direction mandates that

$$F_{BAR} = \sigma_X A_{BAR} \,, \tag{9.43}$$

and consideration of kinematic equivalency between the classical model and the SPLM model in the X-direction mandates that

$$S_1 = \varepsilon_X \,. \tag{9.44}$$

For the case of uniaxial elastic behavior in the classical model

$$\sigma_X = E\varepsilon_X \,. \tag{9.45}$$

Substituting Eqs. 9.43 and 9.44 into 9.45 gives

$$F_{BAR} = EA_{BAR}S_1 \,. \tag{9.46}$$

Therefore, for a uniaxial bar, we get elastic deformation response matching the classical deformation response if we assign the microelastic stiffness parameters in Eqs. 9.40-9.42 values of $a = EA_{BAR}$, $b = 0$, and $c = 0$. Thus,

$$\begin{Bmatrix} F_1 \\ F_2 \end{Bmatrix} = EA_{BAR} \begin{bmatrix} 1 & 0 \\ 0 & 1 \end{bmatrix} \begin{Bmatrix} S_1^e \\ S_2^e \end{Bmatrix} . \tag{9.47}$$

The out-of-axis bond forces, bonds 3-18, are not needed in calculating the axial motion of the particle, because by symmetry they exert no force in the axial (X) direction. However, for computing the 18 plastic stretches and the damage parameter, discussed further in Chapters 10 and 11, it is convenient to perform an auxiliary calculation to determine all 18 bond stretches for an equivalent 3D particle.

With stretches S_1 and S_2 known, the rest of the stretches S_3 through S_{18} are found by first obtaining the full elastic strain matrix, $\{\varepsilon^e\}$, by inverting Eq. 9.29, then finding the average stretch matrix $\{\bar{S}^e\}$ via Eq. 9.14, and finally expanding the average stretch matrix, $\{\bar{S}^e\}$, to the full elastic stretch matrix, $\{S^e\}$. We obtain, for the uniaxial case:

$$\lfloor S^e \rfloor =$$
$$\Big\lfloor 1 \quad 1 \quad \left(\frac{1-3\nu}{4}\right) \quad \left(\frac{1-3\nu}{4}\right) \quad \left(\frac{1-3\nu}{4}\right) \quad \left(\frac{1-3\nu}{4}\right) \quad \left(\frac{1-3\nu}{4}\right) \quad \left(\frac{1-3\nu}{4}\right) \quad \left(\frac{1-3\nu}{4}\right) \quad -\nu \quad -\nu \quad -\nu \quad -\nu \quad \left(\frac{1-\nu}{2}\right) \quad \left(\frac{1-\nu}{2}\right) \quad \left(\frac{1-\nu}{2}\right) \quad \left(\frac{1-\nu}{2}\right) \Big\rfloor S_1^e. \tag{9.48}$$

With $\{S^e\}$ now known, the full 3D force matrix $\{F\}$ is found via Eq. 9.28, using the 3D values of the elastic parameters a, b, and c given by Eqs. 9.40–9.42. Following this, if desired, we can find the stress matrix $\{\sigma\}$ using Eq. 9.24.

The reader might well be wondering why we went to the trouble of determining the 3D stretch and 3D force matrices for a simple uniaxial bar. The answer is that, when plasticity and damage are considered, it is efficacious to consider just the 3D case for determining stress, strain, yield, and damage, rather than to have to develop completely independent theories of plasticity and damage for each special case (uniaxial, plane stress, plain strain, etc.). We present a three-dimensional plasticity theory in Chapter 10, and a three-dimensional damage theory in Chapter 11.

Next, we consider the case of plane stress.

9.7 Plane stress case

In the case of plane stress, we assume that the problem to be analyzed lies in the XY plane, with thickness t_b in the Z direction and we model the problem as a 2D hexagonal lattice body $\mathcal{L}_{\mathcal{R}}$ in Λ^2. The bonds that drive particle motion are the in-plane bonds, B_1 through B_6, as shown in Fig. 6.7. We assume that all externally- applied forces are in the plane of the body. Out-of-plane bond *forces*, even though not necessarily zero, in consideration of symmetry, can have no resultant effect upon the motion of the particle. However, out-of-plane bond *stretches*, as with the uniaxial case, can have an effect upon in-plane bond forces.

As with the 3D case, we equate, for both the classical and the SPLM models, for a single lattice particle P_i, the internal stored elastic energy for equivalent virtual deformations. However, for the SPLM particle, we initially assume all internal energy is stored only in the six in-plane bonds.

For plane problems, the volume of material associated with a single lattice particle is the lattice covolume $d(\Lambda^2)$ times the thickness t_b

$$\Delta V = \frac{\sqrt{3}t_b L^2}{2}. \tag{9.49}$$

For plane stress, the classical constitutive equation is

$$\begin{Bmatrix} \sigma_{XX} \\ \sigma_{YY} \\ \tau_{XY} \end{Bmatrix} = \frac{E}{1-\nu^2}\begin{bmatrix} 1 & \nu & 0 \\ \nu & 1 & 0 \\ 0 & 0 & \frac{1-\nu}{2} \end{bmatrix}\begin{Bmatrix} \varepsilon_{XX} \\ \varepsilon_{YY} \\ \gamma_{XY} \end{Bmatrix}, \text{ or } \{\sigma\} = [D]\{\varepsilon\}. \tag{9.50}$$

The relationship between equivalent in-plane SPLM stretches and in-plane strains is

$$\begin{Bmatrix} S_1 \\ S_2 \\ S_3 \\ S_4 \\ S_5 \\ S_6 \end{Bmatrix} = \begin{bmatrix} 1 & 0 & 0 \\ 1 & 0 & 0 \\ \frac{1}{4} & \frac{3}{4} & \frac{\sqrt{3}}{4} \\ \frac{1}{4} & \frac{3}{4} & \frac{\sqrt{3}}{4} \\ \frac{1}{4} & \frac{3}{4} & \frac{-\sqrt{3}}{4} \\ \frac{1}{4} & \frac{3}{4} & \frac{-\sqrt{3}}{4} \end{bmatrix}\begin{Bmatrix} \varepsilon_{XX} \\ \varepsilon_{YY} \\ \gamma_{XY} \end{Bmatrix} \text{ or } \{S\} = [N]\{\varepsilon\}. \tag{9.51}$$

The relationship between equivalent in-plane bond forces and in-plane stresses is:

$$\{\sigma\} = [M]\{F\}. \tag{9.52}$$

with

$$[M] \equiv \frac{1}{2\Delta V}[N]^T L. \tag{9.53}$$

Similar to the 3D case, we assume

$$\{F\} = [K]\{S^e\} \tag{9.54}$$

with

$$[K] = \begin{bmatrix} (a+b) & b & b & b & b & b \\ b & (a+b) & b & b & b & b \\ b & b & (a+b) & b & b & b \\ b & b & b & (a+b) & b & b \\ b & b & b & b & (a+b) & b \\ b & b & b & b & b & (a+b) \end{bmatrix} \tag{9.55}$$

As with the 3D case, energy equivalence for arbitrary virtual deformations mandates that

$$[D] = [M][K][N]\,. \tag{9.56}$$

Solving Eq. 9.56 for a and b, we obtain:

$$a = \frac{2ELt_b}{\sqrt{3}(1-\nu)}\,,\; b = \frac{2ELt_b(1-3\nu)}{6\sqrt{3}(\nu^2-1)}\;. \tag{9.57}$$

Inserting a and b into Eqs. 9.54 and 9.55, the microelastic stiffnesses necessary to determine the forces acting upon the particle are known. Note that when Poisson's ratio ν is equal to one-third, in Eq. 9.57, the parameter b is zero, in which case the model reduces to the Silling's bond-based model of 2000.

As with the one-dimensional case, it is efficacious, once the in-plane stretches are determined, to compute the equivalent three-dimensional stretch matrix and force matrix. This information is necessary for the plasticity and damage models described in Chapters 10 and 11. We can calculate the out-of-plane bond forces using Eqs. 9.28, and 9.37, assuming 3D microelastic parameters given by Eqs. 9.40-9.42.[h]

---

[h] Note that even though this is a plane-stress problem, the out-of-plane bond forces need not be null; only the resultant of all bond forces on each side of the XY plane must be null.

9.8 Plane strain case

In the case of plane strain, we assume that the out-of-plane strain is zero. The classical elasticity relation for plane strain is

$$\begin{Bmatrix} \sigma_{XX} \\ \sigma_{YY} \\ \tau_{XY} \end{Bmatrix} = \frac{E}{(1+\nu)(1-2\nu)} \begin{bmatrix} (1-\nu) & \nu & 0 \\ \nu & (1-\nu) & 0 \\ 0 & 0 & \frac{1-2\nu}{2} \end{bmatrix} \begin{Bmatrix} \varepsilon_{XX} \\ \varepsilon_{YY} \\ \gamma_{XY} \end{Bmatrix}, \tag{9.58}$$

$$\text{or } \{\sigma\} = [D]\{\varepsilon\}.$$

Following the same procedure as was outlined in Section 9.7 for plane stress, but using $[D]$ for plane strain defined by Eq. 9.58, we obtain the microelastic parameters

$$a = \frac{2ELt_b}{\sqrt{3}(1+\nu)}, \; b = \frac{ELt_b(1-4\nu)}{6\sqrt{3}(2\nu-1)(1+\nu)} . \tag{9.59}$$

When a and b from Eq. 9.59 are inserted into Eq. 9.55, the plane strain micro-elastic stiffness matrix $[K]$ is obtained. Note that when Poisson's ratio is equal to one-fourth, in Eq. 9.59, the parameter b goes to zero, in which case the model reduces to the Silling's bond-based model.

Similar to the plane stress case, with the in-plane particle positions known, one can compute the in-plane bond stretches, and then one can obtain the out-of-plane bond stretches. This information is necessary for the plasticity and damage models described in Chapters 10 and 11. One can calculate the out-of-plane bond forces using Eqs. 9.28 and 9.37 and the three-dimensional microelastic stiffness parameters are given by Eqs. 9.40–9.42.

9.9 Summary

In this chapter, we have developed a set of micro-elastic linear force-stretch stiffness relations for the linear elastic state-based peridynamic lattice constitutive model, including uniaxial, plane stress, plane strain, and fully three-dimensional behavior.

We specialized the continuum state-based peridynamic theory developed in (Silling, S.A., Epton, M., Weckner, O., Xu, J., and Askari, E., 2007) to the state-based peridynamic lattice model (SPLM) in Chapter 8. The SPLM described in Chapter 8 is more general than what we need for a mobile linear elastic state-based lattice model, and in this chapter, we have specialized the theory further to enable practical computational implementation.

We emphasize that the SPLM is not simply a truss model. This is because the force acting between two particles is a function not only of the displacements of the two particles, but is also a function of the displacements of other particles in the neighborhood. There are no finite elements in the model; only bond forces between neighboring particles.

Likewise, the SPLM is not simply a discrete element model (DEM). The theoretical structure provided by the lattice model of the material body described in Chapter 6 provides much more theoretical unity than is provided by a traditional discrete element model. The SPLM is similar to a continuum mechanics model, except without the assumption of continuity.

The derivations in this chapter assume that a full complement of neighboring particles surrounds each particle (thus, each particle has eighteen neighbors in 3D). But in any body containing a finite number of lattice particles, some lattice particles will lie on a domain boundary, and thus some of the bonds will not have neighboring particles. When, for a particular bond, the neighboring particle does not exist, we assume that the total stretch of that bond is equal to that of its opposing bond, and the force in the boundary bond is null. If no opposing bond exists, then we assume that no force acts in either of the bonds, and that the elastic stretch within both bonds is null.

If the magnitude (square root of the sum of the squares of the components) of the stretch matrix $\{S\}$ is small (say, less than 1%), then we have found that the SPLM linear elastic model produces results virtually identical to the classical Navier-Cauchy elasticity model. However, the SPLM is more general than the Navier-Cauchy model because it allows for large rotations without theoretical difficulty. In addition, the SPLM is much more directly implementable on a digital computer than the Navier-Cauchy model, which one must first discretize in in some way.

The SPLM is much easier to extend into the nonlinear realm (elastic or inelastic) than the Navier-Cauchy model, because no assumptions of continuity (essential for the various definitions of strain) nor of geometric or material linearity have been built into the model at the outset.

While the SPLM continues to be a reasonable model as the structure size scale approaches the material size scale, the continuum model certainly is not applicable under such conditions. The requirement that lattice spacing L be larger than the material size scale prevents the modeler from misusing the SPLM. No such inherent limitation prevents unwitting misuse of the continuum model.

In the next two chapters, we present plasticity and damage models appropriate for use with the linear elastic SPLM developed in this chapter.

9.10 Exercises

9.1 (a) Write an exact expression for the stretch S defined by Eq. 9.2 between two particles initially located at $\boldsymbol{X}_a = \begin{Bmatrix} 0 \\ 0 \end{Bmatrix}$ and $\boldsymbol{X}_b = \begin{Bmatrix} 2 \\ 1 \end{Bmatrix}$, and currently located at $\boldsymbol{x}_a = \begin{Bmatrix} 0 \\ 0 \end{Bmatrix}$ and $\boldsymbol{x}_b = \begin{Bmatrix} 2+u \\ 1 \end{Bmatrix}$ in $\mathbb{R}^2$. (b) Approximate in terms of displacement component u the exact expression for stretch S developed in part (a) using a Taylor-series expansion, including all terms up to and including third order. (c) for the exact, first-order, second-order, and third-order Taylor-series approximations plot S versus u, for $-2 \leq u \leq 2$.

9.2 Prove that, assuming small deformation conditions, the stretch of a bond is given by Eq. 9.10: $S = {n_X}^2 \varepsilon_{XX} + {n_Y}^2 \varepsilon_{YY} + {n_Z}^2 \varepsilon_{ZZ} + n_X n_Y \gamma_{XY} + n_Y n_Z \gamma_{YZ} + n_Z n_X \gamma_{XZ}$.

9.3 Using MatLab, show that the components of $[N]$ in Eq. 9.13 are correct for the 3D FCC lattice shown in Fig. 6.7.

9.4 Using MatLab, show that the components of $[M]$ in Eq. 9.27 are correct for the 3D FCC lattice shown in Fig. 6.7.

9.5 Verify the formulas for the microelastic SPLM parameters a, b, and c given in Eqs. 9.40–9.42.

9.6 Show that if the $[K]$ matrix in $\{F\} = [K]\{S^e\}$ includes only the upper left 12×12 matrix $\begin{bmatrix} a+b & b & \cdots & b \\ b & a+b & \cdots & b \\ \vdots & \vdots & \ddots & \vdots \\ b & b & \cdots & a+b \end{bmatrix}$ in Eq. 9.37, thus relating only the forces in the twelve nearest neighbors to the stretches in the twelve nearest neighbors, it is not possible to represent an isotropic elastic continuum using the SPLM linear elastic model unless $b = 0$. Use MatLab.

9.7 Show that using the plane stress microelastic SPLM parameters a and b given in Eq. 9.57 and the microelastic stiffness relation $\{F\} = [K]\{S^e\}$ given by Eq. 9.55, for a given strain matrix $\{\varepsilon\}$, the elastically-stored energy per unit volume is identical regardless of the proper lattice rotation matrix $[Q]$ in Eq. 6.1.

9.8 Show that using the microelastic SPLM parameters a, b, and c given in Eqs. 9.40–9.42 and the microelastic stiffness relation $\{F\} = [K]\{S^e\}$ given by Eq. 9.37, for a given strain matrix $\{\varepsilon\}$, the elastically-stored energy per unit volume is identical regardless of the proper lattice rotation matrix $[Q]$ in Eq. 6.1.

9.9 Represent a uniaxial bar as a plane stress SPLM lattice, with bonds B_1 and B_2 lined up with the X_1-axis, as shown in Fig. 6.7. The lattice is very large, so that boundary effects upon the overall behavior are negligible. The lattice spacing is $L = 0.01\,m$, and thickness is $t_b = 0.1m$. If the axial stress in the bar is $\sigma_X = 1$ Pa, then what are the three-dimensional bond forces F_1 through F_{18} of a typical interior lattice particle?

9.10 Repeat Problem 9.9, but this time, calculate the three-dimensional bond forces F_1 through F_{18} of a typical exterior lattice particle on the left-hand boundary of the bar.

9.11 Write a MatLab program to implement the one-dimensional linear elastic SPLM. Test your program by simulating a uniaxial steel bar with length of 0.1 m, lattice spacing of 1 cm, and cross-sectional area of 0.0001 m^2. Fix the leftmost particle, and apply a load of 1000 N to the rightmost particle. Assume static loading, so that you can solve the peridynamic equations by solving a set of linear equations.

9.12 Represent a uniaxial bar as a plane stress SPLM lattice, with bonds B_1 and B_2 lined up with the X_1-axis, as shown in Fig. 6.7. The lattice spacing is $L = 0.02\,m$, and thickness is $t_b = 0.1m$. The length of the bar L_b is 10 times longer than its width W_b. Assuming the axial stress in the bar is $\sigma_X = 1$ Pa, $E = 1000$ GPa , and $\nu = 0.3$, calculate the stretch of the bar $S_{bar} = {}^{\Delta L}/_{L_b}$ as a function of ${}^{W_b}/_{L}$, for $1 \leq {}^{W_b}/_{L} \leq 50$. Plot ${}^{S_{bar}}/_{S_{classical}}$ versus ${}^{W_b}/_{L}$. What modeling decisions did you make? What conclusions can you draw? (Write an SPLM computer program using MatLab or FORTRAN to solve this problem.)

Chapter 10

Plasticity

Chapter objectives are to:

- review the history of the development of the theory of plasticity;
- describe the continuum plasticity model;
- describe non-ordinary continuum state based peridynamic plasticity;
- develop a state-based peridynamic lattice (SPLM) mobile[a] plasticity model.

The theory of plasticity was developed in recognition of the fact that many materials do not "spring back" after the load has been removed. At sufficiently high deviatoric stress levels, solid materials flow almost like fluids. Plasticity models of solids, used in conjunction with elasticity models, replicate the fluid-like behavior seen at sufficiently high levels of deviatoric stress. Fig. 10.1 shows a clay body plastically deforming.

In this chapter, we first review the history and the content of the continuum plasticity model. Engineers have implemented continuum plasticity models within the finite element framework with moderate success. However, with the commonly used displacement-based finite elements, "element locking", which is often poorly understood, can occur due to the spurious constraints of the displacement interpolation functions. Thus, at present, the finite element modeling of plastic solid bodies requires considerable experience and judgment.

In this chapter, we develop a mobile SPLM plasticity model, for use in conjunction with the elastic SPLM model developed in Chapter 9. In the lattice body context, with no displacement interpolation functions artificially constraining the deformation field, we do not expect the

---

[a] See Definition 8.21 for the definition of "mobile".

"element locking" seen with displacement-based finite elements to be an issue.[b]

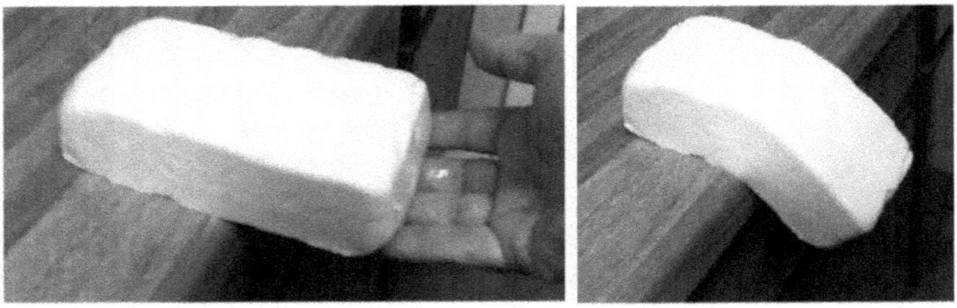

(a) before deformation (b) after deformation

Figure 10.1. Photographs of clay body plastically deforming under its own weight. The body deformed gradually over the course of two minutes in going from state (a) to state (b).

10.1 History of theory of plasticity

The theory of plasticity has a long history, documented for example in (Osakada, 2008). Briefly, the major contributors are Coulomb in 1784, and then in the late 1800s, Tresca, Saint-Venant, Levy, Bauschinger, Maxwell, and Mohr. In the early twentieth century, the researchers Huber, von Mises, Nadai, and Bridgman developed the theory further. Nadai published the first book on plasticity in German in 1927, and the book was translated into English in 1950 under the title "Theory of Flow and Fracture of Solids".

Prandtl and Hencky developed plastic slip line theory in the 1920's, and Hill further developed the theory in the mid-to-late 1900's.

Yamaguchi, Taylor, Polanyi, and Orowan developed dislocation theory of crystals as an explanation for plastic behavior in the 1930s.

Brown University became the center of the plasticity research starting in the 1950's and onward, with contributions by Drucker, Prager, Hodge, Marcal, Rice, Needleman, and others.

---

[b] One can use the SPLM plasticity model developed in this chapter in conjunction with any (linear or nonlinear) elasticity model.

Truesdell and Noll put continuum mechanics on a firm mathematical foundation in the 1950s and 1960s, including notions of plasticity and flow.

Marcal and King, in the 1960s, applied the finite element method to the solution of elasto-plasticity problems. Many researchers have developed numerical implementations of elasto-plasticity since then.

Hutchinson and Rice, in the 1960s and onward, explored elasto-plastic fracture mechanics.

Since the 1970s, the automobile industry has funded research in elasto-plastic finite element simulation as a tool in designing cold-formed steel auto bodies and chassis.

This brief sketch of the history of plasticity is by no means complete, but it provides a start in understanding the history and development of the topic. We next review the basic concepts of continuum plasticity.

10.2 Continuum plasticity

The theory of plasticity, in contrast to the theory of elasticity, is a constitutive model that describes the deformation of a solid that flows under sufficiently intense shear stress. Under plastic flow, unlike with elasticity, the body does not return to its original shape after one removes the load. For example, if a mild steel paper clip is bent to a ninety-degree angle with a small-enough radius of curvature, so that the outermost fibers of the paper clip yield, upon removal of the applied load the paper clip springs back slightly, but remains bent. It is important to recognize that plasticity is a phenomenological (as opposed to mechanistic) engineering theory that usefully describes observed behavior of solids like ductile metals, but that the physics underlying plasticity is complex and various mechanisms at the microscale can lead to plastic behavior at the macroscale. Elastic-plastic behavior of solids, involving both plastic flow and elastic deformation, describes a regime somewhere between the "springiness" of solids and flow of fluids. Aside from the broad categories of elastic deformation and plastic deformation, the other main category of deformation of solids is damage, discussion of which we defer to Chapter 11. One observes plastic behavior in many engineering materials, but one must understand that theories of plasticity only grossly approximate the behavior of real solids. It would be unusual for an engineering model to be able to predict plastic deformation, under an arbitrary multi-axial state of stress, with an accuracy of better than twenty percent. The plastic behavior of many materials is temperature- and time-dependent, and strong coupling usually exists between temperature and plastic flow rates. Additionally, continuum plasticity involves an internal state variable called the plastic strain, whose initial state is not easy to measure. Thus, one cannot expect a plasticity model that neglects time, temperature, and initial plastic strain to provide very accurate predictions.

One observes plastic deformation in almost all materials, including metals, rock, concrete, and soil. In contrast to metal plasticity, in rock and concrete, plastic behavior becomes dominant only under states of high hydrostatic confinement. This is because cementitious materials tend to be weak in tension, and so they fracture (or become damaged) well before plasticity can commence.

Fig. 10.2 shows an idealized stress-strain relation for a uniaxial metal coupon in tension, and Fig. 10.3 shows a further-simplified stress-strain relation called the elastic-perfectly-plastic model.

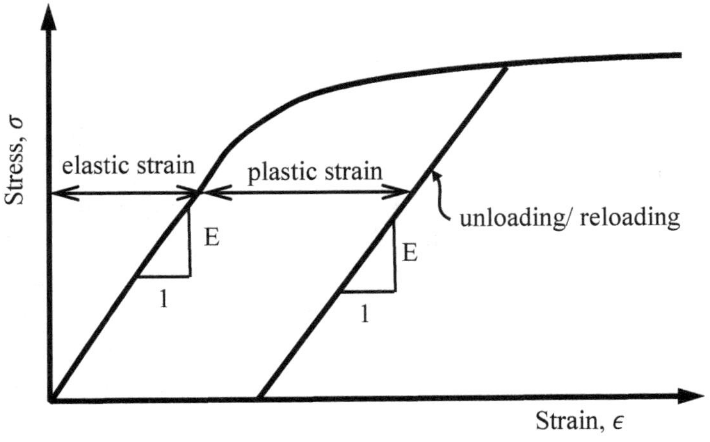

Figure 10.2. Typical idealized uniaxial stress-strain relation involving both elastic and plastic behavior.

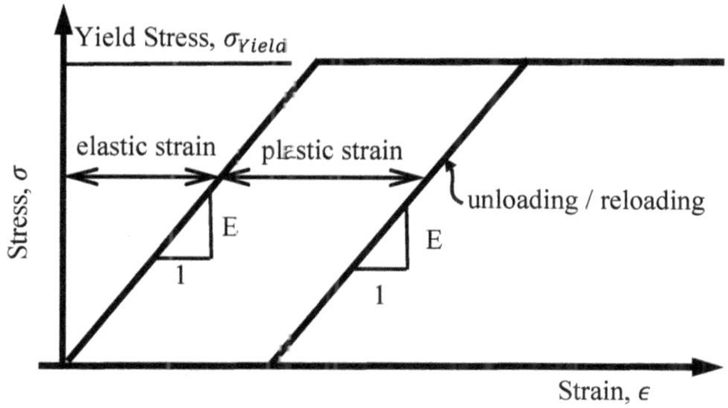

Figure 10.3. Idealized elastic-perfectly-plastic uniaxial stress-strain relation.

At the scale of crystals, which, at a small-enough scale, make up most materials (except for glasses and polymers), plasticity is usually a consequence of dislocations enabling slip planes between layers of atoms within a single crystal. In polymeric organic materials (generically called plastics), plastic-like behavior can arise if polymers unfold upon loading and do not return to their original configuration upon unloading. In cementitious materials (like concrete and rock), the behavior in regimes of compressive and shear deformation is often simulated using a plasticity model, even though the mechanism at a smaller scale might be dominated by complex interacting mechanisms of damage and friction.

Structural engineers, for example, typically model reinforced concrete beams in pure bending using plasticity models. The engineer models the deformation of the concrete on the compression side of the beam as elastic-perfectly-plastic, and likewise models the deformation of the reinforcing steel on the tension side of the beam as elastic-perfectly-plastic. In current engineering practice, engineers assume the tensile strength of concrete is nil in the axial direction of a concrete beam. However, in (difficult-to-defend) design practice, engineers usually assume that in transverse directions, the concrete beam has sufficient cohesion to engage the reinforcing bars; otherwise, the beam would collapse as a heap of sand on the floor, and the reinforcing bars would have no cohesion to the concrete. The American Concrete Institute Building Code Requirements for Structural Concrete (ACI318, 2014) currently use semi-rational formulas for beams, columns, and plates, and as such give little guidance for computational analysis of reinforced concrete structures. Such is the level of sophistication of engineering practice in our day!

Whatever the cause at the micro level, plastic-like behavior often emerges at the macro level, and if it is a dominant mechanism of deformation, engineers must account for it in their engineering models.

One of the features of pure plasticity is that the plastic component of volumetric strain does not change under plastic deformation[c]. A constant-volume process is termed "isochoric". However, some types of plasticity models nonetheless allow for plastic volume change.

Engineers, at least up to now, have thought of plasticity at the macro level as almost exclusively a nonlinear relationship between the rate of change of strain, with respect to stress, as shown in Figs. 10.1 and 10.2. With the introduction of peridynamic states, we have an opportunity to frame plasticity not in terms of strain and stress, but in terms of peridynamic force states and deformation states.

One of the key characteristics of classical plasticity is the notion that, at a given material point, plastic flow begins when the state of stress reaches a critical condition. It has been found that plastic flow does not occur in response to hydrostatic tension or compression; rather, plastic flow occurs in response to the magnitude of the deviatoric (or maximum shear) component of the stress tensor. The deviatoric component of the stress tensor is the total stress tensor minus the hydrostatic component.

The Tresca criterion for yield states that plastic flow will commence when the *maximum shear stress* reaches a prescribed limit, called the shear strength τ_{Yield}. In principal stress space, the Tresca yield criterion is a hexahedral cylinder whose axis lies on the hydrostatic axis as shown in Fig. 10.4.

The Von Mises yield criterion states that yield commences when the deviatoric *strain energy density* reaches a certain level. As shown in Fig 10.4, the resulting Von Mises yield surface is now a circular cylinder, whose axis again coincides with the hydrostatic axis. The Tresca and the Von Mises yield criteria predict the same yield stress for uniaxial and some biaxial stress states, but predict a maximum difference of 15.5% in

---

[c] The elastic component of the strain, however, is not in general, isochoric.

the case of pure shear. The Von Mises yield criterion, being a smooth analytical surface in principal stress space, is easier to implement both analytically and computationally, and is thus usually preferred. For most materials, laboratory testing, producing rather scattered yield strengths, does not definitively determine which of the two yield theories is more accurate; neither model is highly accurate at replicating the onset of yield of real materials in all stress conditions.

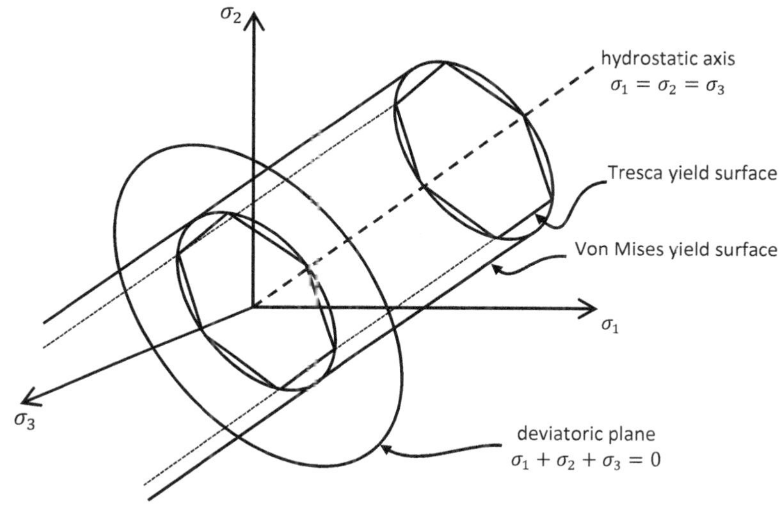

Figure 10.4. Tresca and Von Mises yield criteria in principal stress space.

In the plasticity model, once the stress reaches the yield surface, then the material begins to flow, in what we call *plastic yield*. Neither the Tresca nor the Von Mises yield criteria say anything about how the material flows subsequent to yield.

In a perfectly plastic rate-independent material, the plastic flow commences immediately, and continues instantaneously until the stresses are sufficiently relieved to bring the stress state back inside the yield surface. In a rate-dependent plasticity model, termed viscoplasticity, on the other hand, plastic flow is not instantaneous, and the rate of plastic flow is dependent upon how far outside the yield surface the stress state is located.

One of the common assumptions of plasticity theory is that the total strain $\boldsymbol{\varepsilon}$ can be considered to be the sum of the elastic strain $\boldsymbol{\varepsilon}^{\boldsymbol{e}}$ and the plastic strain $\boldsymbol{\varepsilon}^{\boldsymbol{p}}$:

$$\boldsymbol{\varepsilon} = \boldsymbol{\varepsilon}^{\boldsymbol{e}} + \boldsymbol{\varepsilon}^{\boldsymbol{p}} \,. \tag{10.1}$$

Another characteristic of plasticity is that as plastic flow occurs, the yield surface may evolve. As plastic flow takes place, due to entanglement of dislocations at the microscale, the configuration of the molecules changes, resulting in increased yield strength. Engineers call this "strain hardening" or "work hardening". In one model of work hardening, called *isotropic hardening*, the radius of the yield cylinder shown in Fig. 10.4 grows during plastic flow. On the other hand, in *kinematic hardening*, the yield cylinder retains its initial radius, but the yield cylinder translates in principal stress space. Kinematic hardening accounts for the *Bauschinger effect*, in which it is observed that the yield strength in tension increases as the specimen yields in tension (plastic hardening), while simultaneously the magnitude of the yield stress in compression decreases.

In the following three subsections, we review the content of one of the most common plasticity models, J_2 plasticity.

10.2.1 *Yield condition*

J_2 plasticity uses the Von Mises yield criterion, at least for linear elastic-perfectly plastic materials under small deformation. The literature also refers to the theory as the "maximum distortion strain energy theory of yield". It is perhaps the simplest multi-axial yield criterion. For materials undergoing large deformation and having nonlinear elastic behavior, J_2 plasticity is apparently not applicable.

The Cauchy stress tensor $\boldsymbol{\sigma}$ at a point in a solid can be decomposed into a deviatoric part $\boldsymbol{\sigma_D}$ and a hydrostatic part $\boldsymbol{\sigma_H}$:

$$\boldsymbol{\sigma} = \boldsymbol{\sigma_D} + \boldsymbol{\sigma_H} \,. \tag{10.2}$$

The hydrostatic part of the stress tensor is simply the average of the normal components of stress. This average normal stress is invariant under coordinate rotation:

$$\boldsymbol{\sigma_H} = \frac{1}{3}(\sigma_{XX} + \sigma_{YY} + \sigma_{ZZ})\boldsymbol{I} = \frac{1}{3}(\sigma_1 + \sigma_2 + \sigma_3)\boldsymbol{I} . \tag{10.3}$$

where σ_1, σ_2, and σ_3 are the principal stress components of the stress tensor $\boldsymbol{\sigma}$.

We express the deviatoric component of the stress as

$$\boldsymbol{\sigma_D} = \boldsymbol{\sigma} - \boldsymbol{\sigma_H} = \boldsymbol{\sigma} - \frac{1}{3}trace(\boldsymbol{\sigma})\boldsymbol{I}, \tag{10.4}$$

where $trace(\boldsymbol{\sigma})$ is sum of the stress components located on the diagonal of the stress matrix, and the trace is often called the first invariant, I_I, of the stress tensor.

The norm of the deviatoric stress $\|\boldsymbol{\sigma_D}\|$ is the square root of the sum of the squares of the nine components. We define the invariant parameter J_2 as

$$J_2 = \frac{1}{2}\|\boldsymbol{\sigma_D}\|^2. \tag{10.5}$$

J_2 can also be expressed in terms of the principal stresses as

$$J_2 = \frac{1}{6}[(\sigma_1 - \sigma_2)^2 + (\sigma_2 - \sigma_3)^2 + (\sigma_3 - \sigma_1)^2] . \tag{10.6}$$

The Von Mises yield criterion states that a material will yield when the strain energy density of the deviatoric component of the stress reaches a critical value. For a linear elastic material, J_2 is proportional to the strain energy density of the deviatoric component of the stress. Thus, for linear elastic materials, J_2 plasticity is equivalent to Von Mises plasticity.

The J_2 plasticity theory says that the material will yield if

$$J_2 = k^2 , \tag{10.7}$$

where k is a material constant. On the other hand, if $J_2 < k^2$, the material deforms only elastically, and no plastic flow takes place. The yield surface in stress space is the set of all stress states that satisfies the yield criterion, Eq. 10.7. The yield surface is the cylinder in principal stress space, of radius $\sqrt{2}k$, shown in Fig. 10.4.

Consider Eqs. 10.6 and 10.7 in the case of uniaxial stress, and let σ_{Yield} be called the "uniaxial yield stress". In this case, we know that yield occurs when $\sigma_1 = \sigma_{Yield}$ and σ_2 and σ_3 are null. Thus, $J_2 = \frac{1}{6}[(\sigma_1 - \sigma_2)^2 + (\sigma_2 - \sigma_3)^2 + (\sigma_3 - \sigma_1)^2] = k^2$, from which we find that $k = \frac{\sigma_{Yield}}{\sqrt{3}}$.

A similar analysis, except under pure in-plane shear, with $\tau_{Yield} = \sigma_1 = -\sigma_2$, and $\sigma_3 = 0$, shows that $k = \tau_{Yield}$. τ_{Yield} is termed the "shear yield stress". Thus, we can express the value of the material parameter k in terms of the uniaxial yield stress σ_{Yield} or in terms of the shear stress in pure shear, τ_{Yield} as

$$k = \tau_{Yield} = \frac{\sigma_{Yield}}{\sqrt{3}}, \tag{10.8}$$

and the yield stress in pure shear τ_{Yield} is $\frac{1}{\sqrt{3}} = 57.7\%$ of the uniaxial yield stress σ_{Yield}.

The Von Mises yield criterion and J_2 plasticity theory only indicate when plastic flow commences, but they say nothing about how the material flows. The next section describes how the material flows plastically.

10.2.2 *Plastic flow rule*

So far, we have presented a model for linear elastic material behavior in Chapter 9, and a model that indicates when plastic flow commences in Subsection 10.2.1. In this subsection, we address how the material plastically flows once it has yielded.

For an isotropic material, it is reasonable to assume that the principal plastic strain increments are proportional to the principal components of the *deviatoric* stress:

$$d\varepsilon_1^p = d\lambda\sigma_{D_1},\ d\varepsilon_2^p = d\lambda\sigma_{D_2},\ d\varepsilon_3^p = d\lambda\sigma_{D_3}, \tag{10.9}$$

where $d\lambda$ is a proportionality constant. Eqs. 10.9 are called the "Levy-Mises" flow equations.

Note that the Levy-Mises equations ensure isochoric (constant volume) plastic behavior because

$$d\varepsilon_{volumetric} = d\varepsilon_1^p + d\varepsilon_2^p + d\varepsilon_3^p = d\lambda(\sigma_{D_1} + \sigma_{D_2} + \sigma_{D_3}) = 0\,. \tag{10.10}$$

Algebra shows that in terms of Cartesian stress components,

$$\frac{d\varepsilon_{XX}^p}{\sigma_{D\,XX}} = \frac{d\varepsilon_{YY}^p}{\sigma_{D\,YY}} = \frac{d\varepsilon_{ZZ}^p}{\sigma_{D\,ZZ}} = \frac{d\varepsilon_{XY}^p}{\sigma_{D\,XY}} = \frac{d\varepsilon_{YZ}^p}{\sigma_{D\,YZ}} = \frac{d\varepsilon_{XZ}^p}{\sigma_{D\,XZ}} = d\lambda\,, \tag{10.11}$$

or, using indicial notation,

$$d\varepsilon_{ij}^p = d\lambda\sigma_{D\,ij}\,. \tag{10.12}$$

In terms of the stress components of the total stress tensor $\boldsymbol{\sigma}$ (not of the deviatoric stress tensor $\boldsymbol{\sigma_D}$), we find that

$$\begin{aligned}
d\varepsilon_{XX}^p &= \tfrac{2}{3}d\lambda\left[\sigma_{XX} - \tfrac{1}{2}(\sigma_{YY} + \sigma_{ZZ})\right];\\
d\varepsilon_{YY}^p &= \tfrac{2}{3}d\lambda\left[\sigma_{YY} - \tfrac{1}{2}(\sigma_{XX} + \sigma_{ZZ})\right];\\
d\varepsilon_{ZZ}^p &= \tfrac{2}{3}d\lambda\left[\sigma_{ZZ} - \tfrac{1}{2}(\sigma_{XX} + \sigma_{YY})\right];\\
d\varepsilon_{XY}^p &= d\lambda\sigma_{XY}\,;\\
d\varepsilon_{YZ}^p &= d\lambda\sigma_{YZ}\,;\\
d\varepsilon_{XZ}^p &= d\lambda\sigma_{XZ}\,.
\end{aligned} \tag{10.13}$$

We call this plastic strain rate versus stress relationship a *flow rule*. We cannot have a flow rule that gives the plastic strain increments as explicit functions of the stress, as the yield criterion might not be met; we must include the to-be-determined scalar plastic multiplier $d\boldsymbol{\lambda}$. The plastic multiplier $d\lambda$ is determined by ensuring the stress-state lies on the yield surface during plastic flow.

The "Prandtl-Reuss" equations, on the other hand, combine the linear elastic response predicted by the Navier-Cauchy equations with the plastic flow predicted by the Levy-Mises equations.

The yield surface can evolve as plastic flow takes place, as discussed next.

10.2.3 *Evolution of the yield surface*

As plastic flow occurs, many metals work-harden due to dislocation entanglement. In *isotropic hardening*, at the macroscale, the yield surface grows in radius. Stress-strain behavior under strain-hardening plasticity is generally stable and determinate if the yield surface is convex and growing.

On the other hand, in cementitious materials and soils, subsequent to yield, the material may strain-soften, which may cause an unstable condition leading to strain and stress singularities. In soil mechanics, these conditions are sometimes termed "shear bands". To prevent indeterminate shear-banding behavior, some sort of a localization limiter is required. The SPLM inherently avoids these singularities.

We do not go into detail in describing models for various yield surfaces and their evolution in this book. We simply mention that there are many plasticity models for different types of materials, including among many others, the Mohr-Coulomb, the Drucker-Prager, the Bresler-Pister, the Willam-Warnke, the Bigoni-Piccolroaz, and the Altenbach-Bolchoun-Kolupaev yield surfaces.

The Drucker stability postulates refer to a set of criteria that restrict the possible nonlinear stress-strain relations that can be satisfied by a solid material. A classical material that does not satisfy these criteria is unstable in that stress at a material point can lead to arbitrary deformations and to infinite strain.

Drucker's first stability postulate, motivated by thermodynamic considerations, states that internally stored strain energy density must be non-negative:

$$d\boldsymbol{\sigma} \cdot d\boldsymbol{\varepsilon} \geq 0 \,. \tag{10.14}$$

Drucker's second postulate states that the work done by plastic strains must always be non-negative:

$$d\boldsymbol{\sigma} \cdot d\boldsymbol{\varepsilon}_{\boldsymbol{P}} \geq 0 \,. \tag{10.15}$$

This postulate means that the plastic strain must always remove, or dissipate, energy from the system.

Drucker's postulates, based upon considerations of conservation of energy, are subject to debate, as some materials may convert chemically-stored energy to mechanical energy as plastic flow occurs, thus acting as energy sources, at least until their chemical potential is exhausted.

We next consider how we might implement plasticity theory in the context of the state-based peridynamic model.

10.3 Adaptation of classical material models to continuum state-based peridynamic model

Any classical (stress-strain) constitutive model can be adapted, at least approximately, to the state-based peridynamic model.

As described in Chapter 8, if at a given particle P_i a peridynamic deformation state $\underline{\boldsymbol{x}_R}$, the corresponding reference state $\underline{\boldsymbol{X}_R}$, and a shape tensor $\boldsymbol{K}$ are given, then these states can be reduced to a "corresponding" classical continuum mechanics deformation gradient tensor $\overline{\boldsymbol{F}}$ using the reduction technique of Definition 8.15 and Eq. 8.46:

$$\overline{\boldsymbol{F}}\left(\underline{\boldsymbol{x}_R}\right) = \mathcal{R}\left\{\underline{\boldsymbol{x}_R}\right\} = \left(\underline{\boldsymbol{x}_R} * \underline{\boldsymbol{X}_R}\right)\boldsymbol{K}^{-1} \,.\text{[d]} \tag{10.16}$$

This approximation is exact for spatially homogeneous deformation states not too close to the boundary of the domain.

Note that this reduction depends upon the choice of influence function ω, upon which the shape tensor $\boldsymbol{K}$ depends.

---

[d] The over bar in $\overline{\boldsymbol{F}}$ represents "approximation of $\boldsymbol{F}$".

With this approximation of the deformation gradient $\overline{\boldsymbol{F}}\left(\underline{\boldsymbol{x}_R}\right)$ an approximation of the Cauchy small strain tensor $\boldsymbol{\epsilon}$ (Eq. 3.38), the Euler-Almansi strain tensor $\boldsymbol{e}$ (Eq. 3.31), or the Lagrangian strain tensor $\boldsymbol{E}$ (Eq. 3.29), can be determined through kinematic equations of continuum mechanics derived in Section 3.3. One must calculate this total strain tensor at each discretized particle. If we want to model plasticity, then the plastic strain components $\boldsymbol{\epsilon}^{\boldsymbol{p}}$ must be stored, (as also must be stored the damage parameters) at each discretized particle. Using Eq. 10.1 to compute the elastic strain component $\boldsymbol{\epsilon}^{\boldsymbol{e}}$, we can find the corresponding stress $\boldsymbol{\sigma}$ through the classical elastic constitutive relation:

$$\boldsymbol{\sigma} = \boldsymbol{D}\boldsymbol{\epsilon}^{\boldsymbol{e}} \,. \tag{10.17}$$

With the stress $\boldsymbol{\sigma}$ known at particle P_i, J_2 can be calculated using Eq. 10.6, the yield condition can be checked using Eq. 10.7, the Levy-Mises flow rule (Eq. 10.13) followed, and the evolution of the yield surface traced as time progresses.[e]

Finally, with the stress $\boldsymbol{\sigma}$ now known, we can expand the peridynamic vector force state from the stress tensor, $\boldsymbol{\sigma}$ using Eq. 8.47:

$$\underline{\boldsymbol{T}}\langle\boldsymbol{\xi}\rangle = \underline{\omega}\langle\boldsymbol{\xi}\rangle\boldsymbol{\sigma}\boldsymbol{K}^{-1}\boldsymbol{\xi}\,, \quad \text{or} \quad \underline{\mathbf{T}} = \underline{\omega\mathcal{E}}(\boldsymbol{\sigma}\boldsymbol{K}^{-1}). \tag{10.18}$$

This approach has been applied using an elasto-plastic von-Mises isotropic linear hardening model with an associated flow rule and radial return algorithm, as described by Warren et al. (Warren, T.L., Silling, S. A., Askari, A., Weckner, O., Epton, M. A., and Xu, J., 2009), although the computational implementation is not described in detail. The model is *non-ordinary* (as presented in Definition 8.20), because the bond forces are not aligned with the bond directions. The approach has the advantage that it can be implemented using any collection of discrete peridynamic particles; a regular lattice is not required. On the other hand,

[e] Instantaneous plasticity is actually more difficult computationally to implement than rate-dependent viscoplasticity. For instantaneous plasticity, within each time step, we must employ an iterative/incremental algorithm to satisfy the equations of plasticity. For viscoplasticity, this extra loop is not required.

the method is very involved theoretically and is computationally expensive, requiring calculation of $\overline{\boldsymbol{F}}\left(\underline{\boldsymbol{x}_R}\right)$ involving integration (or finite summation if discrete particles are used) and inversion of the $\boldsymbol{K}$ matrix at every particle using Eq. 10.16.

Two simpler and more direct *ordinary* and *mobile* approaches, taking advantage of the spatial regularity and the topology provided by the lattice body $\mathcal{L}_{\mathcal{R}}$, are described in the next section.

10.4 SPLM mobile plasticity models

We can develop elastic-plastic models directly using the SPLM, without having to resort to the classical continuum strain and stress concepts. The advantage of avoiding strain and stress is that singularities do not arise, and discontinuities emerge without theoretical difficulty. In addition, the direct approaches described in this section are simpler, both conceptually and computationally, than the non-ordinary approach described in Section 10.3.

To summarize, with the non-ordinary approach described in Section 10.3, at each particle in the material domain, the peridynamic deformation state $\underline{\boldsymbol{x}_R}$ is first reduced to an approximate continuum deformation gradient $\overline{\boldsymbol{F}}\left(\underline{\boldsymbol{x}_R}\right)$. We calculate the classical strain $\boldsymbol{\epsilon}$ from this continuum deformation gradient, and then we calculate the classical stress $\boldsymbol{\sigma}$ via a classical elastic constitutive model. From the classical stress $\boldsymbol{\sigma}$, the force state $\underline{\mathbf{T}}$ is computed and then finally used to integrate the peridynamic model in space and time using Eq. 8.30.

Let us now leave the classical stress-strain approach behind completely, and develop our plasticity model from basic considerations within the framework of the SPLM.

Consider the interior three-dimensional SPLM particle P_i, shown in Fig. 6.7, having twelve bonds B_j to its nearest neighbors, and six bonds B_j to its second nearest neighbors in lattice body $\mathcal{L}_{\mathcal{R}}$.

For each lattice particle P_i, we endow each of the eighteen bonds, for example bond B_j, with a plastic response attribute, called the "plastic bond stretch" S_j^p. These eighteen plastic bond stretches S_j^p are state variables stored with particle P_i.

Assume that each bond consists of two halves (the half associated with particle P_i, and the half associated with particle P_k) of equal reference length $L/2$ for the twelve near-neighbor bonds, and $\sqrt{2}L/2$ for the six second-nearest neighbor bonds.

For the bond B_j with $\mathcal{N}[P_i]\langle B_j\rangle = P_k$, thus connecting particles P_i and P_k, assume the stretch of each half of the bond is the sum of elastic and plastic components: $\underline{S}[P_i]\langle B_j\rangle = \underline{S^e}[P_i]\langle B_j\rangle + \underline{S^p}[P_i]\langle B_j\rangle$, and $\underline{S}[P_k]\langle B_{j'}\rangle = \underline{S^e}[P_k]\langle B_{j'}\rangle + \underline{S^p}[P_k]\langle B_{j'}\rangle$. Here, B_j and $B_{j'}$ represent opposite collinear, bonds, as shown in Fig. 6.7. We find the total elastic stretch to be

$$\underline{S^e}[P_i]\langle B_j\rangle = \underline{S}[P_i]\langle B_j\rangle - \frac{1}{2}\left(\underline{S^p}[P_i]\langle B_j\rangle + \underline{S^p}[P_k]\langle B_{j'}\rangle\right). \tag{10.19}$$

Thus, each particle P_i stores eighteen plastic stretches ${S^p}_j$, one for each of its eighteen bonds B_j. The elastic stretch ${S^e}_j$, of each bond B_j is thus the total stretch, S_j, calculated from particle spatial positions $\boldsymbol{X}_i$ and $\boldsymbol{X}_k$ using Eq. 9.2, minus the average of the plastic stretches of each of the bonds $\{S^p\}[P_i]\langle B_j\rangle$ and $\{S^p\}[\mathcal{N}[P_i]\langle B_j\rangle]\langle B_{j'}\rangle$.[f]

To reiterate, we can calculate the total stretch matrix $\{S\}$ for any particle P_i at any time step directly from the relative particle spatial positions $\boldsymbol{x_R}$ using Eq. 9.2. With all plastic stretches known for each particle P_i, the elastic portion, $\{S^e\}$, of the stretch matrix, $\{S\}$, can be calculated using Eq. 10.19, and thus the force matrix, $\{F\}$, can be computed using (Eq. 9.28): $\{F\} = [K]\{S^e\}$.[g] The computed force thus indirectly accounts for the average plastic stretch of each bond. The force state is then computed as $\{T\} = \frac{1}{2}\{F\}$, and Eq. 8.30 is finally used to integrate the particle motion using a finite difference approach described further in Chapter 12.

The following subsections outline several possible approaches to defining the yield condition and flow rule directly from the force state matrix $\{F\}$.

---

[f] $\{S^p\}[\mathcal{N}[P_i]\langle B_j\rangle]\langle B_{j'}\rangle$ means "the plastic stretch of the bond $B_{j'}$ of the neighboring particle via bond B_j of particle P_i", where B_j and $B_{j'}$ are complementary bonds.

[g] We do not need actually to form the microelastic stiffness matrix $[K]$ in each time step. Instead, Eqs. 9.34–9.36 can be evaluated in a do-loop.

10.4.1 *First force state plasticity model*

When does plastic stretch commence? In analogy to J_2 plasticity, we assume that plastic flow is caused by the deviatoric component of the force matrix $\{F\}$, which we define as the force matrix minus the average bond force, and so we define the average bond force as

$$F_{avg} \equiv \frac{1}{18}\sum_{j=1}^{18} F_j \, . \tag{10.20}$$

For a given particle locate within a spatially homogeneous deformation field, the average bond force F_{avg} is invariant with respect to lattice rotation $\boldsymbol{Q}$ for the three-dimensional SPLM elasticity model developed in Chapter 9, but F_{avg} does depend upon the Poisson's ratio ν. For uniaxial stress σ_{XX} and lattice spacing L, we can show that

$$F_{avg} \equiv \sigma_{XX} L^2 \frac{2\nu(2\sqrt{2}-1)+(1+\sqrt{2})}{36(1+\nu)} \, . \tag{10.21}$$

The average force F_{avg} is analogous to the hydrostatic component of stress $\boldsymbol{\sigma_H}$. Similar to the deviatoric component of the stress tensor in Eq. 10.4, we define the deviatoric part of the peridynamic force matrix as

$$\{F_{dev}\} \equiv \{F\} - F_{avg} \, . \tag{10.22}$$

We can show that the sum of the components of $\{F_{dev}\}$ is equal to zero.

The norm of the deviatoric part of the peridynamic force matrix is the square root of the sum of the squares of its components:

$$\|\{F_{dev}\}\| \equiv \sqrt{\sum_{j=1}^{18}\left({F_{dev}}_j{}^2\right)} \, . \tag{10.23}$$

When $\|\{F_{dev}\}\|$ reaches a critical value, call it $F_{devYield}$, we assert that the yield surface has been reached; thus, the condition for SPLM yield is

$$\|F_{dev}\| \geq F_{devYield} \, . \tag{10.24}$$

In terms of the classical uniaxial yield stress σ_{Yield}, lattice spacing L, and Poisson's ratio ν, under homogeneous deformation, one can show that

$$F_{devYield} = \frac{L^2 \sigma_{Yield}}{12(1+\nu)} \sqrt{(93 - 4\sqrt{2}) + (174 - 8\sqrt{2})\nu + (135 + 32\sqrt{2})\nu^2} \,. \quad (10.25)$$

Eq. 10.25 for $F_{devYield}$ has been determined by symbolically computing (using MatLab) the norm of the deviatoric part of the force matrix for a single three-dimensional SPLM particle under the condition of uniaxial yield stress: $\sigma_{XX} = \sigma_{Yield}$, assuming homogeneous elastic deformation.

Note that $F_{devYield}$ is a weak function of Poisson's ratio ν, varying by less than five percent as ν increases from 0 to 0.5.

If satisfaction of Eq. 10.24 indicates that a plastic yield condition has been reached, the plastic stretch matrix $\{S_j^p\}$ must evolve to allow for plastic flow. At a given lattice particle P_i, and at a given time step, we assume that the change in the plastic stretch matrix, $\{\Delta S_j^p\}$, is proportional to the deviatoric force matrix:

$$\{\Delta S^p\} = \Delta\lambda \frac{\{F_{dev}\}}{F_{devYield}} \,, \quad (10.26)$$

where $\Delta\lambda$ is a proportionality constant, similar to the classical theory of plastic flow given by the Levy-Mises equations 10.12.

For rate-independent plasticity, it is necessary to iterate within each time step until equilibrium is achieved. Rather than to assume rate-independent plasticity, it is convenient (and probably more physically correct) to assume viscoplastic (time-dependent) flow occurs. Thus, we define

$$\Delta\lambda = \frac{\sigma_{Yield}}{E}\left(\frac{\|F_{dev}\|}{F_{devYield}} - 1\right), \quad (10.27)$$

so that the flow rate is proportional to the uniaxial yield strain and to the distance between the force state and the yield surface. Thus, the rate of plastic flow depends upon how far the force state lies outside the yield surface. In addition, the increment in plastic flow $\{\Delta S^p\}$ is proportional to the uniaxial yield strain. When plastic flow is occurring, it might be necessary to decrease the time step below that needed for a stable elastic solution to capture the details of the dynamic plastic flow. This aspect of the model requires further study and development.

Because the sum of the components of $\sum_{j=1}^{N_B} F_{devj} = 0$, therefore, in consideration of Eq. 10.26, $\sum_{j=1}^{N_B} \Delta S^p{}_j = 0$. For small stretches, so that, say, $|S_j| < 0.01$ for all S_j, the change in plastic volume ΔV^p of the particle can be shown to be approximately zero (as is to be shown in Exercise 10.9). However, if any of the stretches are large, so $|S_j| > 0.1$ then plastic volume change ΔV^p of the particle will in general be nonzero. In this case, to satisfy the condition of isochoric plastic behavior, a more exact (and likely more computationally expensive) plastic flow theory is warranted.

For perfectly plastic behavior, the yield criterion $F_{devYield}$, given in Eq. 10.24, does not evolve with $\{S^p\}$. However, for some materials, for example strain-hardening steels, the yield criterion evolves as deformation progresses. In this case, $F_{devYield}$ in Eq. 10.24 would be a specified hardening function of $\{S^p\}$. This would describe the case of isotropic hardening. The case of kinematic hardening would be somewhat more complex, but there does not seem to be a reason why we could not develop any number of SPLM theories for plastic yield surface evolution (hardening and softening), just as is the case with classical plasticity models described in Section 10.2.3.

10.4.2 *Second force state plasticity model*

In the previous section, we assumed, somewhat arbitrarily, that plastic flow commences when the norm of the deviatoric component of the force matrix reaches a specified value $F_{devYield}$. We found that $F_{devYield}$ is a weak function of Poisson's ratio. Although the approach is simple and computationally efficient, this does not match the Von Mises theory, in which the yield criterion depends *only* upon stress, and is only indirectly dependent upon the elastic parameters. Thus, we seek an alternative SPLM plasticity model that is equivalent to the Von Mises theory of yield, at least in regions of spatially homogeneous strain and stress.

We can express the stress in terms of the bond forces using Eq. 9.24.

Then we can express J_2 using Eq. 10.5 in terms of the force matrix $\{F\}$. This gives the J_2 directly in terms of the eighteen bond forces for the three-dimensional particle shown in Fig. 6.7:

$$\begin{gathered}
J_2 = \frac{1}{36L^4}(A + B + C + D + E + F), \text{ where} \\
A = \frac{1}{2}\left(\sum_{i=7}^{10} F_i - 2\sum_{i=11}^{12} F_i + 2\sqrt{2}\sum_{i=13}^{14} F_i - \sqrt{2}\sum_{i=15}^{18} F_i\right)^2; \\
B = \left(\sum_{i=1}^{6} F_i - \sum_{i=7}^{12} F_i\right)^2; \\
C = \frac{3}{32}\left(3\sqrt{2}(F_3 + F_4 - F_5 - F_6) + \sqrt{2}(F_7 + F_8 - F_9 - F_{10}) + 4(F_{15} + F_{16} - F_{17} - F_{18})\right)^2; \\
D = \frac{3}{2}\left(F_7 + F_8 - F_9 - F_{10} + \sqrt{2}(-F_{15} - F_{16} + F_{17} + F_{18})\right)^2; \\
E = \frac{1}{32}\left(4\sqrt{2}\sum_{i=1}^{2} F_i - 5\sqrt{2}\sum_{i=3}^{6} F_i + 3\sqrt{2}\sum_{i=7}^{10} F_i - 8\sum_{i=13}^{14} F_i + 4\sum_{i=15}^{18} F_i\right)^2; \\
F = \frac{1}{32}\left(-8\sqrt{2}\sum_{i=1}^{2} F_i + \sqrt{2}\sum_{i=3}^{10} F_i + 4\sqrt{2}\sum_{i=11}^{12} F_i + 8\sum_{i=13}^{14} F_i - 4\sum_{i=15}^{18} F_i\right)^2;
\end{gathered} \tag{10.28}$$

With J_2 at each particle now computed from the bond forces, we can check the yield function using the classical yield criterion, Eq. 10.7, in each time step. If particle yields, then the flow $\{\Delta S^p\}$ can be computed either using Eqs. 10.26 and 10.27, or more in line with the continuum theory, by computing the increment in plastic strain using Eq. 10.12 followed with computing the increment in plastic stretch using Eq. 9.14.

The approach described in this section is precisely Von Mises plasticity, as long as the particle is in an environment of spatially homogeneous deformation. On the other hand, if the deformation is not homogeneous, one can still use the method, but it is not directly comparable to Von Mises plasticity. This SPLM plasticity model is neither better nor worse than continuum Von Mises plasticity – just different, and perhaps easier to implement on a computer. The primary advantage of the SPLM model is that it allows one easily to combine damage and fracture with plasticity as described in the next Chapter.

10.5 Summary

In this chapter, we began by describing, in broad outline, classical plasticity models. Then we focused on the classical J_2 plasticity model, also called the distortional strain energy model, including the Levy-Mises and Prandtl flow rules and various models for the evolution of the yield surface in stress space.

Then, in Section 10.3, we presented a method for adapting classical constitutive equations to the continuum state-based peridynamic model. The method requires reduction of the vector deformation state to the classical deformation gradient, calculation of the strain, use of the classical constitutive equations to determine the stress, and then expansion of the stress to determine the vector force state, which we can then plug into the peridynamic equations of motion. Although this method has the advantage that one can port any classical constitutive model to a peridynamic framework, and that a regular lattice is not required, its principal disadvantage is that the method is complex and appears to be computationally more involved than necessary. In addition, the method requires the selection of an ad-hoc influence function and does not produce a mobile peridynamic model.

Then, in Section 10.4, we introduced two mobile SPLM models for elasto-plasticity. The first of these models completely avoids the use of concepts of strain and stress. Instead, we relate the bond forces to the elastic components of the bond stretches, computing the eighteen components of plastic bond stretch and storing them as particle attributes. The second of these models makes use of classical continuum plasticity, by providing a formula for J_2 directly in terms of bond forces. If the yield condition is satisfied, then we compute the plastic flow either directly from the bond forces, or by first computing the plastic strain from which we can obtain the plastic bond stretches.

Because we compute the bond forces directly from elastic bond stretches and from current bond directions, which we obtain from the current relative particle positions, which are vectors, the method is independent of coordinate system, and is thus frame-indifferent. In addition, the elasticity model developed in Chapter 9 is isotropic and independent of lattice orientation, for homogeneous deformation states. Thus the two SPLM models developed in Section 10.4 appear to be legitimate material models (frame-indifferent and isotropic), which are easy to implement computationally and are theoretically simple. However, the yield surface in the first model depends weakly upon Poisson's ratio, making it slightly different from the Von Mises yield criterion.

In Section 10.4.2, we developed an alternative SPLM plasticity model that is identical to Von Mises plasticity (under homogeneous deformation) but is somewhat more difficult to implement and is not as efficient as the more direct SPLM plasticity model presented in Section 10.4.1.

As long as all bond stretches are reasonably small, the SPLM plasticity models presented in this chapter produce essentially isochoric plastic flow. If large bond stretches are to be allowed, a more involved plastic flow rule is required if isochoric plasticity is mandated.

One can compare the SPLM plasticity model to the classical models, but the SPLM model does not depend for its legitimacy upon the classical models, and it stands on its own as a valid computational model for the deformation of elasto-plastic solids. Its ultimate validity depends upon whether it is useful in simulating the physical behavior of solids; not on whether it exactly matches the classical plasticity model.

In the next chapter, we extend the elasto-plastic SPLM model to also include damage, and therefore, fracture.

10.6 Exercises

10.1 Based upon the definition of J_2, show that Eq. 10.6 is correct:: $J_2 = \frac{1}{6}[(\sigma_1 - \sigma_2)^2 + (\sigma_2 - \sigma_3)^2 + (\sigma_3 - \sigma_1)^2]$, where σ_1, σ_2, and σ_3 are the principal stresses. You may use MatLab for this exercise.

10.2 Show that Eq. 10.8, $k = \tau_{Yield} = \frac{\sigma_{Yield}}{\sqrt{3}}$ is correct.

10.3 Plot the J_2 yield surface in principal (σ_1, σ_2) space for a plate in plane stress. Show τ_{Yield} and σ_{Yield} on the plot.

10.4 Show that if the Levy-Mises flow equations are correct, then Eq. 10.12: $d\varepsilon_{ij}^p = d\lambda\sigma_{D_{ij}}$ is true for any Cartesian coordinate system.

10.5 Verify that the component form of the Levy-Mises equations, Eqs. 10.13, are correct.

10.6 Show that $F_{avg} \equiv \sigma_{XX} L^2 \frac{2\nu(2\sqrt{2}-1)+(1+\sqrt{2})}{36(1+\nu)}$ (Eq. 10.21) is true regardless of proper lattice rotation $\boldsymbol{Q}$.

10.7 Show that the sum of the components of $\{F_{dev}\}$, defined by Eq. 10.22, is equal to zero.

10.8 Show that Eq. 10.25:

$$F_{devYield} = \frac{L^2\sigma_{Yield}}{12(1+\nu)}\sqrt{(93 - 4\sqrt{2}) + (174 - 8\sqrt{2})\nu + (135 + 32\sqrt{2})\nu^2}$$

is correct.

10.9 Write a MatLab script that calculates the volumetric plastic strain ϵ_V^p of a plane strain hexagonal lattice particle as a function of the deformation state $\underline{\boldsymbol{x}_R}$. Assuming that the sum of the components of the $\{S^p\}$ matrix is zero, what is the maximum volumetric plastic strain if no component of plastic strain matrix $\{S^p\}$ is greater than (a) 0.01; (b) 0.1; (c) 0.5?

10.10 Verify expressions given for A through F in the equation $J_2 = \frac{1}{36L^4}(A + B + C + D + E + F)$ given by Eq. 10.28.

10.11 Considering a case of uniaxial stress, show that if the plastic yield surface shrinks with increasing plastic strain (thus strain softens), strain localization will occur. Consider the cases of axial tension and axial compression.

10.12 If a rod is made of the strain-softening material shown in Fig. 10.5, describe the deformation behavior as load is gradually applied in (a) tension; (b) compression.

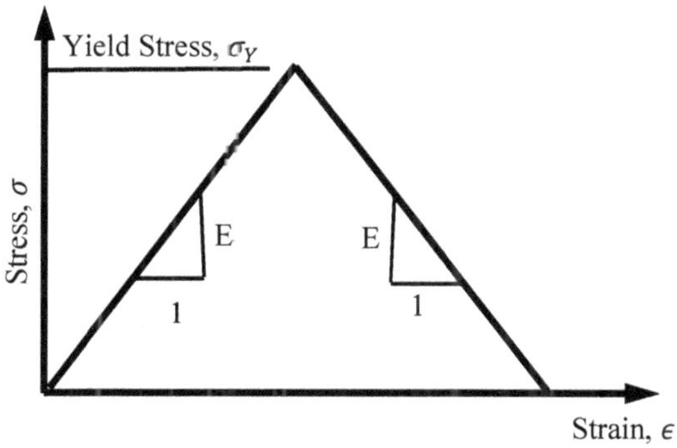

Figure 10.5. Strain-softening plasticity model.

Chapter 11

Damage

Chapter objectives are to:

- review the history of continuum damage mechanics;
- present the theory of continuum damage mechanics;
- discuss damage localization;
- develop a bond-based theory of peridynamic damage;
- develop a state-based peridynamic theory of damage;
- investigate isotropic and anisotropic damage models;
- study the relationship between damage and dynamical behavior;
- and to explain the relationship between damage and fracture.

The subject of this chapter is damage mechanics, which goes under various names: the "smeared crack model", "continuum damage mechanics", and "nonlocal continuum damage mechanics". We question under what conditions damage can be, or should be, thought of as a continuum phenomenon. The bulk of the literature has treated damage mechanics as a branch of continuum mechanics, and as such, engineers have modelled damage primarily using the finite element method.

Damage is different from plasticity in that damage results in a reduction of material stiffness, with no permanent deformation upon unloading, as shown in Fig. 11.1. In contrast, plasticity involves permanent deformation but no reduction in stiffness, as shown in Figs. 10.2, 10.3, and 11.1. Many materials demonstrate features of both damage and plasticity.

However, damage fits only uncomfortably into a continuum mechanics model, because damage, involving breakage of bonds, is essentially a fracture-like phenomenon and thus often leads at some point to discrete cracking. "Distributed microcracking" fits with the continuum

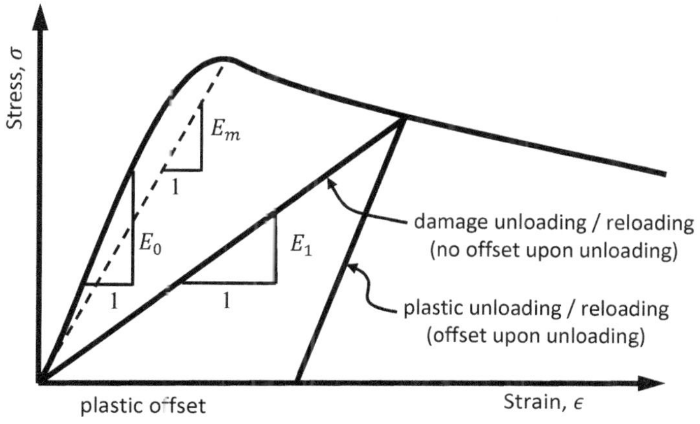

Figure 11.1. Damage and plasticity stress-strain behavior.

approach, as long as one can describe the phenomenon as a continuous distribution of microcracks evolving under increasing load. However, if the damage and strain are increasing while the stress is decreasing, as shown in Fig. 11.1, the microcracking domain coalesces dynamically into a localized macrocrack, which is essentially discontinuous. The post-peak portion of the stress-strain relation in Fig. 11.1 is actually a fiction, because strain is undefined as it localizes.[a] Thus, either an auxiliary localization limiter is required if continuum mechanics is to be employed, or a discontinuous material model should be used instead. We take the latter approach, in the form of the state based peridynamic lattice model (SPLM).

After reviewing the history and theory of continuum damage mechanics, we introduce both a bond-based micropolar damage model and a state-based peridynamic lattice damage model, both similar to the simplest of continuum damage models, while avoiding the issues of lack of objectivity associated with strain localization.

---

[a] Any plot of stress versus strain in which stress decreases with increasing strain is thus nonsense. The softening branch of the curve in Fig. 11.1 is included only as a means of starting the discussion about strain localization.

11.1 History of damage mechanics

The discipline of damage mechanics essentially originated in 1958 with the work of Kachanov (Kachanov, 1958), who studied the softening behavior of solid visco-elastic materials. Kachanov imagined that internally within the material, a volumetrically distributed collection of "microcracks" was both nucleating and propagating in response to applied stress at the macro level, causing the effective stiffness of the continuum to decrease as damage evolved. He represented the growth of these microcracks, or micro voids, which he supposed would be highly random and morphologically varied at the micro level, by a single spatially continuously-varying parameter at the macro level, called the damage parameter ω, which varies continuously with time from $\omega = 0$ (no damage) to $\omega = 1$ (complete damage). Thus, Kachanov postulated that the uniaxial stress-strain (σ versus ε) relationship was of the form

$$\sigma = (1 - \omega)E\varepsilon \, . \tag{11.1}$$

where E is Young's modulus. The time-evolution of the damage parameter was assumed to be a prescribed function of stress or of strain, or perhaps, of some other "energetically conjugate" variable such as the "energy density release rate (with respect to damage)". Note that when damage is complete ($\omega = 1$), the stress is zero for any level of strain.

Throughout the 1960's and 1970's, researchers wrote many papers on the damage mechanics of metals. For example, Rabotnov applied the ideas of Kachanov to creep problems (Rabotnov, 1971). He proposed a coupled set of differential equations that describe how damage rate $\dot{\omega}$ and strain rate $\dot{\varepsilon}$ evolve with respect to stress σ, temperature T, and damage ω:

$$\dot{\varepsilon} = \phi(\sigma, T, \omega) \quad \text{and} \quad \dot{\omega} = \psi(\sigma, T, \omega) \tag{11.2}$$

These equations, however, were difficult to extend into three dimensions.

Among many others, Lemaitre and Chaboche (Lemaitre, J. and Chaboche, J.-L., 1974) extended the theory into three dimensions, assuming isotropic damage, which assumes that microcracks evolve equally in all directions regardless of the stress state. Most of the papers up to the mid-1970s applied damage mechanics to the ductile creep of metals.

Material instability had been studied as early as 1903 by Hadamard, and then later on by (Thomas, 1961), (Hill, 1962), and (Rudnicki J.W. and Rice J. R., 1975). Review articles on damage localization are provided by (Jirásek, 1998), (Bazant, Z.P. and Jirasek, M., 2002), and by (Bessuelle P. and Rudnicki, J. W., 2004).

In the late 1970's, continuum damage mechanics was applied to quasibrittle structures like concrete using tensor approaches by (Krajcinovic, 1983) and by (Mazars, J., and Lemaitre, J., 1984).

These continuum damage mechanics theories were often applied using the finite element method, but considerations of material stability and strain localization were not well-understood, at least by the facture mechanics of concrete and rock community, until the 1980's and 1990's with the work of Bazant (Bazant, Z. P. and Planas, J. P., 1998) and many others.

It seems that researchers have proposed all of these theories in terms of the fundamental objects of continuum mechanics: strain and stress. Due to Silling (Silling, S.A., Epton, M., Weckner, O., Xu, J., and Askari, E., 2007), we now have a model for solids more fundamental than the strain and stress model: deformation states and force states. (Bazant, Z.P. and Jirasek, M., 2002), in the conclusion to their comprehensive review paper, wrote:

> "The last two decades of research gave birth to a wide variety of nonlocal models, with differences that are not quite justified by diversity in the types of materials and practical application. One may now expect a period of crystallization in which many artificially complex or oversimplified models will fade, being recognized as superfluous, and only a few will gain a permanent pedestal in the pantheon of knowledge."

With engineering, and even with science, the "pantheon of knowledge" can rapidly shift in response to computational and other practical realities. The entire category of nonlocal continuum damage models may be more complex than necessary, and the more direct peridynamic model possibly could supersede them.

11.2 Continuum damage mechanics theory

Damage mechanics starts with the idea that at the material meso-level, many volumetrically distributed microcracks can develop in response to an applied stress. As a simplified model of this situation, Fig. 11.2 illustrates a cable consisting of many parallel wires loaded in tension with applied stress σ. As some of the weaker wires break at random locations along the length of the cable, a reduction of the stiffness and strength of the overall cable occurs. (We assume that no shear stress exists between the wires in this discussion.) If we define the damage to be the ratio of the cross-sectional area of broken wires $\bar{A}$ to the cross-sectional area A of the total number of wires, $\omega \equiv \bar{A}/A$ then the "microstress $\bar{\sigma}$" in the remaining wires increases as

$$\bar{\sigma}(A - \bar{A}) = \sigma A \quad \text{or} \quad \bar{\sigma} = \frac{\sigma}{\left((A-\bar{A})/A\right)} = \frac{\sigma}{(1-\omega)} \,. \qquad (11.3)$$

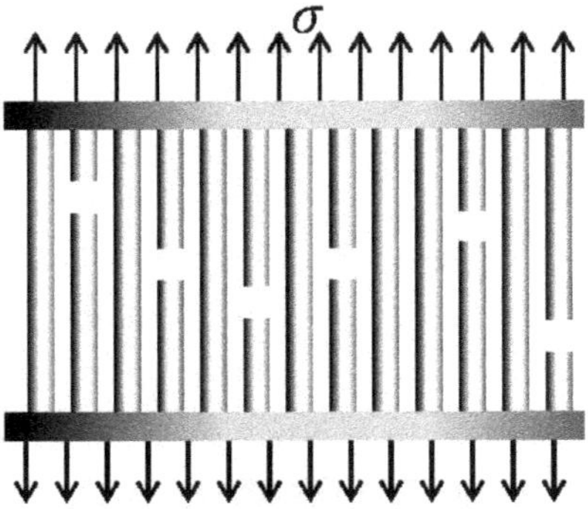

Figure 11.2. Parallel axial members at the meso-level.

Assuming that the unbroken wires remain linearly elastic, then at the macro-level,

$$\sigma = (1 - \omega)\bar{\sigma} = (1 - \omega)E_0\epsilon = E_1\epsilon \,, \qquad (11.4)$$

which illustrates one of the essential ideas of damage mechanics. The evolution of damage, in the case shown in Fig. 11.2, depends upon the distribution of the tensile strengths of the wires. In this case, the damage would be a function of the micro stress $\omega = \omega(\bar{\sigma})$, with the understanding that damage can never decrease.[b]

Damage is not limited to one-dimensional linear elastic structures – it also occurs in three-dimensional nonlinear elastic, viscous, and plastic bodies.

Of course, in real isotropic materials, the microstructure is not nearly as simple as with the system of parallel wires shown in Fig. 11.2. Even so, the concept of damage, to a degree of approximation, can carry over to general bodies if one considers the damage to be some form of spatially distributed bond-breakage, or perhaps microcracking.

Many have attempted to treat the propagation of microcracks associated with damage using linear elastic fracture mechanics (LEFM) concepts. This approach is almost never successful, because the conditions for LEFM, (with the fracture process zone being tiny compared with grain dimensions) simply do not occur at the meso level.

One can express a three-dimensional isotropic continuum damage model, equivalent to Eq. 11.4 as

$$\boldsymbol{\sigma} = (1-\omega)\boldsymbol{D}\boldsymbol{\epsilon} \quad \text{or} \quad \sigma_{ij} = (1-\omega)D_{ijkl}\epsilon_{kl}\,, \tag{11.5}$$

where a suitable damage evolution function must also be prescribed. For example the damage at a material point might depend upon the highest maximum tensile principle stress previously reached at that material point.

For an anisotropic continuum damage model, the damage parameter may be a tensor quantity $\boldsymbol{\omega}$. For example, one might assume

$$\boldsymbol{\sigma} = \boldsymbol{D}(\boldsymbol{\omega})\boldsymbol{\epsilon} \quad \text{or} \quad \sigma_{ij(mnop)} = D_{ijkl}(\omega_{mnop})\epsilon_{kl}\,, \tag{11.6}$$

---

[b] Actually, damage *can* decrease in certain situations, in which a material might "heal" through chemical and thermal processes. Such is the case with ice cubes in a glass of water, and asphalt concrete on a hot afternoon.

where now the stiffness $\boldsymbol{D}$ is an anisotropic function of an anisotropic damage function $\boldsymbol{\omega}$, both being fourth-order tensors.

However, it hardly seems worthwhile to develop highly sophisticated continuum damage models, except as a theoretical exercise, because most of the significant deformation usually occurs subsequent to damage localization to a surface (macrocrack), after which the problem is really one of fracture mechanics, not of continuum damage mechanics.

One observes that, regardless of the nature of the damage at the micromechanical level, materials do suffer damage when loaded, even at low levels of applied load. For example, in concrete cylinders loaded in compression, researchers detect acoustic emissions indicative of micro damage at loads well below the compressive strength of the cylinder. In most materials, however, pre-peak damage (damage that occurs while stresses are still increasing) has only a minor effect upon the observed elastic modulus of the material. For example, in Fig. 11.1, the area between the pre-peak curve and the dotted line going from the origin to the point of peak load is negligible compared to the entire area under the stress-strain curve. (These areas represent the strain energy density dissipation due to damage.) In the same vein, in Fig. 11.1, the value of E_m is only slightly different from that of E_0. It is only after the loading reaches the strength of the material that damage begins to have a major effect upon the structural stiffness. However, by this time, in most cases, the damage has already localized to form one or more discrete cracks, and the damage is thus no longer a continuum phenomenon, as is further discussed in the next section.

11.3 Damage localization

Consider the uniaxial bar shown in Fig. 11.3(a). Parts M and N are made of identical materials, except let us assume that N is just slightly weaker in tension than M. The dotted lines in Fig. 11.3(b) show the hypothetical post-peak stress-strain relations for both parts. Assume a displacement-controlled test. As displacement Δ is applied, initially both parts have identical strains. However, as the stress in part N reaches its peak strength, it begins to unload while elongating further. Assuming the situation is static, part M and part N, loaded in series, both carry identical load P and thus have identical stress σ. However, we have never stressed part M to its strength. Therefore, part M remains undamaged and accordingly shortens, while damage in part N is increasing as it lengthens.

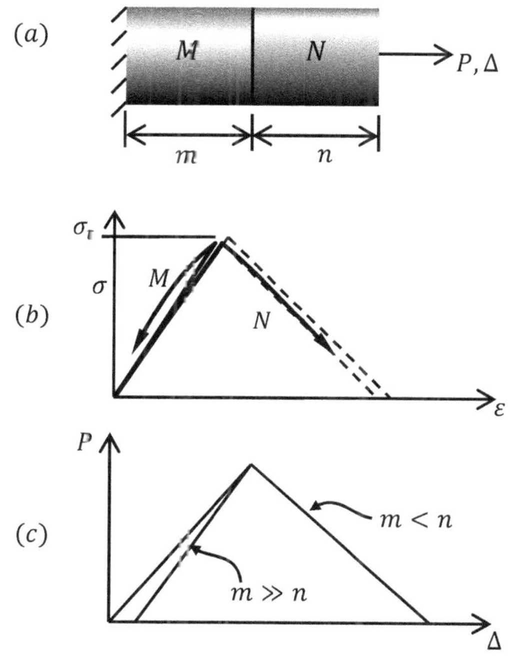

Figure 11.3. Bar showing localization of strain after reaching peak strength.

Let us consider part N now. We can divide part N into two parts with slightly differing strengths, just as the original bar was subdivided. The damage will localize within one of the parts, while the other part will unload elastically.

Thus, in a recursive process, we can think of each damaged part, no matter how small, as being composed of two parts with slightly differing strengths. We can carry out this recursion process an infinite number of times, and thus we see that the damage will ultimately localize to a part of zero length and thus of zero volume. We call this "damage localization". If the damage localizes to a domain of zero volume, then the energy dissipated (assuming that it is equal to the dissipated strain energy density times the volume of damaged material) must be zero, as shown next.

The "fracture energy", or the energy dissipated in creating surface area, is equal to the area under the stress-strain curve shown in Fig. 11.3(b) integrated over the volume of the damaged portion, N, of the structure:

$$W_{fracture} = \int_{vol\,N} \int_{\varepsilon=0}^{\infty} \sigma d\varepsilon \, dvol \,. \tag{11.7}$$

The integral of the stress-strain curve is finite, while the volume of part N approaches zero as reasoned above. Thus, assuming no pre-peak damage, the work of fracture must be zero according to the model shown in Fig. 11.3.

Another way to see that the work of fracture is null is through the overall load-deformation behavior of the bar, shown in Fig. 11.3(c). As can be seen, depending upon the lengths of the two parts of the bar, the overall load-deformation behavior subsequent to achieving the peak load is dependent upon the lengths of the two parts. The fracture energy, or energy lost in the process of damage is equal to the area under the P versus Δ curve. As can be seen from Fig. 11.3(c), the fracture energy depends upon the length n of the damaged portion of the bar. As the length n of the damaged part goes to zero, the energy dissipated due to fracture also goes to zero.

The situation is actually even more serious than just the fact that damage localizes. The part must fail in a dynamic manner, in a process called "snap back". Just as when one stretches a rubber band too far it will break and snap back with considerable dynamic energy, in any body where the post-peak energy dissipation rate with respect to damage formation is less than the rate at which elastic energy is released from the body, dynamic energy will be released. As we have seen, the post-peak energy dissipation rate is zero due to strain localization to a domain of zero volume.

As long as the stress-strain curve continues to slowly rise (prior to reaching the peak strength), volumetrically distributed damage will evolve in a quasi-static process. However, as soon as the stress-strain curve acquires a negative slope, material instability ensues. Therefore, we see that any post-peak stress-strain curve (with negative slope) is meaningless, because post-peak damage immediately localizes into a region of zero material volume and infinite strain, regardless of the nature of the post-peak stress-strain relationship.[c]

Interestingly, stress-strain curves showing a peak strength followed by post-peak softening regime appear throughout the engineering literature. How did researchers develop these relationships from experiments, if we say that post-peak deformation is necessarily discontinuous, and thus strain is undefined? The answer is that in a laboratory test, using a very stiff tensile testing machine, applied displacement Δ is applied to a tension specimen of length L and cross-sectional area A, while the load P is measured using a load cell. Researchers often then plot stress defined as $\sigma = P/A$ versus strain defined as $\epsilon = \Delta/L$. However, after peak load is reached, the strain in the specimen is nonuniformly distributed along its length, localizing at one cross-section along the length of the specimen, and thus the calculation of strain, $\epsilon = \Delta/L$, is wrong. Invariably, regardless of the stiffness of the testing machine, at some point snap-back occurs, the specimen suddenly snaps into two pieces, and thus the stress-strain curve terminates before the load P reduces to zero. Had the researchers used a longer specimen, the post-peak stress-strain curve would have been lower. Thus, the post-peak stress-strain curve is non-objective with respect to specimen length.

However, solids made of real materials exhibit finite fracture energy. The unrealistic situation of damage localization with consequent infinite strain and predicted null fracture energy dissipation results in a discontinuous deformation solution that violates the fundamental assumption of continuum mechanics.

To repair this situation, researchers have devised various strategies to avoid localization in an effort to maintain the continuum paradigm. The most common method is to assume that the stress and the damage at a material point depend not upon the strain at that point, but rather upon the strain (and perhaps other variables) averaged over a finite,

---

[c] Indeed, the very concept of "material volume" breaks down when considering the mesoscale, because at this scale the concept of "material" breaks down.

prespecified neighborhood of the point. Researchers call methods of this type "nonlocal continuum damage mechanics". (Bazant, Z. P. and Planas, J. P., 1998), in Chapter 13, present in detail various nonlocal continuum damage models. These nonlocal continuum damage models require the finite element mesh to be fine enough to represent a smooth variation of damage (and all other responses) across the width of the fracture process zone, which might be very small. So, even though the concept of strain breaks down within the fracture process zone, nonetheless many finite elements are required to capture the imaginary smoothly varying strain and damage fields within this zone. This need for a very fine finite element mesh becomes particularly burdensome when one considers that engineers should conduct finite element convergence studies to assure a converged numerical solution to the mathematically posed continuum mechanics problem. Additionally, most engineers are not concerned with details of what happens within the fracture process zone; rather, they just want to be certain that they have adequately characterized the macroscopic structural behavior, including the energy release rate of discrete cracks.

Implementing a localization limiter within a finite element program is nontrivial. The stress at each Gauss point within a finite element is now a function not just of the strain at that Gauss point, but also of the strains at other Gauss points within the element as well as of the strains at Gauss points within adjacent finite elements. The stress at a Gauss point may possibly even depend upon responses within non-adjacent finite elements (depending upon the size of the localization limiter compared to the size of the finite elements in the region). In addition, the constitutive behavior at Gauss points close to material domain boundaries requires special treatment.

On the other hand, (Bazant, Z. P. and Oh, B.-H., 1983) have developed the "blunt crack band model", in which the material constitutive relation for a given finite element is made a function of the size of the finite element. This approach suffers from several problems. Firstly, in a notorious problem with displacement-based finite elements, element locking may occur when the crack does not line up with the finite element it traverses. Secondly, if the finite element is large compared to the realistic size (width or length) of the fracture process zone, then snap-back within the finite element can occur, in which it is impossible to adequately represent the correct fracture energy dissipation rate. In Chapter 8 of (Bazant, Z. P. and Planas, J. P., 1998), various remedies to these problems are tentatively suggested, but these remedies are ad-hoc and unreliable.

There is a simpler way to model distributed damage while still avoiding spurious localization and while adequately characterizing the fracture energy of propagating discrete cracks. In the next two sections, we develop two damage models; the first is suitable for bond-based micropolar peridynamic lattices, and the second is suitable for the state-based peridynamic lattice model (SPLM) described in Chapter 8.

11.4 Damage model for micropolar peridynamic lattice

One of the simplest approaches to model damage in a lattice model was taken by (Schlangen, E. and Van Mier, J.G.M., 1992), in which linear elastic frame finite elements are simply removed from the (linear elastic frame element) lattice when the maximum stress (including both axial and bending stresses) in an element exceeds a specified tensile strength. This approach has the negative consequence of unphysically removing mass from the structure, as well as of manifesting a reasonably strong directional lattice bias. Additionally, with this approach it is difficult to control the fracture energy dissipation rate, except via changing the lattice spacing.

In describing this approach, (Man H.-K and Van Mier, J.G.M., 2007) are interested in comparing the behavior of random frame lattices with that of regular lattices. Using these lattices, they simulate the meso-structure of the material, explicitly modeling cement matrix, stone aggregate, and an "interfacial transition zone". They did observe a size effect, but their simulation results, being quasistatic and leaving out the time dimension, are only somewhat physically accurate. This approach of modeling the meso-structure explicitly using lattice networks has been used by many investigators in the 1990's and onward, and is philosophically different from the approach taken in this book, in which every effort is made to *avoid* explicit modeling of the mesoscale structure of the material.

Gerstle and his students (Gerstle, W. H. and Sau, N., 2004), (Gerstle, W., Sau, N., and Silling, S., 2005), (Gerstle, W. H., Sau, N., and Aguilera, E., 2007), (Gerstle, W., Sakhavand, N., Rahman, A., and Tuniki, B.K., 2011), (Gerstle, W., Honarvar Geitanbaf, H., and Asadollahi, A., 2013) have written about peridynamic modeling of concrete structures. In their early work, they assumed a tensile bond-based damage model of the type shown in Fig. 11.4. In this simple model, the bond is essentially a linear elastic spring with a sudden loss of force upon attaining a specific stretch. This is very similar to the early damage models of van Mier (Schlangen, E. and Van Mier, J.G.M., 1992). While the model is appealing for its simplicity, it suffers from the following shortcomings:

1. There is no way, other than by altering lattice spacing, independently to control the fracture energy of a propagating crack.
2. The model is not state-based. Damage in a bond depends only upon the states of the two particles interacting through that bond, and does not depend upon the surrounding material environment.
3. The method is not sensitive to volumetric deformation.
4. The model does not accurately represent compressive failure.
5. The method is nonobjective with respect to lattice orientation.

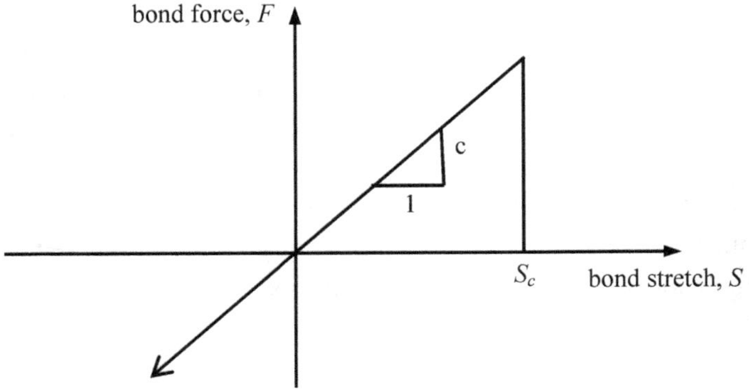

Figure 11.4. Micro elastic peridynamic model with damage. This model governs the forces between two particles situated within the material horizon, δ, of each other.

Nonetheless, the model showed promise in that it could simulate complex tensile cracking behavior, as demonstrated by Fig. 11.5.

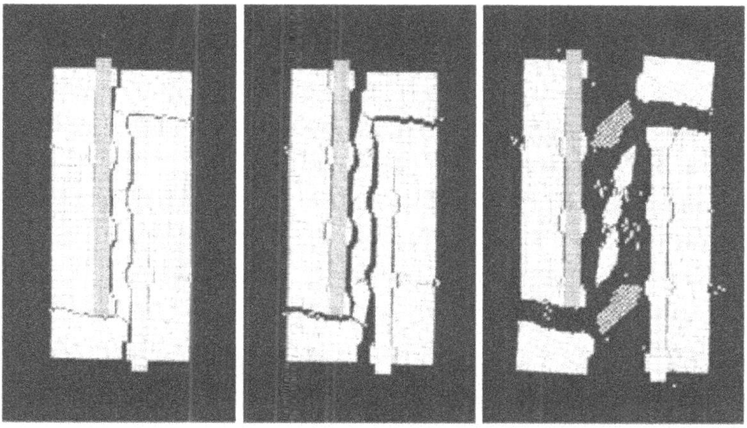

(a) (b) (c)

Figure 11.5. Magnified deformed shapes of splice of reinforcing bars in concrete at three stages (lattice spacing is 0.03 m). Ref: (Gerstle, W. H. and Sau, N., 2004).

(Gerstle, W., Honarvar Geitanbaf, H., and Asadollahi, A., 2013) proposed a reasonably simple, micropolar peridynamic damage model with two damage parameters, one for tension and one for compression, of the form

$$\{f\} = (1 - \omega_t)\,[K]\{d\} \quad \text{in the tensile regime, and} \tag{11.8}$$

$$\{f\} = (1 - \omega_c)\,[K]\{d\} \quad \text{in the compressive regime,} \tag{11.9}$$

where $\{f\}$ is the force vector (including micropolar axial forces, shear forces, and moments) acting between particles P_i and P_j, $[K]$ is the linear elastic stiffness matrix, $\{d\}$ is the vector of particle deformations, and ω_t and ω_c are the two damage parameters, for tension and for compression, respectively, associated with the interaction between the particles. Because there are many interactions per particle, this form allows damage to be anisotropic.

With reference to Fig. 11.6, the micropolar axial stretch of bond ij is

$$\epsilon_a \equiv \frac{d_x^j - d_x^i}{d}. \tag{11.10}$$

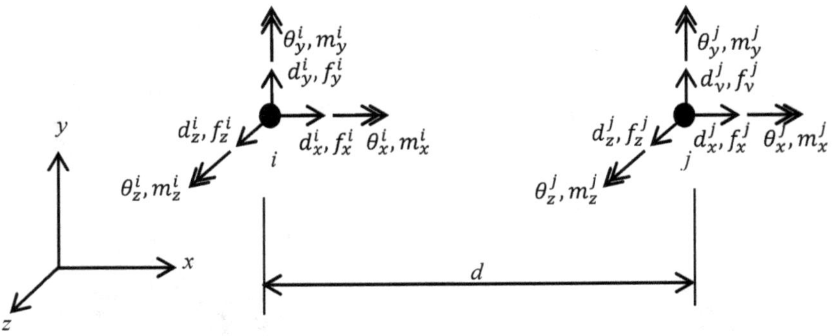

Figure 11.6. Generalized displacement and force components, in local coordinates, acting between particles P_i and P_j, separated by reference distance, d.

Similar to Euler-Bernoulli beam theory, with reference to Fig. 11.6, the maximum micropolar "curvatures" about the local z, y and x axes, respectively, of interaction $i\,j$ are:

$$\begin{gathered}
\psi_z \equiv max\left[\left|\frac{2}{d}\left(2\theta_z^i + \theta_z^j - \frac{3}{d}(d_y^j - d_y^i)\right)\right|, \left|\frac{2}{d}\left(2\theta_z^j + \theta_z^i - \frac{3}{d}(d_y^i - d_y^j)\right)\right|\right],\\
\psi_y \equiv max\left[\left|\frac{2}{d}\left(2\theta_y^i + \theta_y^j - \frac{3}{d}(d_z^j - d_z^i)\right)\right|, \left|\frac{2}{d}\left(2\theta_y^j + \theta_y^i - \frac{3}{d}(d_z^i - d_z^j)\right)\right|\right],\\
\psi_x \equiv \left|\frac{\theta_x^j - \theta_x^i}{d}\right|.
\end{gathered} \tag{11.11}$$

The model specifies measures of micropolar tensile and compressive deformation as

$$\epsilon_{mp+} \equiv \epsilon_a + \beta d\sqrt{{\psi_x}^2 + {\psi_y}^2 + {\psi_z}^2}, \text{ and } \epsilon_{mp-} \equiv \epsilon_a - \beta d\sqrt{{\psi_x}^2 + {\psi_y}^2 + {\psi_z}^2}, \tag{11.12}$$

where β is a dimensionless model parameter (whose value is determined below). Thus, both axial deformation and bending deformation of the bond are included in these measures.

The damage parameters ω_t and ω_c, are defined in terms of these deformation measures, with reference to Fig. 11.7, as follows:

$$
\begin{aligned}
&\text{For tension damage:} \\
&\quad \text{for } \epsilon_{mp+} \leq \epsilon_t, && \omega_t = max(0, \omega_{tprev}) \\
&\quad \text{for } \epsilon_t \leq \epsilon_{mp+} \leq \alpha_t\epsilon_t, && \omega_t = max(\Omega_t(\epsilon_{mp+}), \omega_{tprev}) \\
&\quad \text{and for } \alpha_t\epsilon_t \leq \epsilon_{mp+}, && \omega_t = 1, \\
&\text{and for compression damage} \\
&\quad \text{for } \epsilon_{mp-} \leq \alpha_c\epsilon_c, && \omega_c = 1 \\
&\quad \text{for } \alpha_c\epsilon_c \leq \epsilon_{mp-} \leq \epsilon_c, && \omega_c = \max(\Omega_c(\epsilon_{mp-}), \omega_{cprev}) \\
&\quad \text{and for } \epsilon_c \leq \epsilon_{mp-} && \omega_c = \max(0, \omega_{cprev}),
\end{aligned}
\tag{11.13}
$$

(a)

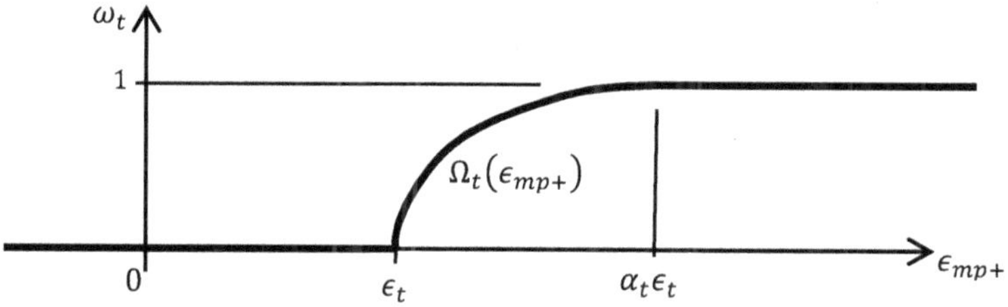

$$\Omega_t(\epsilon_{mp+}) = 1 - \frac{\epsilon_t}{(\epsilon_t - \alpha_t\epsilon_t)^3}\left(\epsilon_t^3(3\alpha_t^2 - \alpha_t^3)/\epsilon_{mp+} - 6\alpha_t\epsilon_t^2 + 3\epsilon_t(1+\alpha_t)\epsilon_{mp+} - 2\epsilon_{mp+}^2\right)$$

(b)

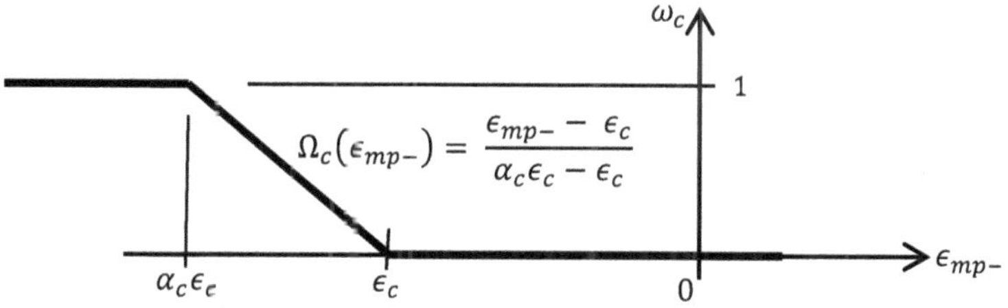

Figure 11.7. (a) Damage, ω_t, versus the micropolar strain measure, ϵ_{mp+}. (b) Damage, ω_c, versus the micropolar strain measure, ϵ_{mp-}. (ω_t and ω_c never decrease with time.)

where ω_{tprev} and ω_{cprev} are the values of the tensile and compressive damage parameters, respectively, for the bond in the immediately preceding time step. Fig. 11.7 defines functions $\Omega_t(\epsilon_{mp+})$ and $\Omega_c(\epsilon_{mp-})$. Additionally, for the plane strain and three-dimensional models, the authors assume that nonzero compression damage ω_c can develop in interaction ij only after nonzero tensile damage attached to any bond attached to either particle P_i or particle P_j has first occurred. This precludes the development of unrealistic hydrostatic crushing.

It may be possible to dispense with the compressive damage parameter ω_c completely. If an interaction is in compression, there is a possibility of geometric instability at the peridynamic (particle) level. This instability can only arise if the stiffness matrix is dependent upon current particle geometry. After enough transverse interactions fail in tension (due to Poisson's ratio transverse strain and consequent tensile damage), the interaction in compression will fail at some critical load, by geometric instability. This mechanism – transverse tensile damage followed by compressive instability may be a somewhat realistic model for compressive and shear failure in quasibrittle materials.

The constitutive model presented by (Gerstle, W., Honarvar Geitanbaf, H., and Asadollahi, A., 2013) has eight parameters: peridynamic lattice spacing parameter L, two micro-elastic stiffness parameters c^*, and d^* given in Eqs. 7.40-7.41, and the five parameters governing damage evolution: ϵ_t, α_t, ϵ_c, α_c, and β.

An engineer must calibrate all of these damage parameters against laboratory test data. The parameter ϵ_t is calibrated to reproduce the tensile strength, f_t, of the concrete: $\epsilon_t \approx \frac{f_t}{E}$. The parameter, $\beta \approx 0.1$, is chosen to replicate the ratio of uniaxial compressive load to uniaxial tensile load, usually around ten, as is observed empirically for normal-strength concrete. The parameter $\epsilon_c \approx 0.001$ replicates the strain at which uniaxial compressive failure commences, and $\alpha_c \epsilon_c \approx 0.003$ represents the ultimate compressive strain. Finally, the parameter α_t replicates the tensile fracture energy G_F of the material.

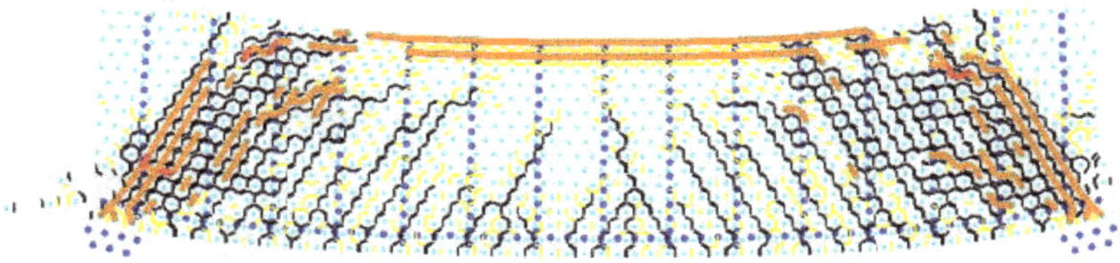

Figure 11.8. Magnified deformed shape of a uniformly loaded simply supported reinforced concrete beam with flexural steel and stirrups. Ref: (Gerstle, W., Honarvar Geitanbaf, H., and Asadollahi, A., 2013).

The model described appears to give reasonable results for the behavior of concrete beams, as demonstrated by (Gerstle, W., Honarvar Geitanbaf, H., and Asadollahi, A., 2013), and as shown by the promising crack pattern showed in Fig. 11.8. However, this model seems somewhat ad-hoc, and the micropolar aspect of the model is more complex than necessary to capture the essential behavior of concrete. Additionally, one cannot readily add isochoric plasticity to this micropolar model.

In the next section, we present a conceptually simpler isotropic damage model within the framework of the state-based peridynamic lattice model.

11.5 Isotropic SPLM damage mechanics

Softening tensile damage, unlike hardening plasticity and hardening damage, is essentially a localized phenomenon. At the mesoscale, tensile damage results when grains and chemical bonds of one form or another, (ionic, covalent, or Van der Wahl's bonds), break. Because these grains and bonds are random and anisotropic at the mesoscale, it is impossible precisely to predict the direction of a broken bond. Indeed, in actual materials, the character of the damage can be both quite random and quite morphologically complex, and this is what causes the observed tortuosity of cracked surfaces. The morphology of these surfaces can indicate a lot about the nature of the damage at the mesoscale. However, structural engineers are not generally interested in the details of crack morphology, and instead attempt to hide these details using the regularizing simplifications of fracture mechanics and damage mechanics.

In local continuum damage mechanics, we have assumed that the damage parameter at a point depends upon either the stress tensor or the strain tensor at that point. In nonlocal continuum mechanics, the damage at a point depends upon the stress or strain averaged over a neighborhood of the point.

In the micropolar damage model described in the previous section, damage was associated with the bond and not with the particle. As there are many more bonds than particles (eighteen more, in three-dimensional FCC lattice models including second-nearest neighbors), this requires perhaps more damage parameters to be stored in the computational model than are necessary. If instead we associate a single damage parameter ω with a lattice particle (rather than with each bond), we need compute and store less data, and the essential features of damage behavior might still perhaps be adequately represented.

With the SPLM, we use a three-dimensional close-packed FCC lattice body $\mathcal{L}_{\mathcal{R}}$. Although the lattice has some symmetry, it still has an orientation, and we would like to avoid directional bias that might reflect this lattice orientation in describing the onset of damage. In addition, once damage has started, cracks propagate, and we would like the crack propagation direction to be objective with respect to lattice orientation. Similarly, both the energy release rate and the energy dissipation rate of a propagating crack should be independent of lattice orientation.

The simplest way to ensure that the initiation of damage is independent of lattice rotation is to assume that the damage parameter is a scalar function of some scalar quantity that is independent of lattice rotation. At a material lattice particle P_i, when we assume scalar damage, we assume that the stiffness of all bonds connected to that point will soften identically as damage evolves.

One scalar measure of the deformation state is the average elastic stretch in all bonds connected to a given lattice particle P_i:

$$S^e_{Avg} = \frac{1}{N_b}\sum_{j=1}^{N_b} S^e{}_j, \tag{11.14}$$

where N_b is the number of bonds connected to the lattice particle, and $S^e{}_j$ is the elastic component of the stretch of bond i. It can be shown that for a close-packed FCC lattice modeling a linear elastic material subjected to a specified spatially homogeneous deformation, S^e_{Avg} is independent of lattice rotation. It is physically reasonable to assume that when S^e_{Avg} is sufficiently high damage ω will initiate and evolve.

In the case of uniaxial stress, using the three-dimensional FCC SPLM, we can show that the uniaxial strain ϵ_{axial} relates to the average elastic stretch S^e_{Avg} as

$$\epsilon_{axial} = S^{e}_{Avg}\left(\frac{3}{1-2\nu}\right) . \tag{11.15}$$

We now define the *critical average stretch* $S_{AvgCrit}$ as

$$S_{AvgCrit} = \epsilon_t\left(\frac{1-2\nu}{3}\right) = \frac{\sigma_t}{E}\left(\frac{1-2\nu}{3}\right) . \tag{11.16}$$

The critical average stretch $S_{AvgCrit}$ provides a measure for deciding when damage commences. S^{e}_{Avg} provides a single quantity similar to the definition of the Von Mises stress: a one-dimensional simplification of a more complex stress state.[d]

Furthermore, to enable damage evolution once damage has initiated, we define an "equivalent crack opening displacement" $COD_{eq} = S_{TotMax}L$, where S_{TotMax} is the maximum total stretch (elastic plus plastic) of any bond connected to the particle under consideration, and L is the lattice spacing.

We assume that tensile damage ω evolves as follows:

$$\begin{aligned}
&\text{if } S^{e}_{Avg} < S_{AvgCrit} \quad \omega = \max(0, \omega_{prev}) \\
&\text{else if } COD_{eq} \leq COD_c \quad \omega = \max\left(1 - \frac{\gamma\frac{\sigma_t L}{E\,COD_{eq}}(COD_c - COD_{eq})}{COD_c},\ \omega_{prev}\right) \\
&\text{else } \omega = 1 ,
\end{aligned} \tag{11.17}$$

where the parameters σ_t, E, COD_c, and γ are illustrated in Fig. 11.9. Eqs. 11.8, 11.16, and 11.17 define an "elastic/cohesive" SPLM damage model. This model describes the linear softening relation shown in Fig. 11.9.

In addition to the parameter γ shown in Fig. 11.9, the linear cohesive softening model is defined in terms of tensile strength σ_t, and the critical crack opening displacement COD_c. Thus,

---

[d] To clarify, we are specifying that damage commences when the average elastic stretch S^{e}_{Avg} reaches a critical value $S_{AvgCrit}$, which we have calibrated using the classical uniaxial tensile strength.

$$\epsilon_t = \frac{\sigma_t}{E} \text{ and } COD_{eq} = S^e{}_{max}L \,. \tag{11.18}$$

In our model, the damage parameter ω immediately jumps to a finite value once damage commences – damage "pops in". Prior to the initiation of damage, the force state is given by the state-based elastic peridynamic model, Eq. 9.28 $\{F\} = [K]\{S^e\}$, and immediately after reaching the tensile strength, the force in each bond B_j is determined by a bond based damage model based upon Eq. 7.2 : $F_j = (1 - \omega_{avg})cS_j^e$, where ω_{avg} is the average damage of the two particles connected by the bond.

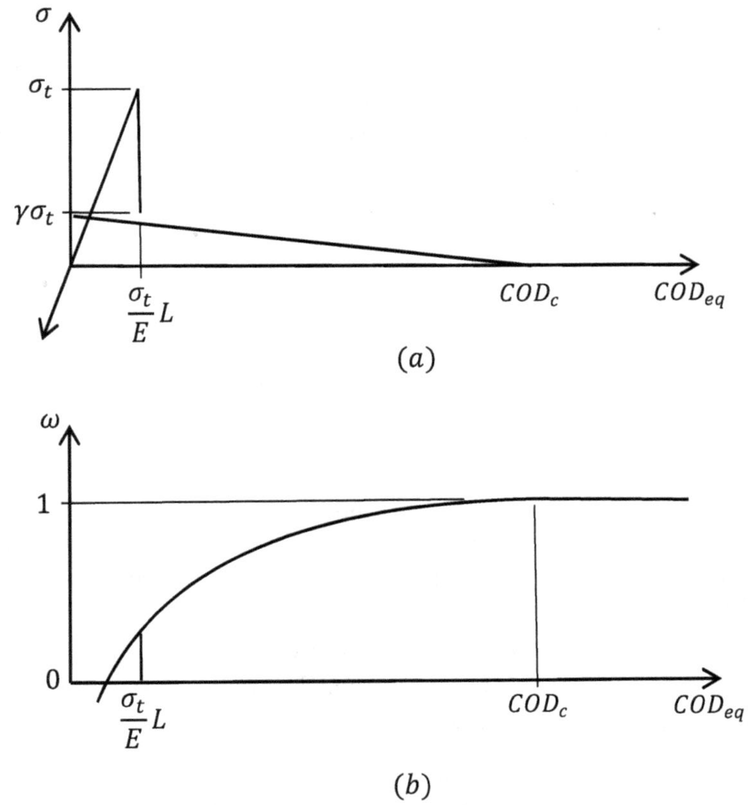

Figure 11.9. (a) Stress versus equivalent crack opening displacement; (b) damage versus equivalent crack opening displacement.

Of course, it is possible that in a very spatially non-uniform deformation field, individual bonds connected to a given particle might be highly stretched and others highly compressed, resulting in a low average elastic stretch S^e_{Avg}. Such would be the case for a point load quasistatically applied to a single interior lattice particle within a large lattice fixed at its outer boundaries. [e] Having zero average elastic stretch S^e_{Avg} (due to antisymmetry of the applied loading), the loaded lattice particle would have null S^e_{Avg} and would therefore not be damaged, but on the other hand, the neighboring lattice particle on the tension side of the applied point load would suffer high S^e_{Avg}, and would be damaged at some level of applied load. Thus, making the initiation of damage ω a function of S^e_{Avg} appears to make sense even for the case of highly non-uniform deformation-state and force-state fields.[f] After damage commences, further damage evolution depends upon the maximum elastic stretch $COD_{eq} = S^e{}_{max}L$.

Notice that with this model, initiation of tensile damage depends upon the volumetric tensile elastic strain (and associated hydrostatic stress), but not upon the maximum principal strain (and associated maximum principal stress). Because S^e_{Avg} is invariant with respect to lattice rotation under homogeneous deformation environments, the initiation of damage is also invariant with respect to lattice rotation. However, subsequent to damage initiation, there will unfortunately be some sensitivity to lattice orientation. This is the cost in choosing to include only eighteen neighbors in our peridynamic constitutive model.

To repeat, the tensile damage model described in this section is perfectly isotropic in a homogeneous deformation field only up to the load at which damage initiates. After damage initiates, microcracks will tend to coalesce and a macrocrack will eventually propagate in a direction that is perpendicular to the general direction of maximum principal stress. The direction of crack growth is constrained by the lattice orientation. This level of non-objectivity may be acceptable in

---

[e] This is analogous to the Kelvin problem of a point load applied to an infinite elastic continuum.

[f] Rather than S^e_{Avg}, F_{Avg} defined by Eq. 10.20 could instead have been chosen as an indicator of the onset of damage. As damage evolves in post-peak situations, it seems simplest to use $COD_{eq} = S^e{}_{max}L$ as the indicator of damage evolution.

many engineering situations. If one seeks more objectivity in crack propagation direction, then one would need to increase the peridynamic horizon with respect to lattice spacing. However, then the method would require many more force computations per time step – one must strike a judicious balance between simplicity and accuracy.

With this model, if the damage parameter ω associated with a particle P_i is equal to unity, particle P_i becomes completely detached from the material body, and no longer interacts with its neighbors through material bond forces.[g]

11.6 Relationship to fracture mechanics

Linear elastic fracture mechanics (LEFM), described in Chapter 4, is the discipline which considers the extreme situation where all damage and plasticity is localized within a propagating fracture process zone (FPZ) of negligible size compared to other dimensions of the material body. In other words, LEFM considers the behavior and the propagation of a crack in an infinitely large linear elastic body.

Assume that a given elastic material has a damage model. The linear elastic SPLM was described in Chapter 9, and is represented by the material parameters a, b, and L. The parameters of the SPLM damage model are tensile strength σ_t, the parameter γ, and the critical crack opening displacement COD_c. These parameters must be chosen in such a manner that the LEFM fracture energy release rate G_F is replicated correctly in a problem for which the FPZ is small compared to other problem dimensions. So if the fracture energy is given for a material, the

---

[g] However, a re-contact force may come into play if a damaged particle comes close enough to its neighboring particles, damaged or undamaged. A simple way to simulate particle re-contact, once damage is complete, is to assume that if the elastic bond stretch is positive, the bond force is null. On the other hand, if the elastic bond stretch is negative, a prescribed force function of bond stretch prevents the particles from interpenetrating.

values of σ_t, COD_c, and γ must be properly chosen to produce the correct fracture energy G.[h]

Even with these constraints, there are many combinations of COD_c and γ that can reproduce the correct behavior of a linear elastic LEFM material. The length of the FPZ might be another physical feature that one might want faithfully to replicate. This puts one more constraint upon the damage parameters, but it appears that the chosen parameters are sufficient accurately to represent even nonlinear cohesive materials like concrete.

An engineer can certainly create more involved damage models within the framework of SPLM as necessary to model unusual materials.

11.7 Damage due to excessive plastic deformation

If the average tensile stretch is low enough, tensile damage will not occur, but plastic flow may occur if the norm of the deviatoric component of the force state is sufficiently high, as described in Chapter 10. However, plastic flow cannot continue forever; at some point, damage will occur due to excessive plastic flow. A simple model for damage due to excessive plastic flow is to assume that damage evolution commences once some measure of the plastic stretch, say $\|S^p_{tot}\|$, reaches a critical value, call it S_{1plast} and is total when the plastic stretch reaches the critical value S_{2plast}. This type of damage has not yet been implemented by the author.

11.8 Summary

In this chapter, we have presented damage mechanics. We started with a brief outline of the history of the development of damage mechanics, and then presented the theory of continuum damage mechanics. We found that when the material strength is decreasing as damage increases, damage is an unstable phenomenon that tends to

[h] If a plasticity model is also included, the FPZ may also include the localized plasticity at the crack tip, and the fracture energy G must reflect plastic energy dissipation as well as damage energy dissipation.

localize as far as the mesomechanical structure of the material will allow. With a lattice-based description of the material body, damage cannot localize to a level of detail that is smaller than the lattice spacing, and damage localization is thus limited to the lattice spacing. In this sense, the lattice provides an intrinsic localization limiter to the damage model.

We then presented a damage model for a bond-based peridynamic lattice, focusing our attention on a micropolar lattice model. This damage model allowed damage to evolve as a function of bond stretch, which included bond "curvature" as well as axial stretch. Although this damage model appears to have some predictive capability, it is more complicated than appears to be necessary, and is an inappropriate model for implementation together with plasticity.

Finally, we presented a damage model for the state-based peridynamic lattice model. Our operating principle was to keep the model as simple as possible while still providing reasonable predictive power. Damage initiates when the average elastic stretch S^{e}_{Avg} reaches the critical value $S_{AvgCrit}$ described by Eq. 11.16. Subsequent to the initiation of damage, further evolution of damage depends upon an "equivalent crack opening displacement" COD_{eq} described by Eq. 11.17. This simple model appears to provide a reasonable basis for predicting damage initiation and subsequent tensile cracking. Note that, using this model, damage "pops in" suddenly, causing localized dynamic snap back events similar to the acoustic emissions observed in the laboratory.

Damage can also occur in response to excessive plastic deformation, and can potentially be modeled as described in Section 11.7.

The simulation examples of reinforced concrete structures presented in Chapter 14 illustrate the behavior of the SPLM damage models developed in this chapter.

11.9 Exercises

11.1 A beam of rectangular cross-section of depth d and width w is made of a material that is linear elastic with axial tensile damage behavior as shown in Fig. 11.10. Determine the ultimate bending strength M_u of the beam if (a) $E_1 = 0.1E$; (b) $E_1 = 0$; (c) $E_1 = -E$; (d) $E_1 = -\infty$. (Hint: assume plane sections remain plane.)

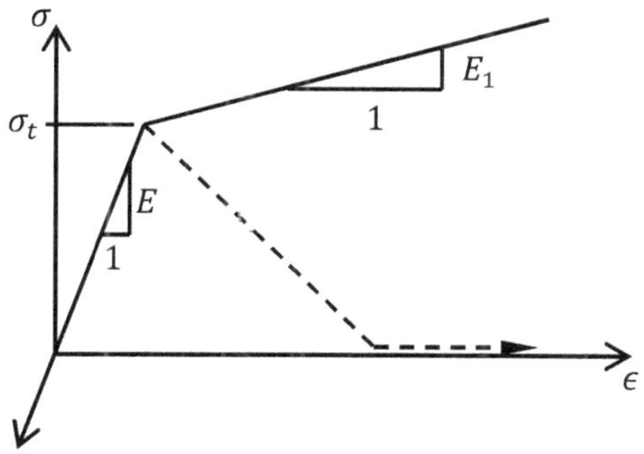

Figure 11.10. Stress-strain relation for Problem (11.1).

11.2 A plane-stress plate of thickness t_b and of infinite size has a long straight central crack of length $2a$ sawn into it. Assume that the fracture process zone is a band of width w_c in the plane of the plate. A remote tensile stress σ_0 acting transversely to the crack is applied. Inside the fracture process zone, assume that the strain is uniform in the direction transverse to the crack. Outside the fracture process zone, assume that the material exhibits linear elastic behavior. What will be the amount of energy dissipated per unit area of crack growth G_F if the material behavior is as shown in Fig. 11.10? Assume (a) $E_1 = 0$; (b) $E_1 = -E$; (c) $E_1 = -\infty$.

11.3 Show that $S^e_{Avg} = \frac{1}{N_b}\sum_{j=1}^{N_b} S^e{}_j$ defined by Eq. 11.14 for a close-packed FCC lattice modeling a linear elastic material subjected to a specified spatially homogeneous deformation is independent of lattice rotation.

11.4 For a close-packed FCC lattice modeling a linear elastic material subjected to a specified spatially homogeneous deformation, show that $\epsilon_{axial} = S^e_{Avg}\left(\frac{3}{1-2\nu}\right)$ (defined by Eq. 11.15).

11.5 Write a MatLab function that plots ω versus COD_{eq} as shown in Figs. 11.9(b) given the material parameters in Eqs. 11.17.

11.6 Using the damage model described in Section 11.5, and using Eqs. 11.17, what is the approximate energy required to completely break a bond of length L?

11.7 For a plane stress hexagonal lattice, using the damage model described in Section 11.5, what is the minimum number of bonds that must be broken per particle to create a discrete crack running through a lattice body?

11.8 For a plane stress hexagonal lattice, using the damage model described in Section 11.5, what is the approximate work of fracture per unit crack area, G_F as a function of crack angle θ with respect to the X axis?

11.9 Apply the damage model described by Eqs. 11.17 and Fig 11.9 to a one-dimensional rod of cross sectional area A_{rod} and of length L_{rod}. Qualitatively describe the deformation behavior of the rod subsequent to initiation of damage if quasistatic conditions apply up to the point of initiation of damage. Assume the initial damage occurs at (a) the middle of the rod; (b) at one end of the rod.

Chapter 12

Particle Dynamics

Chapter objectives are to:

- study the equations of particle motion;
- discretize the equations of particle motion;
- develop a method to integrate the equations of motion in time;
- derive the critical time step for a material;
- present damping models; and to
- investigate dynamic relaxation methods to arrive at a static solution.

The purpose of this chapter is to present methods for solving the peridynamic equations of particle motion, Eqs. 5.5 and 8.30. These second-order differential equations represent the application of Newton's second law, $\boldsymbol{F} = m\boldsymbol{a}$, to each particle in each lattice body $\mathcal{L}_{\mathfrak{R}}$.

We do not assume that the equations of motion are linear, and thus we use an explicit time integration approach[a] to solving the equations of motion. This requires a small-enough time step, and we address the question of determining the largest permissible time step, or *critical time step* necessary to obtain a valid dynamical solution to the problem.

Every material exhibits internal damping, in which energy is dissipated, and we address the question of how to model this internal material damping realistically. In addition, vibrating structures lose energy by radiating sound waves into the surrounding medium, be it air or water. We discuss how to model this type of energy loss as well.

Finally, we are often primarily interested not in the dynamic behavior, but in the final resting state, or the static solution to the

[a] In contrast to an implicit time integration approach, in which one may use much larger time steps, at the cost that the simulation may miss important behaviors. The implicit approach involves simultaneous solution of many linear equations within each time step.

problem. In such cases, we can use dynamic relaxation techniques to arrive at the static solution using as few time steps as possible. We must be careful when using artificial damping, however, as this artificiality can have an effect upon the static solution, and produce incorrect results for nonlinear problems.[b]

Let us begin by presenting a basic algorithm for explicit particle dynamics.

12.1 Basic algorithm for explicit particle dynamics

According to Newton's second law, the acceleration $\ddot{\boldsymbol{x}}$ of a single particle in an SPLM lattice $\mathcal{L}_{\mathcal{R}}$ is

$$\ddot{\boldsymbol{x}} = \frac{1}{m}\left(\boldsymbol{F}^{int} + \boldsymbol{F}^{ext}\right), \tag{12.1}$$

where m is the mass of the particle, $\boldsymbol{F}^{int}$ is the internal force due to interactions with neighboring particles in $\mathcal{L}_{\mathcal{R}}$, and $\boldsymbol{F}^{ext}$ is the externally applied force. Let us assume that the time step Δt and the mass m of the particle do not change with time. Also, at the beginning of the time step, time t_i, let us assume that we know the position $\boldsymbol{x}(t_i)$ and the velocity $\dot{\boldsymbol{x}}(t_i)$ of the particle. Finally, let us assume that during the time step, the forces $\boldsymbol{F}^{int} + \boldsymbol{F}^{ext}$ acting upon the particle do not change appreciably.

Integrating Eq. 12.1 with respect to time once, we obtain the velocity after the time step Δt has elapsed:

$$\dot{\boldsymbol{x}}(t_i + \Delta t) = \int_{t=t_i}^{t_i+\Delta t} \frac{1}{m}\left(\boldsymbol{F}^{int} + \boldsymbol{F}^{ext}\right) dt + \dot{\boldsymbol{x}}(t_i)\,, \tag{12.2}$$

and if we assume a very small time step, so that the integrand changes little during the time step Δt, the velocity at the end of the time step is approximately

$$\dot{\boldsymbol{x}}(t_i + \Delta t) \approx \frac{1}{m}\left(\boldsymbol{F}_i^{int} + \boldsymbol{F}_i^{ext}\right)\Delta t + \dot{\boldsymbol{x}}(t_i)\,, \tag{12.3}$$

[b] Nonlinear problems, unlike linear problems, may have multiple solutions. The artificial damping may have an effect upon which, of many, static solutions is found.

Similarly, because the time step is very small, we assume that the velocity changes very little during the time step, so that the particle position at the end of the time step is

$$\boldsymbol{x}(t_i + \Delta t) = \int_{t=t_i}^{t_i+\Delta t} \dot{\boldsymbol{x}}\, dt + \boldsymbol{x}(t_i) \approx \dot{\boldsymbol{x}}(t_i + \Delta t)\Delta t + \boldsymbol{x}(t_i)\,. \tag{12.4}$$

In other words, we are using a forward Euler integration scheme to obtain the velocity, and a backward Euler integration scheme to obtain the position. This is an explicit time integration scheme, because we compute the updated position and velocity directly from the previous position, velocity, and acceleration without having to solve any simultaneous equations, or to iterate within a time step. The solution proceeds, or is integrated, in a purely incremental fashion.

Many variants of this time integration scheme exist, including for example the Newmark beta methods (Newmark, 1959), the Runge-Kutta methods (Kutta, 1901), and the Verlet methods (Verlet, 1967). However, it appears that the method that we have described in this section is stable and accurate, and is as simple as any other explicit method.[c]

In the next section, we investigate how large the time step Δt may be to obtain an approximately correct dynamical solution.

12.2 Critical time step for a single degree of freedom system

If the time step Δt used in an explicit time integration algorithm applied to a linear system is too large, the resulting solution will not even be approximately correct; instead, the solution will "blow up" and diverge rapidly to infinity. In this section, we determine the maximum stable time step Δt_{crit} for a single degree-of-freedom linear system; then in the next section, we interpret this result in light of the fact that our problems will in general have many degrees of freedom and be nonlinear.

---

[c] Other explicit time integration methods may make additional assumptions about the smoothness of the solution, and may use states at additional earlier time steps to predict the state at the next time step. We do not use these methods, as we want to be as general as possible and we do not want to require that the states at previous time steps be stored in memory.

The equation for the unforced vibration of a single degree of the freedom linear system shown in Fig. 12.1 with mass m, damping coefficient c, and stiffness k is:

$$m\ddot{x} + c\dot{x} + kx = 0\ , \tag{12.5}$$

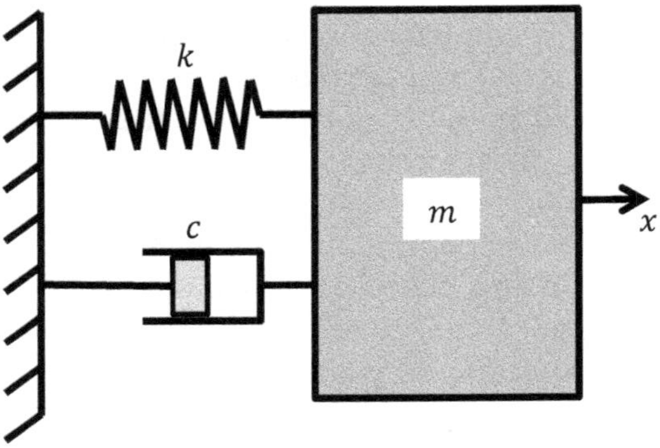

Figure 12.1. Single degree-of-freedom linear system.

where x is the position of the mass with reference to its undeformed position. Most books express this equation as:

$$\ddot{x} + 2\xi\omega_n\dot{x} + \omega_n^2 x = 0\ , \tag{12.6}$$

where $\omega_n^2 = \frac{k}{m}$ and $c = 2m\xi\omega_n$. One calls the parameter ξ the damping ratio and the parameter ω_n the natural frequency of the system. Thus the value of acceleration at time t is

$$\ddot{x}(t) = -2\xi\omega_n\dot{x}(t) - \omega_n^2 x(t)\ , \tag{12.7}$$

and the velocity $\dot{x}$ at $t + \Delta t$ is approximated using forward Euler integration from the velocity and acceleration known at time t, as described in Section 12.1, as

$$\dot{x}(t + \Delta t) = \dot{x}(t) + \ddot{x}(t)\Delta t\ , \tag{12.8}$$

and the displacement is approximated using backward Euler integration from the now known velocity at time $t + \Delta t$ as

$$x(t + \Delta t) = x(t) + \dot{x}(t + \Delta t)\Delta t\ , \tag{12.9}$$

To find the critical time step, we next express x at time $t + \Delta t$ in terms of x at time t and x at time $t - \Delta t$. To do this, first we obtain $\dot{x}(t + \Delta t)$ in terms of $x(t)$ and $\dot{x}(t)$ by substituting $\ddot{x}(t)$ from Eq. 12.7 into Eq. 12.8

$$\dot{x}(t + \Delta t) = \dot{x}(t) + \big(-2\xi\omega_n\dot{x}(t) - \omega_n^2 x(t)\big)\Delta t\ , \tag{12.10}$$

and then by substituting $\dot{x}(t + \Delta t)$ from Eq. 12.10 into Eq. 12.9, we obtain

$$x(t + \Delta t) = x(t) + \big(\dot{x}(t) + \big(-2\xi\ \omega_n\dot{x}(t) - \omega_n^2 x(t)\big)\Delta t\big)\Delta t\ . \tag{12.11}$$

Next, using Eq. 12.9 evaluated at time $t = t - \Delta t$, we find $\dot{x}(t)$ in terms of $x(t)$ and $x(t - \Delta t)$

$$x(t) = x(t - \Delta t\ ) + \dot{x}(t)\Delta t\ \ , \tag{12.12}$$

or

$$\dot{x}(t) = \frac{x(t) - x(t - \Delta t)}{\Delta t}\ . \tag{12.13}$$

Now, we substitute $\dot{x}(t)$ from Eq. 12.13 into Eq. 12.10, and then $\dot{x}(t + \Delta t)$ from 12.10 into Eq. 12.9

$$\begin{aligned} x(t + \Delta t) = x(t) + \left(\frac{x(t) - x(t - \Delta t)}{\Delta t}\right) \\ + \left(-2\xi\omega_n\left(\frac{x(t) - x(t-\Delta t)}{\Delta t}\right) - \omega_n^2 x(t)\right)\Delta t)\Delta t\ , \end{aligned} \tag{12.14}$$

and rearranging Eq. 12.14,

$$x(t+\Delta t) = (2-\omega_n^2\Delta t^2 - 2\xi\omega_n\Delta t)x(t) \\ +(-1+2\xi\omega_n\Delta t)x(t-\Delta t). \tag{12.15}$$

We write Eq. 12.15 in matrix form as

$$\begin{bmatrix} x(t+\Delta t) \\ x(t) \end{bmatrix} = \begin{bmatrix} 2-\omega_n^2\Delta t^2 - 2\xi\omega_n\Delta t & -1+2\xi\omega_n\Delta t \\ 1 & 0 \end{bmatrix} \begin{bmatrix} x(t) \\ x(t-\Delta t) \end{bmatrix}, \text{ or}$$

$$\begin{bmatrix} x(t+\Delta t) \\ x(t) \end{bmatrix} = [A] \begin{bmatrix} x(t) \\ x(t-\Delta t) \end{bmatrix}, \text{ where} \tag{12.16}$$

$$A = \begin{bmatrix} (2-\omega_n^2\Delta t^2 - 2\xi\omega_n\Delta t) & (-1+2\xi\omega_n\Delta t) \\ 1 & 0 \end{bmatrix}.$$

We can express the solution after $n+1$ time steps recursively in terms the initial conditions $x(0)$ and $x(\Delta t)$ as

$$\begin{aligned} \begin{bmatrix} x((n+1)\Delta t) \\ x(n\Delta t) \end{bmatrix} &= [A] \begin{bmatrix} x(n\Delta t) \\ x((n-1)\Delta t) \end{bmatrix} \\ &= [A]\left([A]\begin{bmatrix} x((n-1)\Delta t) \\ x((n-2)\Delta t) \end{bmatrix}\right) \\ &\ \ \vdots \\ &= [A]\left([A](\ldots\left([A]\begin{bmatrix} x(\Delta t) \\ x(0) \end{bmatrix}\right)\ldots)\right) \\ &= [A]^n \begin{bmatrix} x(\Delta t) \\ x(0) \end{bmatrix}. \end{aligned} \tag{12.17}$$

For $x(n\Delta t)$ to be bounded as n goes to infinity, matrix A^n must have bounded elements. We can use the eigenvalues of $[A]$ to evaluate the boundedness of its elements when we raise $[A]$ to the nth power as n tends to infinity. The eigenvalues λ and eigenvectors v of matrices $[A]$ and $[A]^n$ are determined by using the following equations

$$[A]\{v\} = \lambda\{v\} \text{ and we can show that } [A]^n\{v\} = \lambda^n\{v\}\,. \tag{12.18}$$

Matrices $[A]$ and $[A]^n$ thus have the same eigenvectors, but the eigenvalues of matrix $[A]^n$ are equal to the eigenvalues of A raised to the nth power.

The *spectral radius* $\rho([A])$ of a matrix $[A]$ is defined as the maximum of the absolute value of its eigenvalues, so

$$\rho([A]) = \max_i(|\lambda_i|)\text{ , and therefore } \rho([A]^n) = \max_i(|\lambda_i{}^n|). \tag{12.19}$$

To have bounded values for the elements of $[A]^n$ as n goes to infinity, the spectral radius $\rho([A])$ must be less than one.

To find the eigenvalues of the matrix $[A]$, we must solve the eigenvalue problem $\big[[A] - \lambda[I]\big]\{v\} = \{0\}$. To obtain a nontrivial solution for $\{v\}$, by a theorem of linear algebra, we require that $det\big[[A] - \lambda[I]\big] = 0$. This is called the characteristic equation of the matrix $[A]$. Evaluating the characteristic equation of $[A]$ given by the last of Eqs. 12.16, we obtain

$$\lambda^2 - (2 - \omega_n^2\Delta t^2 - 2\xi\omega_n\Delta t)\lambda + 1 - 2\xi\omega_n\Delta t = 0\,. \tag{12.20}$$

Solving this equation for the two eigenvalues using the quadratic formula, we obtain

$$\lambda_{1,2} = \frac{(2-\omega_n^2\Delta t^2 - 2\xi\omega_n\Delta t)}{2} \pm \sqrt{\frac{(2-\omega_n^2\Delta t^2-2\xi\omega_n\Delta t)^2}{4} - (1 - 2\xi\omega_n\Delta t)}. \tag{12.21}$$

The spectral radius is thus

$$\rho(A) = max\left(\left|\frac{(2-\omega_n^2\Delta t^2-2\xi\omega_n\Delta t)}{2} \pm \sqrt{\frac{(2-\omega_n^2\Delta t^2-2\xi\omega_n\Delta t)^2}{4} - (1 - 2\xi\omega_n\Delta t)}\right|\right). \tag{12.22}$$

To obtain the value of the critical time step, we set $\rho(A) = 1$. Inserting the right-hand side of Eq. 12.22 into $\rho(A) = 1$ and solving for Δt_{cr}, we obtain four possible answers

$$\Delta t_{cr} = \left(\frac{2}{\omega_n}\right)\left(\pm\sqrt{1+\xi^2} - \xi\right) \text{ and } \Delta t_{cr} = \pm 0. \tag{12.23}$$

Only one of these solutions is positive, and because the critical time step must be positive, we choose

$$\Delta t_{cr} = \left(\frac{2}{\omega_n}\right)\left(\sqrt{1+\xi^2} - \xi\right). \tag{12.24}$$

In conclusion, for a linear single degree of freedom system with natural frequency ω_n and damping ratio ξ, the time step Δt must be less than Δt_{cr} given by Eq. 12.24, to provide a bounded, and thus physically acceptable, solution.

12.3 Critical time step in terms of speed of sound

The critical time step Δt_{cr} can also be determined by considering the speed of a sound pulse of amplitude U_0 traveling in the X_1 direction at the sound speed c_0 through a linear elastic SPLM lattice body $\mathcal{L}_{\mathcal{R}}$. The shortest spatial pulse length that can be represented in a lattice body with particle spacing L is of length $\alpha_0 L$, where α_0 depends upon the length of the pulse, but is a small number; but $\alpha_0 \geq 2$, as shown by Fig. 12.2(a). A particle P_i will see a displacement of $U(t)$ as the pulse travels by it, as shown in Fig. 12.2(b).

In order to provide a stable solution, the explicit time integration method requires the time step Δt to be sufficiently small to approximate the shape of the pulse in time. We need at least 2π time steps accurately to describe the pulse in time as it travels past the particle in Fig 12.2(b). The duration of the pulse as it travels past particle P_i is $\frac{\alpha_0 L}{c_0}$. The stability limit for a linear elastic solid without damping is thus approximately

$$2\pi\Delta t_{cr} \leq \frac{\alpha_0 L}{c_0}, \text{ or setting } \alpha_0 = 2,\ \Delta t_{cr} = \frac{2L}{2\pi c_0} = \frac{L}{\pi c_0}, \tag{12.25}$$

where L is the lattice spacing and c_0 is the sound speed in the material.

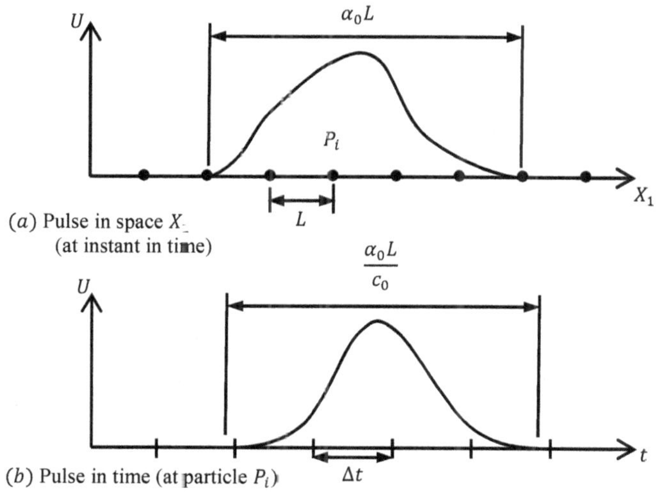

Figure 12.2. Pulse traveling through a lattice body.

In practice, we find that the critical time step to obtain a stable solution also depends upon the damping parameter, and may need to be somewhat smaller than the value given by Eq. 12.25. As indicated by Eq. 12.24, to account for the damping effect, we can set the critical time step to

$$\Delta t_{cr} = \frac{L}{\pi c_0}\left(\sqrt{1+\xi^2} - \xi\right). \tag{12.26}$$

The sound speed in a linear elastic solid is

$$c_0 = \sqrt{\frac{K}{\rho}}, \tag{12.27}$$

where K is the bulk modulus and ρ is the mass density. The bulk modulus of an elastic solid in terms of Young's modulus E and Poisson's ratio ν is

$$K = \frac{E}{3(1-2\nu)}. \tag{12.28}$$

If, during the simulation, the solution becomes unstable, the discretized body "explodes" and the simulation displays unphysically large deformations – quickly overflowing the size limit of a floating-point number. You will know it when you see it.

Even if one chooses a stable time step smaller than Δt_{cr} given by Eq. 12.26, one must realize that the dynamic solution is not exact. If a smaller time step is chosen, the dynamic solution may change, however usually only slightly. If we seek a static solution, we find that setting the time step much smaller than the critical time step typically has only a negligible effect upon the static solution. However, for nonlinear problems, the modeler does need to be aware that the time step size may have a significant effect upon the static solution (due to time-dependent variations in the distribution and sequence of plastic flow and damage), and therefore for nonlinear quasistatic problems, convergence studies (though only in the time dimension) may be necessary.

The time step may need to be even smaller for highly nonlinear problems. For example, in damage mechanics, one may need to specify very tiny time increments to ensure that progressive damage initiates and propagates in the proper sequence. If, due to an insufficiently small time step, damage initiates in a spurious location, the resulting crack propagation may also be spurious. However, this type of sensitivity also raises fundamental issues about the stability and objectivity of the physical problem one wants to solve. In addition, artificially high damping, often used to reduce the number of time steps required to obtain a static solution, may have an effect upon the static solution.

We assume that the stiffness of a material bond never becomes larger than its original stiffness. The stiffness decreases as damage progresses, while the mass remains constant, so the sound speed can never increase. In plasticity, the material stiffness does not change – only the plastic stretch. Therefore, the critical time step will never decrease, and the chosen time step, if smaller than the critical time step, will continue to be valid throughout the simulation. If we presume nothing further about the form of the internal and external forces, then it makes little sense to further refine, and complicate, our time integration method (forward difference for velocity, backward difference for position). Note that in some problems, contact of the particles originally from different bodies can occur and we can handle the contacts using "short-range forces" which are different from the "long-range forces" used to represent a solid body. In this case, the bond stiffness increases with time, so the modeler must ensure that the stiffness due to short-range contact forces does not exceed that of the virgin material (or the modeler must decrease the time step size appropriately to handle the short-range forces).

Next, we consider various types of damping used in SPLM simulations.

12.4 Damping

We may employ damping in a simulation for three distinct reasons:

(1) to represent the actual internal material damping of a solid material;
(2) to represent external damping due to interaction of a solid structure with its surrounding environment; and
(3) as a means to artificially force the dynamic simulation to arrive at a static solution as quickly as possible, in what is called the "dynamic relaxation" method.

We investigate each of these cases in the following subsections.

12.4.1 *Internal damping*

Internal material damping results from the conversion of mechanical energy to thermal energy due to the relative velocity between neighboring particles. As such, it primarily affects the highest natural frequencies ω_0 of the lattice body. The highest natural frequency of vibration in an SPLM lattice body can be determined through simulation, or it can be approximated by considering the behavior of the shortest wavelength that can be physically represented by the lattice, say, $\lambda_0 = 2L$. As such a short wave travels past a single particle at the sound speed c_0, one period of vibration T_0 takes place; thus $T_0 = \lambda_0/c_0 \cong 2L/c_0$. Therefore, the highest natural frequency of vibration of a lattice body is

$$\omega_0 \approx \frac{2\pi}{T_0} = \frac{\pi c_0}{L}. \qquad (12.29)$$

It is useful to describe internal material damping in terms of the damping ratio ξ. From Eqs. 12.5 and 12.6, the damping coefficient for the particle is $c = 2m\xi\omega_0$.

If one wishes to damp the vibration in the direction of two complementary bonds, then, although other bonds also contribute to the damping in a minor way, these two complementary bonds dominate in damping out the motion. We thus assign to each bond a damping coefficient

$$c = \frac{2m\xi\omega_0}{2} = m\xi\omega_0 \,. \tag{12.30}$$

The force matrix due to damping is therefore

$$\{F_{damp}\} = c\{v_{relAxial}\} = m\xi\omega_0[L_i]\{\dot{S}\}\,, \tag{12.31}$$

where $v_{relAxial}$ is the axial component of the relative velocity vector, and $[L_i]$ is the diagonal matrix of bond lengths defined in Eq. 9.20. Each damping force coefficient F_{dampj} acts in the deformed direction $dir\ \boldsymbol{x_r}\langle B_j\rangle$ of bond B_j; thus the material is mobile.

Eq. 12.31 is an approximate formula, but provides at least an order-of-magnitude estimate of the appropriate internal damping model for an SPLM body. When ξ is set equal to unity, critical damping should achieved, and high frequency vibrations should be entirely damped out; to simulate reasonable material internal damping, ξ is perhaps around ten percent for steel and perhaps twenty percent for concrete.[d] This is effective at damping out high frequency behavior, but is not effective at damping out the low-frequency modes of a structure.

Note that if the damping ratio ξ is set to unity, then by Eq. 12.24, the critical time step Δt_{cr} is decreased to approximately 41 percent of its undamped value, and if $\xi = 10$, the critical time step is decreased to approximately five percent of its undamped value.

External damping is effective for damping out the low-frequency vibrations, as discussed in the next section.

---

[d] These damping values are guesses. Material damping values warrant more research.

12.4.2 *External damping*

Unless a vibrating structure is in the vacuum of outer space, it radiates energy in the form of acoustic waves to the surrounding medium, whether gas, liquid, or solid. Instead of actually modeling the surrounding medium, it is often sufficient to assume absolute damping, in which we assume that the damping force acting upon a particle on the surface of the body is proportional to the velocity of the particle.

Thus, for a given particle on the surface of the body,

$$\boldsymbol{F}_{\boldsymbol{damp}} = -c_{ext}\dot{\boldsymbol{x}} = -2m\xi_{ext}\omega_{struct}\dot{\boldsymbol{x}}\,, \tag{12.32}$$

where ω_{struct} is the fundamental natural frequency of the entire structure.

We can estimate the value of the damping coefficient c_{ext} based upon the characteristics, like viscosity and density, of the surrounding medium.

12.4.3 *Artificial damping for dynamic relaxation*

Often, engineers are more interested in the static response than in the dynamic behavior of a deformable structure. In this case, the goal is to damp out the vibration as quickly as possible by judiciously choosing the damping coefficients, both internal and external. The literature calls this process of applying artificial damping "dynamic relaxation".

It is crucial to estimate the maximum, or fundamental, period of vibration of the structure, T_{struct}, which can be found in various ways. One can estimate the fundamental period T_{struct} by developing a one-degree-of-freedom spring-mass model of the structure, or by starting with a linear elastic approximation of the structure, and running the SPLM simulation for sufficient duration to observe the fundamental period and associated deformed shape emerge.

Idealizing the structure as a single-degree-of-freedom system shown in Fig. 12.1 and characterized by Eq. 12.5, the damping coefficient is $c = 2M\xi_{ext}\omega_{struct}$, where M is the total mass of the structure. Thus

$c = 2M\xi_{ext}\left(\frac{2\pi}{T_{struct}}\right) = \frac{4\pi\xi_{ext}N_p m}{T_{struct}}$. We can apply absolute external damping c_{ext} to all particles by dividing c by N_p

$$\begin{aligned} c_{ext} &= \frac{4\pi\xi_{ext}m}{T_{struct}}, \quad \text{so that} \\ \boldsymbol{F}_{damp} &= -\frac{4\pi\xi_{ext}m}{T_{struct}}\dot{\boldsymbol{x}} = -2\xi_{ext}m\omega_{struct}\dot{\boldsymbol{x}}\,. \end{aligned} \tag{12.33}$$

Because we made several approximations to arrive at Eq. 12.33, we may need to adjust ξ_{ext} to obtain appropriate convergence characteristics. A practical way to damp out the fundamental mode of vibration is to start out by setting the damping ratio ξ_{ext} to zero, and observing the period of vibration. Typically, the end time should be set so that there are at least two to eight full periods of vibration. Then, to find a static solution, adjust the value of ξ_{ext} upward toward unity, until you observe critical damping behavior.[e]

12.5 Summary

We have chosen to solve the SPLM equations of motion using an explicit time integration method, which requires many tiny time steps. Why do we not choose an implicit approach (solving a set of simultaneous equations in each time step), which would allow us to use much larger time steps? The answer is that we have decided to avoid making any assumptions about the mechanical behavior. We assume a priori the possibility of nonlinear phenomena like plasticity, damage, geometric nonlinearity and contact. Therefore, tiny time steps are required to ensure that the simulation misses no aspect of nonlinear dynamic behavior, which can of course commence quite suddenly.

This approach is quite different from the more traditional solid mechanics approaches, in which engineers have made many simplifying assumptions to start out, and then over time these simplifying assumptions have been gradually relaxed.

---

[e] Damage may cause sudden microseismic events that cannot be damped out using external damping.

Because we are counting on the presently available powerful computers to be capable of solving problems of interest, there seems to be little need to use more involved solution strategies, which usually involve linearizations and simplifications of various types, and although clever, exhibit limitations in simulating actual physical behavior.

12.6 Exercises

12.1 Derive Eq. 12.5, $m\ddot{x} + c\dot{x} + kx = 0$, and Eq. 12.6, $\ddot{x} + 2\xi\omega_n\dot{x} + \omega_n^2 x = 0$, for an unforced single-degree of freedom linear spring-mass-damper system.

12.2 Solve the equation Eq. 12.6, $\ddot{x} + 2\xi\omega_n\dot{x} + \omega_n^2 x = 0$, for damped unforced vibration of a linear single degree of freedom system, assuming initial conditions $x(0) = 0$ and $\dot{x}(0) = 1$. Plot $x(t)$ for $\omega_n = 1$ and damping ratio $\xi = 0,\ .5,\ 1,$ and 2.

12.3 Explain why, in most materials, one would not expect relative damping to exist between particles that are not close neighbors.

12.4 Explain why external damping might be expected to be nonzero only for particles on the boundary of the body.

12.5 Solve the undamped equation Eq. 12.6, $\ddot{x} + \omega_n^2 x = 0$ for unforced vibration of a linear single degree-of freedom system. Use the numerical integration approach described in Section 12.1. Assume $\omega_n = 1$. Plot $x(t)$ using time step $\Delta t = 0.5\Delta t_{cr}, 0.9\Delta t_{cr}, \Delta t_{cr},$ and $1.1\Delta t_{cr}$.

12.6 Show that, in Eq. 12.18, if $[A]\{v\} = \lambda\{v\}$ then $[A]^n\{v\} = \lambda^n\{v\}$. Also, show that matrices $[A]$ and $[A]^n$ thus have the same eigenvectors but that eigenvalues of $[A]^n$ are equal to the eigenvalues of $[A]$ raised to the nth power.

12.7 Explain why use of the dynamic relaxation approach to obtain a static solution might result in an unphysical solution. Give an example of a situation where use of the dynamic relaxation approach could cause one to obtain an incorrect static solution.

12.8 Determine the critical time step for steel, assuming a 10% damping ratio, assuming the lattice spacing $L = 1\ cm$.

12.9 Determine the critical time step for normal concrete, assuming a 20% damping ratio, assuming the lattice spacing $L = 1\ cm$.

12.10 Assuming a plane-stress hexagonal lattice, determine the highest natural frequency of vibration ω_n, in terms of the classical elastic parameters E and ν, and in terms of the lattice spacing L.

Chapter 13

Computational Implementation

Chapter objectives are to:

- discuss capabilities of modern computers;
- present a parallel particle domain decomposition algorithm;
- indicate how the parallel algorithm can be implemented;
- show how the SPLM is implemented; and to
- investigate the computational efficiency of the parallel particle simulation algorithm.

The purpose of this chapter is to describe how one can implement the state-based peridynamic lattice model (SPLM) either on a single processor (a laptop or desktop computer) or on many processors running in parallel using the MPI (Message Passing Interface) protocol. The molecular dynamics research community has led the way in developing methods for simulating particle dynamics on massively parallel computers.

Experience has shown that for the SPLM, the simulation wall time $T_{simulation}$ is approximately proportional to the number of particles N_{ptcls}, and to the number of time steps $N_{time\ steps}$, and inversely proportional to the number of processors N_{procs}:

$$T_{simulation} = K \times \left(\frac{N_{ptcls} \times N_{time\ steps}}{N_{procs}}\right), \tag{13.1}$$

where the simulation constant $K \approx 2 \times 10^{-5} \frac{processor-seconds}{particle-time\ step}$ for a parallel computer of the 2010 era.[a]

---

[a] This value of K is for "Nano", a linux cluster housed at the University of New Mexico Center for Advanced Research Computing. Nano is rated at 1300 GFLOPS, has one head

To simulate the behavior of a large reinforced concrete beam or column, perhaps one million particles and one million time steps are required. For example, a three-dimensional SPLM model of a simply supported concrete beam one meter deep by one-half meter wide by ten meters long requires a lattice with approximately 625,000 particles spaced at the aggregate size of approximately 2 cm. For concrete, the critical time step size is approximately one-millionth of a second. Because the fundamental period of such a beam is approximately 0.14 seconds per cycle of vibration, at least 140,000 time steps are required per fundamental period. To obtain a quasi-static solution (obtained as a critically damped dynamic relaxation analysis) might require eight fundamental periods, or 1,120,000 time steps. For such a problem, Eq. 13.1 tells us that about 162 days of computer time may be necessary if the problem is running on a single processor (using a typical desktop or laptop computer of 2010 vintage). A 160-processor parallel computer might compute the solution in around one day. The point is that simulations of reasonably large concrete structures are becoming increasingly feasible.

Parallel computers can solve smaller problems in a few hours. Chapter 14 gives several examples along with the computer times necessary to solve them.

Although particle simulation programs can be rapidly prototyped using a high-level scientific computing language like MatLab or Mathematica, computers interpret such languages source-code line by source-code line, and are thus too slow to solve large problems with millions of particles and time steps. In order to produce computer programs that are as computationally efficient as possible, one needs a compiled code like FORTRAN, C, or C++. We favor FORTRAN for particle simulation, as this language is the scientific programming language that is closest to the architecture of a digital computer, and the most computationally efficient. Additionally, FORTRAN is a simpler and lower-level language than C, C++, or JAVA, and the simple data

---

node and 36 compute nodes, with two Intel Xeon 5140 2.33 GHz CPU's per node with two cores per CPU, with a Myrinet interconnect between the compute nodes.

types in a particle simulation application do not require the sophisticated data structures, such as the programmer-created objects, provided by C++ and JAVA.

Although individual computers are becoming increasingly powerful, following Moore's law, computer designers may reach the physical limits of single-processor computers within the next ten years. Hence, parallel computers are even now quite common, and even low-cost workstations may have sixteen or more processors. Massively parallel computers are essentially a rack of computers, each perhaps having its own hard drive, connected together with a high-speed interconnect. Super computers, residing at high-performance computing centers, are accessible over the internet, so engineers can now access high-performance computing "in the cloud".

Parallel programs typically handle interprocessor communication using calls to MPI (Message Passing Interface) subroutines. A group of computer scientists funded by the U.S. National Science Foundation and by the Commission of the European Community developed the MPI standard during the course of the 1990's. MPI provides designers of parallel computers with a clearly defined set of procedures that they must implement as subroutines in software libraries, which then provides scientific programmers with the set of subroutines, called an Application Programmer Interface (API), which they may use to transfer data rapidly back and forth between processors (or cores, on multi-core computers).

In recent years, computer designers have developed another level of parallelism, called Graphics Processing Units (GPUs). A GPU, acting as a parallel floating-point coprocessor, working in coordination with a single computational processor, may offer the capability of speeding up floating-point computations on a single processor by several orders of magnitude. While promising, the author has not yet implemented this level of parallelism, and we do not discuss GPUs further in this book.

13.1 Domain decomposition algorithm

Beginning in 2009 Susan Atlas and Walter Gerstle at the University of New Mexico, along with their graduate students Navid Sakhavand, Vijay Janardhanam, and Hossein Honarvar (Sakhavand, 2011),

(Honarvar Gheitanbaf, 2013) collaboratively developed a parallel particle simulation code called pdQ. pdQ originally was an acronym for "particle dynamics – quantum", but in light of the fact that all of the quantum physics has been removed from the essential particle code, pdQ now stands for "particle dynamics – quickly", and coincidentally, "pretty darn quick". pdQ uses a spatial decomposition algorithm integrating Plimpton's message passing approach and Beazley's cell decomposition approach (Beazley, D.M. and Lomdahl, P.S., 1994), (Plimpton, 1995).

Still not completely devoid of physics, pdQ assumes that all particles reside within a three-dimensional physical space $\mathbb{R}^3$, described by Cartesian coordinates X_i, as shown in Fig. 13.1. Our computational analysis "stage" is a cuboidal domain $\mathcal{D} \subset \mathbb{R}^3$ that includes the set of spatial locations

$$\mathcal{D} = \{\forall \boldsymbol{X} \in \mathbb{R}^3 : \ X_{imin} \leq X_i \leq X_{imax}, \ i = 1,2,3\} .^{b} \tag{13.2}$$

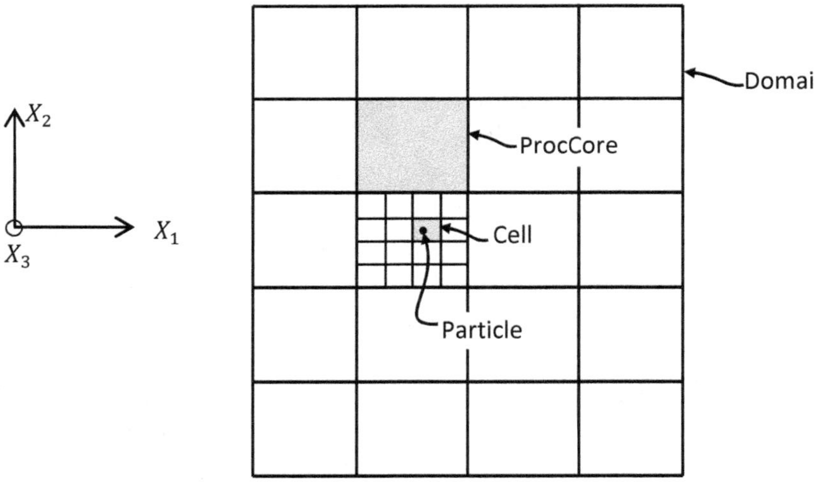

Figure 13.1. Domain $\mathcal{D}$, procCube core $\mathcal{D}_{procCore}$, cell $\mathcal{D}_{cell}$, and a particle.

Depicted in Fig. 13.1, $\mathcal{D}$ includes all of the material bodies, and hence all material particles, within the simulation throughout the time of the simulation. The dimensions of the rectangular domain are

---

[b] Well, actually not $\mathbb{R}^3$, but rather $\mathbb{F}^3$, where $\mathbb{F}$ is the set of floating point numbers.

$$L_{\mathcal{D}i} = X_{imax} - X_{imin}, \ i = 1,2,3\ . \tag{13.3}$$

When using more than one computational processor, we partition the domain $\mathcal{D}$ into a rectangular array of $N_{PC1} \times N_{PC2} \times N_{PC3}$ "procCube cores", or more simply, *procCores*. Each procCore has identical size, given by

$$L_{PCi} = {(X_{imax} - X_{imin})}/{N_{PCi}}, \ i = 1,2,3\ . \tag{13.4}$$

Each procCore is owned by a computational processor Pr_i. Processor Pr_i is responsible for computing the forces acting upon all particles P_i within its procCore. Processor Pr_i is also responsible for updating the states of all particles P_i within its procCore.

In addition to the one-dimensional processor address Pr_i, each procCore has a three-dimensional procCore address $PC(I_1, I_2, I_3)$ with the three indices I_i ranging from one to the number of procCores N_{PCi} in each direction i.

The physical space associated with each procCore $\mathcal{D}_{PC(I_1,I_2,I_3)}$ is given by

$$\mathcal{D}_{PC(I_1,I_2,I_3)} = \left\{ \begin{matrix} \forall \boldsymbol{X} : X_{imin} + (I_i - 1)L_{PCi} < X_i \\ \leq X_{imin} + (I_i)L_{PCi}, \ i = 1,2,3 \end{matrix} \right\}. \tag{13.5}$$

Because for a given particle the force state $\underline{\boldsymbol{T}}$ depends only upon the states of particles closer than the peridynamic horizon δ it makes sense to further partition each procCore into cells.[c] Thus, each procCore is partitioned into $N_{C1} \times N_{C2} \times N_{C3}$ rectangular cells of length L_{celli} in dimension i slightly larger than the material horizon δ:

[c] Actually, the cell size must be at least 2δ for state-based peridynamics, because the force acting on a particle within the procCore may depend upon the force states of particles residing within the procSkin.

$$L_{celli} = {}^{L_{PCi}}/_{N_{Ci}}, \; i = 1,2,3 \,. \tag{13.6}$$

Within a given procCore, the cells are numbered from $I_{celli} = 1$ to N_{Ci} in each of the three directions.

The physical space associated with each cell is

$$\mathcal{D}_{cell(I_{cell1},I_{cell2},I_{cell3})} = \left\{ \begin{array}{l} \forall \boldsymbol{X} : X_{iPCmin} + (I_{celli} - 1)L_{celli} < X_i \\ \leq X_{iPCmin} + (I_{icell})L_{celli}, \; i = 1,2,3 \end{array} \right\}. \tag{13.7}$$

To start a computational simulation, the program reads all material particles P_i from a file called *ptclAttrsInit.pdq* in each computational processor.

Fixed attributes, such as global particle identifier, $GlobalID(P_i)$, reference position $\boldsymbol{X}(P_i)$, particle lattice body $\mathcal{L}_{\mathcal{R}}(P_i)$, and particle mass $m(P_i)$, do not change during the course of the simulation. The minimal set of fixed attributes contains $GlobalID(P_i)$ and initial reference position $\boldsymbol{X}(P_i)$; the rest of the fixed attributes depend upon the physical problem to be solved, and are specified by the user.[d]

On the other hand, alterable attributes, in addition to the global ID $GlobalID(P_i)$ may include the current particle position $\boldsymbol{x}(P_i)$, velocity $\boldsymbol{v}(P_i)$, and temperature $\mathcal{T}(P_i)$. The values of the alterable attributes at the beginning of the simulation represent the initial conditions of the simulation. The minimal set of alterable attributes contains the global identifier $GlobalID(P_i)$, the initial position $\boldsymbol{x}_0(P_i)$, and the initial velocity $\boldsymbol{v}_0(P_i)$.[e] Depending upon the physics involved, the user may specify additional alterable attributes for a particular type of model.

A material particle P_i is read in and stored in memory by a particular computational processor Pr_i if its current position $\boldsymbol{x}$ resides within the physical domain of that processor's procCore; otherwise the processor does not store the particle. The material particle is stored in a single

---

[d] The global ID $GlobalID(P_i)$ is an integer, and no two particles may have the same global ID. However, the global ID's need not be sequential or even positive. They are simply unique integer names, or handles, for the particles.

[e] Although the global particle ID never changes during the simulation, it is included as an element of the alterable attributes simply for convenience.

floating point array on a procCore Pr_i. We call the array *ptclAttrs*, and this array contains all the particle attributes, both fixed and alterable. The *ptclAttrs* array is five-dimensional, with the five indices: I_{attr}, I_{ptcl}, I_{cell1}, I_{cell2}, I_{cell3}. These arrays can be dynamically allocated only after the number of particle attributes N_{attrs}, the maximum number of particles contained by any cell in the domain $N_{maxPtclsCell}$, and the number of cells in each direction N_{cellsi}, are known.

In the spatial decomposition parallel processing algorithm, it is necessary that particles in spatially adjacent cells within different procCores interact. To allow this interaction between particles on different computational processors, a *procSkin* of cells surrounds each procCore. Together, the procCore and the procSkin create what we call a *procCube*. While the cells of the procCore are numbered from 1 to N_{cellsi} in each direction i, the cells of a procCube are numbered from 0 to $N_{cellsi} + 1$ in each direction i. Contiguous arrays of cells on each face of the procCube, called *skinWalls*, within the *procSkin* overlap with identically-spatially-located arrays of cells within the procCores of adjacent procCubes, called *coreWalls*.

In each time step of a computationally parallel simulation, prior to calculation of particle peridynamic states, the program copies the particles within all coreWalls of each procCube into the respective skinWalls of adjacent procCubes using Message Passing Interface (MPI) procedures. Following this interprocessor communication, on each procCube, an up-to-date procSkin surrounds the procCore, and the peridynamic calculations on the particles within the procCore can proceed.

In some types of particle simulations, particles P_i may move large distances compared to the cell size, and in the process interact with new sets of particle neighbors, possibly in new cells. In such simulations, the program reassigns particles to new cells as their spatial positions $\boldsymbol{x}(P_i)$ move into new cells. We call this the *shuffle* process.

Additionally, during the course of the simulation, some particles may move outside the boundary of the simulation domain $\mathcal{D}$, in which case either the particles are removed from the simulation or they are transferred to an appropriate cell on the other side of the domain $\mathcal{D}$, in a process known as *periodic boundary conditions*, or “PBC”.

13.2 Design and implementation of pdQ

In addition to *ptclAttrs.pdq*, pdQ reads two additional input files, *Input.pdq* and *UserInput.pdq*. *Input.pdq* contains several global parameters required by all pdQ simulations (like the number of procCubes in each direction, the inter-particle interaction horizon, and the number of time steps), and *UserInput.pdq* is used for additional user specifications (like particle and material parameters) that may vary from model to model.

pdQ both requires and permits the modeler to write user-specific subroutines in user files that specify the user setup methods (*userSetup.F*), the interaction functions (*userForce.F*), the integration methods (*userIntegrate.F*), and the output methods (*userWriteOut.F*). In addition, the file *userModules.F* declares all user variables that are common to more than one user subroutine. By both requiring and permitting the user to program these five subroutines, pdQ retains vast modeling generality while still providing to the user nontrivial and difficult-to-program functionality concerning the spatial decomposition and parallelization. Thus, pdQ has few preconceived notions about the physics of the model, and handles the complex tasks of parallelization, particle shuffling, and periodic boundary conditions, while the user needs only to program the physical behavior of the particles. We have found that while about a dozen particle simulation codes exist, they are all tied to one or another type of discipline (biology, material science, etc.), and having a "clean" parallel particle simulation program is particularly valuable. The pdQ program architecture allows the user to add higher-level program layers if desired, perhaps including user-defined common data structures (in userModules) that define constraints, links, and even perhaps finite elements. Thus, one can think of pdQ as an Application Programmer Interface (API).

After the program reads in the particles from ptclAttrsInit.pdq, and partitions them between procCubes and cells, the program enters the time integration loop. The user specifies the number of time steps or programs stopping criteria in one of the user files. Each time step includes interaction computation, time integration, interprocessor communication, and, if needed, particle shuffling between cells, and outputting of results, as follows:

prepare the necessary arrays (various pdQ subroutines)
allow user to set up additional arrays (userSetup)
Loop in time until the stopping criterion is satisfied:
  force computation (user routine userForce)
  time integration (user routine userIntegrate)
  exchange particle attributes (pdQ exchangePtclAttrs)
  shuffle particles (pdQ routine Shuffle)
  output the results (user routine userWriteOut)
end time loop

Next, we discuss the details of interprocessor communication.

13.3 Interprocessor communication

To explain the communication between processors, we focus now on a single reference procCube, which we refer to as the *home* procCube, iPr, shown in Fig. 13.2.

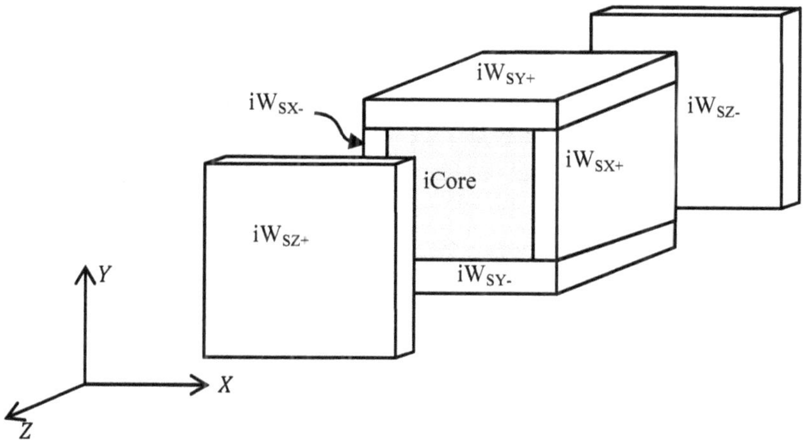

Figure 13.2. The home procCube, its core, and its receiving walls.

Particles residing in a cell on the boundary, but within the procCore of the home procCube can potentially interact with particles that are within its material horizon but which lie in the procCores of adjacent procCubes. The program must send the necessary particle data from procCores of adjacent procCubes to the skin of the home procCube. The Message Passing Interface (MPI) subroutine MPI_SENDRECV executes these communications.

The information that must be passed between processors in each time step consists of particles within an array, *ptclSend* and another array called *nPtclsCellSend*, which tells the receiving processor how many particles reside within each sent cell. (Before entering the time loop, particles have already been loaded into the core and skin cells of each procCube.)

The home procCube core is potentially adjacent to twenty-six adjacent procCube cores. Using Plimpton's method, the program accomplishes the message passing in just six MPI send-receive messages per time step, as depicted in Fig. 13.3. Before proceeding, we explain the naming convention for sending and receiving walls. A wall is a cuboidal array of cells that is responsible for sending and receiving the necessary particle attributes between processors.

Table 13.1. Sending and receiving walls.

Sending walls		Receiving walls	
Identifier	Cell indices	Identifier	Cell indices
iW_{SX-}	$[1, 1{:}N_{CY}, 1{:}N_{CZ}]$	iW_{RX-}	$[0, 1{:}\ N_{CY}, 1{:}\ N_{CZ}]$
iW_{SX+}	$[N_{CX}, 1{:}N_{CY}, 1{:}N_{CZ}]$	iW_{RX+}	$[N_{CX}+1, 1{:}\ N_{CY}, 1{:}\ N_{CZ}]$
iW_{SY-}	$[0{:}N_{CX}+1, 1, 1{:}N_{CZ}]$	iW_{RY-}	$[0{:}\ N_{CX}+1, 0, 1{:}\ N_{CZ}]$
iW_{SY+}	$[0{:}N_{CX}+1, N_{CY}, 1{:}nC_Z]$	iW_{RY+}	$[0{:}\ N_{CX}+1, N_{CY}+1, 1{:}\ N_{CZ}]$
iW_{SZ-}	$[0{:}\ N_{CX}+1, 0{:}\ N_{CY}+1, 1]$	iW_{RZ-}	$[0{:}\ N_{CX}+1, 0{:}\ N_{CY}+1, 0]$
iW_{SZ+}	$[0{:}N_{CX}+1, 0{:}\ N_{CY}+1, N_{CZ}]$	iW_{RZ+}	$[0{:}\ N_{CY}+1, 0{:}\ N_{CY}+1, N_{CZ}+1]$

For cognitive reasons, we now let $X_1 = X$, $X_2 = Y$, and $X_3 = Z$, as shown in Figs. 13.2 and 13.3. The cell indices of the core of the home procCube (iP) are defined as $iCore = [1{:}N_{CX}, 1{:}N_{CY}, 1{:}N_{CZ}]$. Table 13.1 defines the cell indices of sending and receiving walls. Fig. 13.2

shows the receiving walls; Table 13.1 indicates by their cell indices the receiving walls.

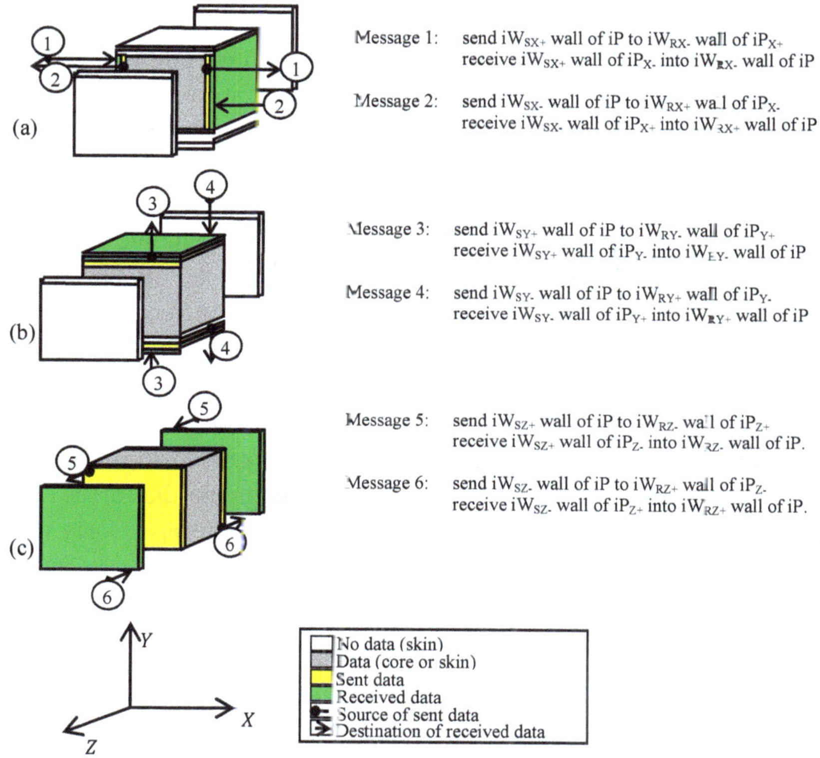
(a)
(b)
(c)
Message 1: send iW_{SX+} wall of iP to iW_{RX-} wall of iP_{X+}
receive iW_{SX+} wall of iP_{X-} into iW_{RX-} wall of iP
Message 2: send iW_{SX-} wall of iP to iW_{RX+} wall of iP_{X-}
receive iW_{SX-} wall of iP_{X+} into iW_{RX+} wall of iP
Message 3: send iW_{SY+} wall of iP to iW_{RY-} wall of iP_{Y+}
receive iW_{SY+} wall of iP_{Y-} into iW_{RY-} wall of iP
Message 4: send iW_{SY-} wall of iP to iW_{RY+} wall of iP_{Y-}
receive iW_{SY-} wall of iP_{Y+} into iW_{RY+} wall of iP
Message 5: send iW_{SZ+} wall of iP to iW_{RZ-} wall of iP_{Z+}
receive iW_{SZ+} wall of iP_{Z-} into iW_{RZ-} wall of iP.
Message 6: send iW_{SZ-} wall of iP to iW_{RZ+} wall of iP_{Z-}
receive iW_{SZ-} wall of iP_{Z+} into iW_{RZ+} wall of iP.
Y
X
Z
No data (skin)
Data (core or skin)
Sent data
Received data
Source of sent data
Destination of received data

Figure 13.3. Six message-passing steps, with focus on home procCube, iP.

Table 13.2. Indices of procCubes adjacent to the home procCube.

Non-periodic boundary conditions		Periodic boundary conditions	
Identifier	ProcCube Indices	Identifier	ProcCube Indices
iP_{X-}	$[iP_X-1, iP_Y, iP_Z]$	$iP_{X-} = 1$	$[N_{PX}, iP_Y, iP_Z]$
iP_{X+}	$[iP_X+1, iP_Y, iP_Z]$	$iP_{X+} = N_{PX}$	$[1, iP_Y, iP_Z]$
iP_{Y-}	$[iP_X, iP_Y-1, iP_Z]$	$iP_{Y-} = 1$	$[iP_X, N_{PY}, iP_Z]$
iP_{Y+}	$[iP_X, iP_Y+1, iP_Z]$	$iP_{Y+} = N_{PY}$	$[iP_X, 1, iP_Z]$
iP_{Z-}	$[iP_X, iP_Y, iP_Z-1]$	$iP_{Z-} = 1$	$[iP_X, iP_Y, N_{PZ}]$
iP_{Z+}	$[iP_X, iP_Y, iP_Z+1]$	$iP_{Z+} = N_{PZ}$	$[iP_X, iP_Y, 1]$

The procCube indices of the six procCubes adjacent to the home procCube $[iP_X, iP_Y, iP_Z]$, are shown in Table 13.2 for non-periodic and periodic boundary conditions.

Using these wall indices and adjacent procCube indices, we next illustrate the six message passing steps. Note that we execute these six steps in sequence. In addition, prior to this step, we assume that particles in the core have been updated in the previous time step by the user-defined subroutines in the *userIntegrate* file.

In the first step, as illustrated in Fig. 13.3(a), the home procCube sends particle attributes from its iW_{SX+} wall into the iW_{RX-} wall of procCube iP_{X+}. At the same time, the home procCube receives particle attributes from the iW_{SX+} wall of procCube iP_{X-} into its iW_{RX-} wall. Steps two through six are similar to the first step, and are illustrated graphically in Fig. 13.3.When the message passing is complete in all six directions, all the necessary particle attributes data have been received in the skins of the home procCube.[f] Now, we can correctly compute the forces on all particles in the core of the home procCube (the home procCore) using particle information, which resides on the home procCube.

We next discuss neighbor lists.

13.4 Neighbor lists

With state-based peridynamics of solid bodies, particles within a given lattice body $\mathcal{L}_{\mathcal{R}}$ never change their neighbors defined by the neighbor list $\mathcal{N}[P_i]\langle B_j \rangle$. There is a caveat, however: if a particle from one portion of the material body encounters a particle from another part of the same body, or meets with a particle from another body altogether, then the program must prevent interpenetration of the particles using a particle contact algorithm. In this case, particles will interact with new neighbors. Let us call these *nonmaterial neighbors*, as opposed to *material neighbors*.

---

[f] We call this the "Plimpton message-passing algorithm". Cleverly, it is not necessary to send or receive messages from procCubes only adjacent through corners or through edges. You might want to convince yourself of this by following the message sequence.

Thus, for some types of problems, we do need to consider the problem of nonmaterial particle contact. If we wanted to simulate large deformation problems, then the nonmaterial neighbor list would need to be recomputed every so often, and probably even in every time step. Repeating, the material neighbor list $\mathcal{N}[P_i]\langle B_j\rangle$ used for the SPLM lattice body $\mathcal{L}_{\mathcal{R}}$ is distinct from the nonmaterial neighbor list representing contact of particles that were not originally material neighbors.

The issue of when and how to compute neighbor lists is problem-dependent, and therefore the designers of pdQ have chosen to leave computation of particle neighbor lists entirely to the user.

For peridynamic simulation of reinforced concrete structures, for example, it is unnecessary to consider interpenetration of particles that are distant from each other in the reference configuration, and thus the neighbor lists need not change during the simulation. However, for simulation of a piece of taffy or modeling clay under large deformations, to account for the creation of new interfaces, the program needs to re-compute nonmaterial neighbor lists every time step.

As the material neighbor list $\mathcal{N}[P_i]\langle B_j\rangle$ never needs to change over the course of the simulation, the subroutine *userSetup* computes $\mathcal{N}[P_i]\langle B_j\rangle$ before entering the time integration loop. For example, for the SPLM model described in Chapters 6 through 12, in *userModules* we have included two integer arrays dimensioned as follows:

```
ptclNeighMyMat(4, numMyMatBonds, maxNumPtclsCell, numCellsX,
                                        numCellsY, numCellsZ)
```

This array gives, for each particle in the procCore:

```
ptclNeighMyMat(1, iBond, iPtcl, iCellX, iCellY, iCellz) = jCellPtcl
ptclNeighMyMat(2, iBond, iPtcl, iCellX, iCellY, iCellz) = jCellX
ptclNeighMyMat(3, iBond, iPtcl, iCellX, iCellY, iCellz) = jCellY
ptclNeighMyMat(4, iBond, iPtcl, iCellX, iCellY, iCellz) = jCellZ
```

Thus, particle(jCellPtcl, jCellX, jCellY, jCellz) is the address of the particle associated with link iBond, numbered as shown in Fig. 6.7.

The other integer array is ptclNeighOtherMat(4, maxOtherMatBonds, maxNumPtclsCell, numCellsX, numCellsY, numCellsZ)

This array gives, for each particle in the procCore:

ptclNeighOtherMat (1, iBond, iPtcl, iCellX, iCellY, iCellz) = jCellPtcl
ptclNeighOtherMat (2, iBond, iPtcl, iCellX, iCellY, iCellz) = jCellX
ptclNeighOtherMat (3, iBond, iPtcl, iCellX, iCellY, iCellz) = jCellY
ptclNeighOtherMat (4, iBond, iPtcl, iCellX, iCellY, iCellz) = jCellZ

(jCellPtcl, jCellX, jCellY, jCellz) is the address of a neighboring particle associated belonging perhaps to another material body.

This information about particle neighbors, both within material bodies and between material bodies, enables the user efficiently to program the state-based force functions $\underline{\boldsymbol{T}} = \underline{\tilde{\boldsymbol{T}}}(\underline{\boldsymbol{Y}})$ in subroutine userForce.

13.5 Shuffle: moving particles from cell to cell

Each particle has a time-varying current spatial location $\boldsymbol{x}$. The spatial locations of procCubes and their cells do not change during the simulation. If we activate "particle shuffling", then within each time step, pdQ raises a flag if a particle's current location (X, Y, Z) is no longer spatially located sufficiently close to its containing cell's cuboid. This flag indicates that it is necessary to move, or "shuffle" the particle to a neighboring cell. (The assumption is made that the magnitude of particle motion in each time step is small compared to the inter-particle interaction horizon δ, so that a particle need only ever be shuffled from its current cell to an immediately adjacent cell.)

Two operations are necessary to shuffle particle P_k from cell $iCell$ to adjacent cell $jCell$ within a procCube: first, delete the particle from $iCell$ and second, add the particle to $jCell$. To delete the particle P_k from $iCell$ with N_{iCell} particles, the $N_{iCell}{}^{\text{th}}$ particle is copied to the location of the $P_k{}^{\text{th}}$ particle in the *ptclAttrs* array, and then the number of particles in $iCell$ is reduced by 1: $N_{iCell} = N_{iCell} - 1$. Thus, gaps in the particle arrays can never develop. To add particle P_k to cell $jCell$ with N_{jCell} particles, first a check is made to ensure that $N_{jCell} < N_{maxNumPtclsCell}$ (if $N_{jCell} = N_{maxNumPtclsCell}$, either an "out of memory" error message is issued, or memory is dynamically allocated allow for more particles). Then particle P_k is copied to the particle address $N_{jCell} + 1$ in the *ptclAttrs* array, and the number of particles is increased by 1: $N_{jCell} = N_{jCell} + 1$.

Next, consider shuffling a particle from procCube $iProc$ to an adjacent procCube $jProc$. Let us assume that at the beginning of the time step, each particle's spatial position $\boldsymbol{x}$ and other attributes are up-to-date and that the particle is located within the geometrical limits of its containing cell.

Within each time step, first we compute the internal interactions acting between particles within the procCube core and skin.[g] These internal interactions depend upon particle states within both core and skin cells in the home procCube. As particle positions, attributes, and cell addresses are all up-to-date, and each particle in the procCore interacts with its full complement of neighbors, the algorithm will correctly compute the interactions acting upon procCube core particles.[h]

Next, the user-defined time integration subroutine, *userIntegrate*, updates the positions of particles in the core (only). Some of these particles may now have locations outside their containing cells, and thus the program needs to shuffle these particles to adjacent cells; perhaps into a skin cell. However, at this stage, the algorithm does not yet shuffle particles to an adjacent cell.

Next, the up-to-date particles contained by the procCube core cells are sent, via message-passing, into the corresponding skins of adjacent procCubes, over-writing obsolete particles in the skin cells, as illustrated in Fig. 13.3. Note that some of these received particles in skin cells may have geometric locations that are not within their containing cells. At the end of this step, all particles in both skin and core cells have correct

[g] Particles in the skin, within δ of the outer procCube boundary may not have correct forces. But this is okay, as these particles are not used in updating particle forces within the procCore.

[h] Even particles in the skin that reside within δ of the core will be up-to-date.

geometric locations, but their cell addresses may be incorrect because the particle geometric locations may no longer be within their containing cells.

Next, if we have activated the shuffle flag, the algorithm performs a shuffle step within each procCube separately. In the shuffle step, the program changes the addresses of all particles no longer spatially located within their containing cells to the appropriate adjacent cells. Particles contained by core cells may move into adjacent skin cells, and vice versa. In addition, particles in skin cells may move completely outside the procCube, in which case they are simply deleted from the procCube (or transferred to another procCube, in the case of periodic boundary conditions). At the end of this step, some particles in skin cells may be missing, but these particles will not, in any case, interact with particles in the core of the procCube, because they will be outside the inter-particle interaction horizon with any particle in the procCube core. Importantly, after the shuffle step, all core and skin cells are guaranteed to have particles with compatible locations and cell addresses as well as up-to-date attributes, and the condition necessary to start the next time step is fulfilled.

If this seems complex, it is. This is why pdQ has value – users can make use of pdQ as an API and not have to worry about the details of parallelizing the particle simulation themselves. Note that absolutely no accuracy is lost in using the parallel processing capability. Multiple processors working in parallel obtain exactly the same solution as is obtained as on a single processor.

13.6 Domain boundary conditions

In certain types of problems, particularly in molecular dynamics simulations, particles may move close to the boundary of the analysis domain $\mathcal{D}$. If this happens, then under normal circumstances, the particle interacts with no particles located outside of $\mathcal{D}$. However, if the *periodic boundary condition (PBC)* option is specified, then the domain is assumed to possess periodic translational symmetry, or cyclic symmetry, in all three directions X_1, X_2, and X_3. If this type of symmetry is specified, then particles close to the boundary of $\mathcal{D}$ interact with particles on the opposite side of $\mathcal{D}$; it is essentially a three-dimensional torus (something very difficult to visualize). In this case, during inter-processor communication, both the reference and the current particle coordinates in the PBC direction of transfer are shifted by L_{Di} in the appropriate direction and sense.

Table 13.2 shows procCube core adjacencies for periodic boundary conditions.

Also with PBC, when shuffling particles, it is important to move the particles to the appropriate procCube.

13.7 Pre- and post-processing

pdQ requires three input files: *ptcAttrsInit.pdq*, *input.pdq*, and *userInput.pdq*. These files are straightforward and can be automatically created by a preprocessing program. We have used either FORTRAN or MatLab to create these files.

The main files that pdQ writes to as it runs are the *timeHistory.pdq* and *restart.timeStep.pdq* files. The *timeHistory.pdq* file contains in each record the alterable attributes of a single chosen particle at a given time step. The *restart.timeStep.pdq* files output all particle data at pre-chosen time step intervals, in precisely the same format as the *ptclAlterAttrs.pdq* file, which provided the initial conditions. A post-processing program can read these files in, and can make pictures and animations of the simulation as it progresses. We have used MatLab to write the post-processing programs, but other commercial post-processing programs, like TecPlot may also be used.

As the simulation progresses, the user may wish to write out other information in user-created files. Subroutine *userSetup* opens these files, and subroutine *userWriteOut* writes to these files during the time stepping procedure.

13.8 Computational performance

Using up to 32 processors, we have found that the time required for domain decomposition and message passing varies compared to the time required for the force calculations. The overhead involved with message passing can be large compared to the force calculation time if a processor has too few particles. On the other hand, if the simulation employs too few parallel processors, then one does not benefit from parallelism to the extent possible. Therefore, there is an art to specifying the optimal number of processors in each spatial direction. This optimal processor configuration is problem-dependent.

Assuming that the problem has a very large number of particles, the time to achieve a solution is roughly inversely proportional to the number of processors used – up to a point. For larger numbers of processors, the message-passing overhead could become burdensome; we have little experience to date with large numbers of processors.

On single processor machines, we have not so far run into memory limitation issues, even for problems with as many as one million particles.[i]

We find, using Eq. 13.1, for SPLM problems of the type described in this book, that the simulation constant K is about $2 \times 10^{-5}\ \frac{processor-seconds}{particle-time\ step}$. This constant, however, depends upon the type of force calculations performed in the simulation, and on the processor speed and interprocessor communication overhead. The reader should use this constant only as a rule of thumb, and with due caution.

(Sakhavand, 2011) provides more detail regarding the computational performance of pdQ.

[i] In post-processing with MatLab, however, we do run into time and memory issues for problems approaching one million particles in size. We can avoid these issues by displaying a pruned set of particles. For example, we might choose to display only the boundary particles of a three-dimensional domain.

13.9 Summary

In conclusion, we believe that pdQ provides an essential, (relatively) easy-to-understand facility for particle simulations on parallel computers, while at the same time allowing the user maximum modeling flexibility to perform simulations of various physical phenomena.

The parallel programming aspect of the code, using MPI, is quite involved. However, we have completed this programming in the application programmer interface, pdQ.

The scientific modeler need not be concerned with the details of the parallel aspect of the code. The user subroutines *userSetup*, *userForce*, *userIntegrate*, *userModules*, and *userWriteOut*, provide the physical modeling simulation interface, and these subroutines are identical whether programming for one serial computer or for many parallel computers.

The interface exposed to the user is in the FORTRAN language, making pdQ an application programmer interface (API), rather than an end-user software application. Despite having to program in FORTRAN, we believe that using an API rather than an end-user software application better serves the engineer or scientist. The modeling possibilities are much wider if the scientific modeler is able to write his or her customized simulation program.

No other such parallel particle dynamics code of which we are aware provides equivalent capability with such a basic interface. With computer software, it is advantageous to use basic, stripped-down, application programmer interfaces, rather than fully developed software applications with many "bells and whistles" perhaps developed for an entirely different discipline. With such higher-level applications, if the source code is available, one spends too much time first trying to understand the application, then stripping it down, and finally adding the code necessary to implement the simulation task. If the source code is not available, aspects of the software that are difficult to understand are often frustrating.

Choosing the number of processors in each direction for a given simulation is problem dependent. Generally, it is best to minimize the number of particles on the interfaces between procCubes, and to maximize the number of particles per procCube, while at the same time employing as many processors as are available. These competing demands mean that the user must experiment with the number of processors for each simulation to see what works best to minimize the simulation time.

13.10 Exercises

13.1 Obtain a copy of pdQ from the author (or write your own parallel simulation program in any language you choose!), and use it to simulate a one-dimensional axial linear elastic SPLM steel bar 1 cm in diameter and 1 m long, fixed at its left end and loaded in tension with 100 kN at its right end. Assume a lattice spacing of $L = 2$ cm. Run this benchmark problem on a serial computer, and display the time history of the x-displacement at the free end of the bar. (Hint: start by running a benchmark problem downloaded from the pdQ website.)

13.2 Repeat Problem 13.1, this time using two processors.

13.3 Determine the simulation constant K for your computer system for a variety of problems of varying size, using the non-optimized (-g) compiler flag.

13.4 Determine the simulation constant K for your computer system for the problems investigated in Problem 13.3, as a function of the compiler optimization flag used.

13.5 Obtain the pdQ source code from the author. Read the FORTRAN code to identify and clarify the algorithms described in this chapter.

13.6 Estimate the pdQ execution time $T_{simulation}$ for a reinforced concrete cantilever column five meters long with a square cross section of dimensions 0.5 m by 0.5 m. Assume one computational processor. Assume a lattice spacing of 2 cm.

Assume $K \approx 2 \times 10^{-5} \frac{processor-seconds}{particle-time\ step}$. Assume (a) A fully three-dimensional SPLM model; (b) A plane stress SPLM model; (c) A one-dimensional axial SPLM model.

Chapter 14

Simulation of Reinforced Concrete

Chapter objectives are to:

- describe an SPLM model for reinforced concrete structures;
- verify the SPLM elasticity, plasticity, and damage models at the particle level;
- examine the behavior of a concrete cylinder loaded axially in tension and in compression;
- simulate the Brazilian split cylinder test; and
- demonstrate SPLM modeling of reinforced concrete beams.

The desire to model reinforced concrete structures has played a major role in driving advances in computational solid modeling. Reinforced concrete is a nonhomogeneous composite involving a concrete medium interacting with embedded steel reinforcing bars. Characterizing bond-slip behavior between concrete and rebar has been less than successful using traditional ideas of classical continuum mechanics (Gerstle, W. H. and Ingraffea, A.R., 1991). Complicating the situation even more, plain concrete is a composite material at the mesoscale, consisting of gravel, sand, cement and water. Featuring nonlinear damage and fracture behavior in the tensile regime and quasi-continuous plastic-like behavior in the compressive regime, concrete has been a challenge to model using finite element methods. With so many size scales, it has been difficult for modelers to settle upon the size scale of interest in modeling concrete structures. The SPLM developed in this book provides a natural framework for specifying the minimum size scale of the model.

In this chapter, we provide dynamic and quasistatic simulations of various plain and reinforced concrete structures based upon the SPLM elasticity, plasticity, and damage models defined in Chapters 9, 10, and 11 respectively.

First, let us unambiguously specify the SPLM models, used for the simulation of the problems in this chapter, for concrete and for steel. In addition, we shall specify how the concrete and steel lattice bodies interact.

14.1 SPLM model for reinforced concrete structures

Computational models of reinforced concrete structures must faithfully simulate both the behavior of the concrete, of the steel reinforcing bars, and of their interaction (bond). We set the lattice spacing for both steel and concrete to be $L = 1.0$ cm. This value reflects a lower limit beyond which the normal concrete is certainly no longer homogeneous, and below which we would need to explicitly define the geometry of the ribs on the surface of a deformed reinforcing bar. We use a hexagonal lattice for two-dimensional problems, and a face-centered cubic lattice for three-dimensional problems.

Because the damage and plasticity models described in this book are rate-dependent, we fix the integration time step at $\Delta t = 1.0 \times 10^{-7}$ s, a value somewhat less than the critical time step required for linear elastic dynamical simulation of the steel (which has a critical time step of approximately $\Delta t = 3 \times 10^{-7}\, s$), which controls for the composite material. Although even smaller time steps would be desirable for capturing materially nonlinear dynamic damage and fracture events, we judge that for efficacious computational modeling, this time step is sufficiently small. Thus, in keeping with the notion of fixed lattice spacing, we also fix the time step.

With fixed time step, fixed lattice spacing, and fixed material models and associated parameters, we find the material models to be reasonably objective, with lattice rotation and translation of each lattice body being the only remaining free modeling parameters. Note that with the SPLM, we are not trying to solve a set of partial differential equations with boundary conditions, known as a boundary value problem; rather we are trying directly to simulate a physical problem modeled as a set of interacting lattice bodies. Hence, no mesh convergence studies are required.

Each reinforced concrete structure has a fundamental period of linear elastic vibration of $T_{fundamental}$. The modeler must approximately evaluate this period before each simulation. Experience shows that to simulate quasistatic loading, the time of the simulation should be approximately $8 \times T_{fundamental}$. The number of time steps required for a quasistatic simulation is thus $t_{end} = {}^{8 \times T_{fundamental}}/_{\Delta t}$.[a] In each simulation in this chapter, we apply smoothly time-varying displacement-controlled loading $\Delta_Y(t) = \frac{\Delta_{max}}{2}\left(1 - cos\left(\frac{\pi t}{t_{rampEnd}}\right)\right)$, with $t_{rampEnd} = 0.8 \times t_{end}$.

To arrive at the static solution as quickly as possible without affecting the static simulation results, we apply external damping as described in Section 12.4.2. For all example problems in this chapter, we fix the external damping ratio at $\zeta_{external} = 0.2$.

Note that our model realistically simulates large geometric rotations and translations, as long as the bond stretches are small. Thus, for example, the possible funicular behavior of the rebar in response to applied loading is included in our simulations. We thus simulate behavior far into the nonlinear post-failure regime.[b]

In the following subsections, we describe the SPLM models for the concrete, the steel rebar, and the bond between them. We also discuss the steel loading plates and the interaction between these plates and the concrete body.

14.1.1 *Plain concrete*

We model plain concrete using the SPLM elasticity models described in Chapter 9, the plasticity model described in Section 10.4.1, and the damage model described in Section 11.5. For a normal-strength concrete,

---

[a] Note that snap-back events due to damage events can occur despite slow rates of loading. These events are fundamentally dynamic, and are objective features of the simulation despite slow loading rates. Some problems simply are not quasistatic – they are inherently dynamic. Think about stretching a rubber band until it snaps.

[b] We use the term "post-failure regime" loosely here; the definition of "failure" is up to the modeler.

with uniaxial compressive strength $F'_c = 4000\ PSI$, used exclusively in this chapter, we assume the material parameters shown in Table 14.1.

Table 14.1. Material parameters used for plain concrete with $F'_c = 4000\ PSI$.

Parameter	English Value	SI Value
Compressive Strength, $F'_c = \sigma_{Yield}$	4000 PSI	27.58 MPa
Young's modulus, E	3605 KSI	24.86 GPa
Poisson's ratio, ν	0.20	0.20
Mass density, ρ	145 PSI	2323 Kg/m^3
Lattice spacing, L	0.3937 inch	1.0 cm
Internal damping ratio, $\zeta_{internal}$	0.2	0.2
Uniaxial tensile strength, $F_t = \sigma_t$	400 PSI	2.758 MPa
Ultimate tensile damage crack opening displacement, w_c	0.008736 inch	0.2 mm
Tensile damage parameter gamma, γ	0.25	0.25
Fracture energy G_F (derived)	0.787 lb./inch	137.9 N/m

Assuming the crack band to be two lattice spacings $2L$ in width, we estimate the fracture energy G_F as the area under the cohesive stress versus crack opening displacement curve shown in Fig. 11.9(a). This area is

$$G_F \approx 2\left(\frac{F_t^2 L}{2E} + \frac{\gamma F_t COD_c}{2}\right) \approx \gamma F_t COD_c\,. \tag{14.1}$$

Using the values given in Table 14.1, we find $G_F = 137.9\,\frac{\text{N}}{\text{m}} = 0.787\ \frac{lb}{in}$. This value is quite reasonable for normal-strength concrete, for which values of G_F between $100\,\frac{\text{N}}{\text{m}}$ and $150\,\frac{\text{N}}{\text{m}}$ are commonly cited values.

14.1.2 *Reinforcing bars*

We model the steel reinforcing bars as linear elastic – perfectly plastic axial members with no bending stiffness. Although we could, we do not in this chapter model tensile damage in the steel bars, as damage in steel bars typically only occurs at plastic stretches well beyond those encountered in realistic concrete structures. This is in keeping with standard structural engineering practice. Table 14.2 shows the material parameters used for the steel reinforcing bars.

The steel loading plates have SPLM parameters identical to those of the rebar, except that, for modeling simplicity, we inactivate both damage and plasticity in these lattice bodies.

Table 14.2. Material parameters used for steel rebar and steel loading plates.

Parameter	English Value	SI Value
Yield Strength, $F_Y = \sigma_{Yield}$	60,000 PSI	413.7 MPa
Young's modulus, E	29,000 KSI	200.0 GPa
Poisson's ratio, ν	0.30	0.30
Mass Density, ρ	490 PSF	7849. Kg/m^3
Lattice spacing, L	0.3937 inch	1.0 cm
Damping Ratio, ζ	0.2	0.2

14.1.3 *Bond between concrete and reinforcement*

We must specify the force interaction between the concrete and the steel lattice bodies. Consideration of the fact that reinforcing bars have deformations on their surfaces dictates that slip cannot occur between the reinforcing bars and the surrounding concrete (Gerstle, W. H. and Ingraffea, A.R., 1991). However, the bars do need to be able to separate from the surrounding concrete if the concrete becomes damaged.

The peridynamic horizon between a potentially interacting steel particle and a concrete particle is assumed to be the maximum lattice spacing of the associated steel and the concrete lattice bodies: $\delta = \max(L_{conc}, L_{steel}) = L$, which is 1 cm in our simulations.

The force F in an individual SPLM bond with stretch S between steel and concrete is assumed to be

$$F = (1 - \omega_{concrete})a_{concrete}\left(\frac{m_{steel}}{m_{concrete}}\right)S, \tag{14.2}$$

where $a_{concrete}$ is the stiffness parameter for the concrete lattice body, given by Eq. 9.40 for three-dimensional or Eq. 9.57 for the case of plane stress concrete domains, $\omega_{concrete}$ is the damage of the concrete particle, and m_{steel} and $m_{concrete}$ are the masses of the steel and concrete lattice particles, respectively. The factor $\left(\frac{m_{steel}}{m_{concrete}}\right)$ ensures that the elastic stiffness of the bonds connecting the steel body to the concrete body will not be so high that numerical instability arises (otherwise, numerical instability might occur if the steel particles were very light relative to the concrete particles).

The steel rebar overlays the concrete domain, leading to a slight over-estimation of the mass of the structure, because concrete does not actually occupy the volume taken by the rebar. This simplification is acceptable and in accordance with typical engineering practice, as the volume of steel is normally approximately one percent of the volume of the surrounding concrete.

To avoid a pathological situation in which the bond force direction might change rapidly with time, possibly causing computational stability issues, if a steel particle and a concrete particle are sufficiently close together in the reference configuration (closer than $0.2L$), then the program assumes the force between the two particles is null.

14.1.4 *Interaction between concrete and loading plates*

The interaction between the concrete structure and the steel loading plates is the same as that between concrete and rebar described in Eq. 14.2 in the previous subsection.

We must take care to ensure adequate overlap in the reference configuration between the concrete beam and the loading plate – otherwise the two bodies may not interact in a stable manner.[c]

In all of the steel loading plates used in simulations described in this chapter, both the damage flag and the plasticity flag are turned off, so the loading plates are simulated as being perfectly elastic.

Next, we verify the SPLM elasticity, plasticity, and damage models at the lattice particle level.

14.2 Particle-level study of SPLM

To verify the SPLM model, let us begin by studying the behavior of a single particle within a spatially homogeneous, gradually applied deformation field. We focus upon a single lattice particle located in this homogeneous deformation environment in order to investigate whether the behavior matches the behavior mandated by the classical elasticity, plasticity, and damage models, especially as we rotate the lattice and alter the loading rate. We assume small stretches of all bonds, (say $S < .01$). We start with a one-dimensional lattice, and then investigate the behavior of two-dimensional plane stress and plane-strain lattices, and finally we examine the behavior of a three-dimensional lattice. We consider the following behaviors: linear elasticity, plasticity, initiation of damage, and subsequent fracture behavior.

14.2.1 *One-dimensional lattice*

Tests using a one-dimensional lattice under applied uniaxial homogeneous deformation (with no transverse radial constraint) show that while the SPLM simulation replicates Young's modulus exactly[d],

[c] We have not attempted simulating a frictionless condition between the loading plate and the concrete structure, although this should be possible. If too few particles interact, stability issues may ensue.

[d] The SPLM represents Young's modulus exactly, except for minor variations due to numerical noise due to the imprecision of floating-point operations. In the remainder of this chapter, when we say "exact", we mean exact except for numerical round-off errors,

regardless of lattice rotation, the plastic yield strength is mildly rate sensitive. This rate sensitivity is as expected according to the plastic flow algorithm given by Eqs. 10.26 and 10.27. Fig. 14.1 shows the plastic flow behavior with the displacement gradually applied over 50 time steps. Notice that the stress is slightly above the yield stress where the rate of applied deformation is high. Under deformation held constant for at least twenty time steps, the simulation almost exactly replicates the specified uniaxial yield stress σ_{Yield}.

$$E = 24.86 \times 10^9\, Pa;\ \nu = 0.2;\ \sigma_{Yield} = 27.58 \times 10^6\, Pa;\ \Delta t = 1 \times 10^{-7} s$$

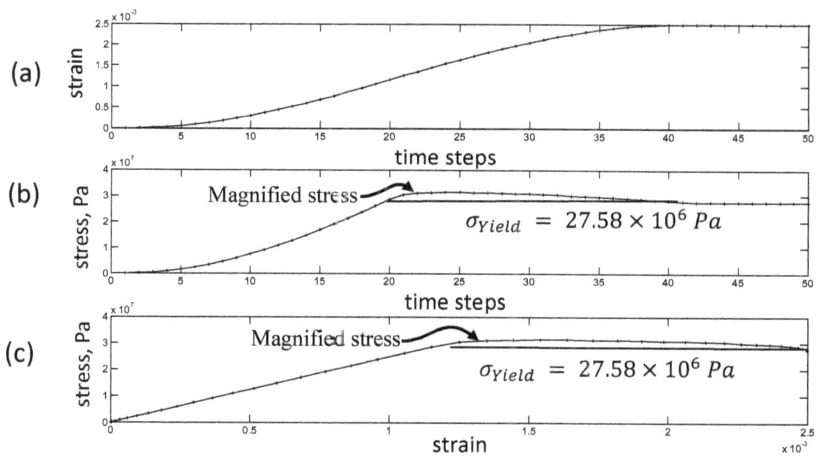

Figure 14.1. Behavior of one-dimensional SPLM model of a uniaxial bar subject to elastic-plastic time-varying axial deformation over 50 time steps. Note that the yield stress is affected by the rate of applied deformation.

Given the fact that the plasticity model itself is a gross approximation of physical reality, we conclude that this elasto-plasticity model, although slightly rate-dependent, is satisfactory. In fact, we may even consider this rate-dependency to be a feature of the model, as we observe in the laboratory qualitatively similar rate-dependency of this type. To implement rate-*independent* plasticity would be much more

---

which are typically in the seventh or eighth decimal place of the fractional component of the double-precision floating-point number, and are minor. In addition, we assume small stretches (and corresponding strains); otherwise, comparison to the classical Young's modulus would not be meaningful computationally and algorithmically intensive. It is computationally beneficial to avoid iterations within a time step, as would be required for a truly rate-independent plasticity model.

Next, let us investigate the *damage* behavior of a lattice particle within the one-dimensional lattice subject to applied homogeneous deformation. Fig. 14.2 shows that with continuously applied loading over fifty time steps, the simulation over-estimates the peak stress σ_t by a factor of about 23%. This is because with the high rate of deformation, the simulation uses only about five time steps to reach the stage at which damage commences.

$E = 24.86 \times 10^9\ Pa;\ v = 0.2;\ L = 0.01\ m\ \sigma_t = 2.758 \times 10^6 Pa;\ w_C = 0.2\text{ mm};\ \Delta t = 1 \times 10^{-7} s$

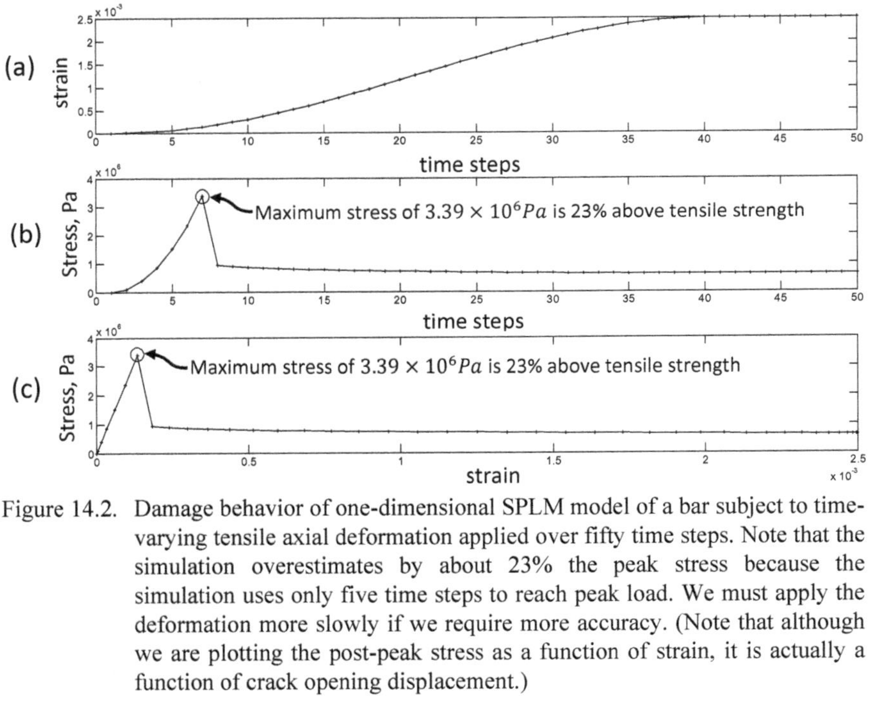

Figure 14.2. Damage behavior of one-dimensional SPLM model of a bar subject to time-varying tensile axial deformation applied over fifty time steps. Note that the simulation overestimates by about 23% the peak stress because the simulation uses only five time steps to reach peak load. We must apply the deformation more slowly if we require more accuracy. (Note that although we are plotting the post-peak stress as a function of strain, it is actually a function of crack opening displacement.)

Fig. 14.3 shows the same bar, but this time with the deformation applied more gradually, over 1000 time steps. Fig. 14.3 shows that now the stress-strain relationship is as we expect, with the peak tensile stress over-estimated by less than 1.7%.

The lesson from these benchmark studies is that if deformations change very quickly from time step to time step, there can be an effect upon plasticity and upon damage behavior. However, the linear elastic internal force state is not affected by the rate of deformation.

If both spatial homogeneity and quasistatic conditions prevail, the SPLM results are identical to the classical elasticity, plasticity, and damage models. However, these conditions hardly ever occur physically when damage occurs. Because we are using dynamic simulations, and such simulations are necessary to model damage realistically, it is important for the modeler to recognize and understand the rate effects upon simulated material behavior of the modelling approach.

$E = 24.86 \times 10^9\ Pa;\ \nu = 0.2;\ L = 0.01\ m\ \sigma_t = 2.758 \times 10^6 Pa;\ w_C = 0.2$ mm; $\Delta t = 1 \times 10^{-7} s$

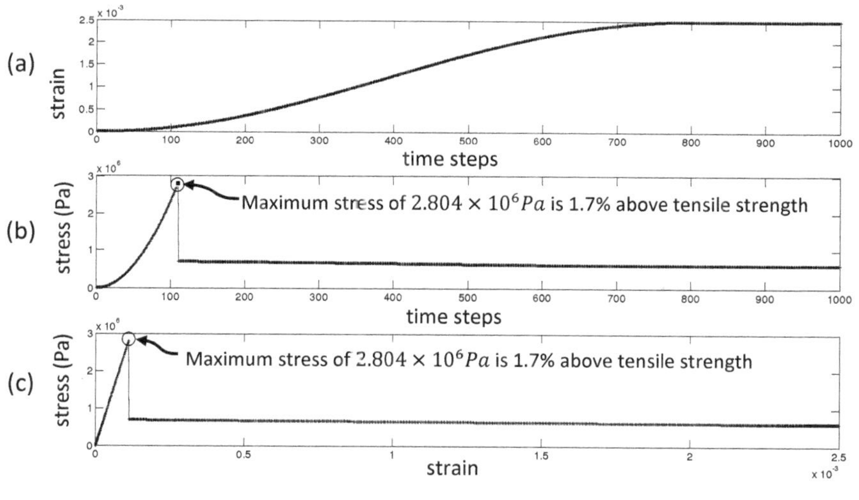

Figure 14.3. Damage behavior of one-dimensional SPLM model of a bar subject to time-varying tensile axial deformation applied over 1000 time steps. The computational simulation overestimates the specified peak stress by about 1.7%.

14.2.2 *Two-dimensional lattice*

Let us now investigate the behavior of a single lattice particle located within a two-dimensional hexagonal lattice, as shown in Fig. 14.4, subject to a spatially homogeneous deformation environment.

We consider four benchmark cases in which we may directly apply homogeneous deformation to the lattice particles:

1. uniaxial strain applied in the Y-direction, with no transverse constraint (plane stress with no constraint in the X-direction);

2. applied axial strain in the Y-direction with full lateral constraint in the X direction (plane stress with full constraint in the X-direction);
3. applied axial strain in the Y-direction with full lateral constraint in the Z direction (plane strain with no constraint in the X-direction);
4. and applied axial strain with full biaxial transverse constraint (plane strain, with full constraint in the X direction).[e]

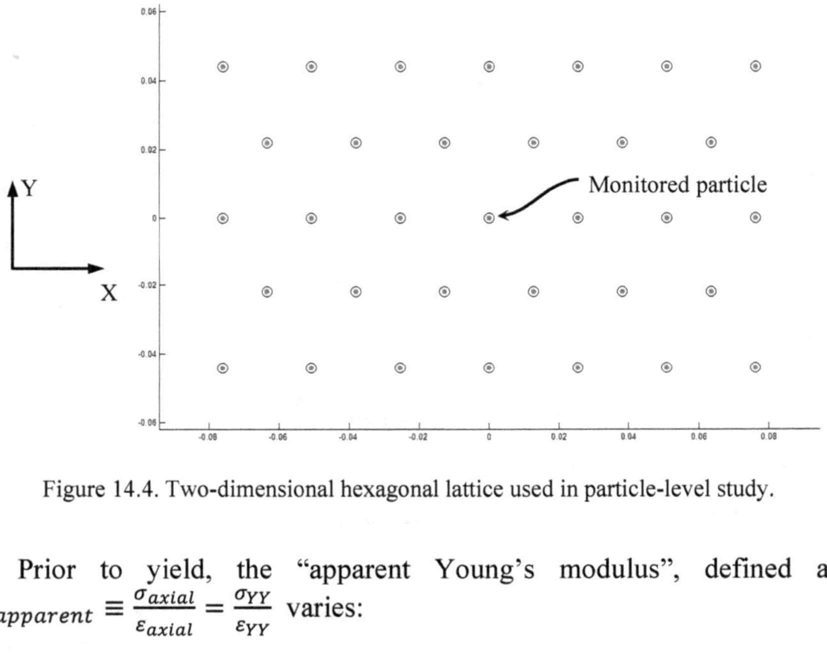

Figure 14.4. Two-dimensional hexagonal lattice used in particle-level study.

Prior to yield, the "apparent Young's modulus", defined as $E_{apparent} \equiv \frac{\sigma_{axial}}{\varepsilon_{axial}} = \frac{\sigma_{YY}}{\varepsilon_{YY}}$ varies:

$$E_{apparent} = E \qquad \text{(no transverse constraint)}$$

$$E_{apparent} = \frac{1}{(1-\nu^2)} E \quad \text{(transverse constraint in one direction only)} \qquad (14.3)$$

$$E_{apparent} = \frac{(1-\nu)}{(1+\nu)(1-2\nu)} E \quad \text{(full transverse constraint)}$$

[e] We apply uniaxial strain, rather than uniaxial stress, as a means to study the stress-strain behavior while avoiding dynamic behavior, which would complicate the response.

Assuming $\nu = 0.2$, $E_{apparent} = 1.0E$ for the case of no lateral constraint; $E_{apparent} = 1.04167E$ for transverse constraint in only one direction; and $E_{apparent} = 1.11\bar{1}E$ for full transverse constraint.

Similarly, for each of these cases, with J_2 plasticity, the axial yield stress $\sigma_{Yield(apparent)}$ in the first principal (Y) direction varies as:

$$\sigma_{Yield(apparent)} = \sigma_{Yield} \quad \text{(no transverse constraint)}$$

$$\sigma_{Yield(apparent)} = \frac{\sigma_{Yield}}{\sqrt{1-\nu+\nu^2}} \quad \text{(transverse constraint in one direction)} \tag{14.4}$$

$$\sigma_{Yield(apparent)} = \frac{\sigma_{Yield}(1-\nu)}{(1-2\nu)} \quad \text{(full transverse constraint)}$$

Assuming $\nu = 0.2$, $\sigma_{Yield(apparent)} = 1.0\sigma_{Yield}$ for no transverse constraint; $\sigma_{Yield(apparent)} = 1.09\overline{109}\sigma_{Yield}$ for transverse constraint in only one direction; and $\sigma_{Yield(apparent)} = 1.3\bar{3}\sigma_{Yield}$ for full transverse constraint.

In the same vein, we can calculate the apparent axial tensile strength $\sigma_{t(apparent)}$ for the three cases as

$$\sigma_{t(apparent)} = F_t \quad \text{(no transverse constraint)}$$

$$\sigma_{t(apparent)} = F_t\frac{(1-\nu)}{(1-\nu^2)} \quad \text{(transverse constraint in one direction)} \tag{14.5}$$

$$\sigma_{t(apparent)} = F_t\frac{(1-\nu)}{(1+\nu)} \quad \text{(full transverse constraint)}$$

Assuming $\nu = 0.2$, $\sigma_{t(apparent)} = 1.0F_t$ for no transverse constraint; $\sigma_{t(apparent)} = 0.83\bar{3}F_t$ for transverse constraint in only one direction; and $\sigma_{t(apparent)} = 0.66\bar{6}F_t$ for full transverse constraint.

Fig. 14.4 shows the lattice used for the benchmark studies, with the simulation monitoring the strain and stress at the central particle indicated. The apparent Young's modulus matches the theoretical prediction given by all three cases of Eq. 14.1 perfectly, regardless of lattice rotation about the Z-axis.

To study the plastic behavior while avoiding the complication of dynamic effects, we consider only the cases of plane stress and plane strain behavior, both constrained in the X-direction, with controlled strain applied in the Y-direction.

With 1000 time steps, the elasto-plastic stress-strain relation for the plane stress case, constrained laterally in plane, is well within one percent of the theoretical prediction for all lattice rotation angles. For the plane strain case, again constrained laterally in plane, the yield stress is also as expected; however, post-yield behavior requires knowledge of the out-of-plane plastic strain component. We have not yet implemented the case of plane strain plasticity even though to do so would not be difficult.

Damage initiation at a particle, as described in Chapter 11, depends upon the average elastic stretch[f] of all bonds attached to the particle, and as this average elastic stretch is independent of lattice rotation, damage initiation (and hence tensile strength) is independent of lattice rotation.

Subsequent to damage initiation at a particle, further damage evolution depends upon the *maximum* elastic stretch (not the *average* elastic stretch), which is somewhat dependent upon the lattice rotation. However, subsequent to damage initiation, the assumption of homogeneous deformation, and indeed of quasistatic conditions, is no longer realistic. Essentially, subsequent to damage initiation, further damage evolution is more of a structural, rather than a material, phenomenon.

We defer further consideration of post-peak damage evolution to Section 14.3.

14.2.3 *Three-dimensional lattice*

We find that the elastic strain and stress (derived from the deformation state and force state fields) at a lattice particle within a three-dimensional SPLM lattice body subject to homogeneous deformation are completely

[f] One can show that, for small stretch magnitudes, the average stretch is identical to the volumetric strain in a homogeneous deformation field.

invariant with respect to lattice rotation. This is unsurprising, as we designed the force state function, given by Eq. 9.28, to exhibit this property of objectivity with respect to lattice rotation.

The plastic behavior of a homogeneously deformed three-dimensional lattice is similarly – by design – independent of lattice rotation. This is borne out by numerical simulations.

We find damage *initiation* in a homogeneously deformed three-dimensional lattice to be – again by design – independent of lattice rotation. However, subsequent damage evolution is mildly dependent upon lattice orientation, as is investigated further in Section 14.3.2.

As before, initiation of damage and plasticity depends somewhat upon rate of loading, as we discussed in Section 14.2.1 for the one-dimensional lattice.

14.3 Standard concrete cylinder under axial loading

We have demonstrated that the pre-peak elastic, plastic, and damage behavior of the SPLM is invariant with respect to lattice rotation, at least at the particle level. Let us now investigate, as test cases, the simulated mechanical behavior of a standard 6-inch diameter by 12-inch long plain concrete cylinder, as shown in Fig. 14.5, loaded first in axial compression and then in axial tension.

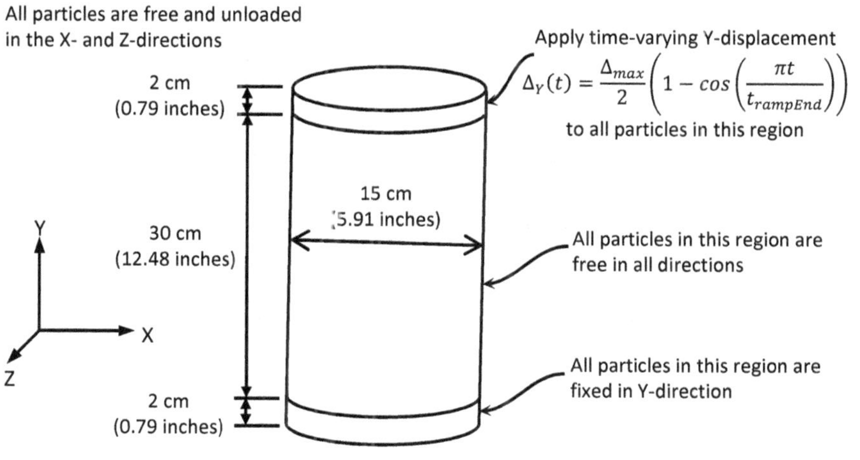
All particles are free and unloaded in the X- and Z-directions
Apply time-varying Y-displacement
$\Delta_Y(t) = \frac{\Delta_{max}}{2}\left(1 - cos\left(\frac{\pi t}{t_{rampEnd}}\right)\right)$
to all particles in this region
2 cm
(0.79 inches)
15 cm
(5.91 inches)
30 cm
(12.48 inches)
All particles in this region are free in all directions
Y
X
Z
All particles in this region are fixed in Y-direction
2 cm
(0.79 inches)

Figure 14.5. Concrete cylinder, showing dimensions and boundary conditions.

In each case, we apply uniaxial displacement-controlled time-varying loading $\Delta_Y(t) = \frac{\Delta_{max}}{2}\left(1 - cos\left(\frac{\pi t}{t_{rampEnd}}\right)\right)$ to the top two-centimeter-long end of the cylinder, leaving the central section unloaded. We set $t_{rampEnd} = 0.8 \times t_{end}$. The entire cylinder is unconstrained in the radial direction. We restrain the bottom two-centimeter long region of the cylinder from movement in the Y-direction. We seek to understand the effects of lattice rotation, defined by θ_Z, and rate of applied loading, defined by Δ_{max} and $t_{rampEnd}$, upon the simulated mechanical behavior.

In the following three subsections, we model the cylinder first as a one-dimensional lattice body, then as a two-dimensional cuboidal plane stress lattice body, and finally as a fully three-dimensional cylindrical lattice body. In the last subsection, we load the cylinder diametrally, in what civil engineers call the Brazilian split-cylinder test.

These simulations provide insight into the elastic, the damage, and the fracture behavior of the SPLM model. Only once we are satisfied with the model's ability to capture the behavior of plain concrete does it make sense to move on to the more complex cases of larger reinforced concrete structures, such as the reinforced concrete beams, presented in Section 14.4.

14.3.1 *One-dimensional lattice*

We first model the concrete cylinder as a one-dimensional lattice body, consisting of 35 particles spaced at 1 cm in the reference configuration, whose axis is oriented in the Y-direction, as shown in Fig. 14.6(d). The fundamental period of axial vibration T, assuming the concrete cylinder is fixed at the base and free at the top, is $T = 0.00042$ s. We set $t_{rampEnd} = 8 \times T = 0.00336$ s. The number of time steps during the ramp loading is $N_{steps} = \frac{0.00336s}{1\times10^{-7}s/step} = 33{,}600\ steps$, and with $t_{rampEnd} = 0.8 \times t_{end}$, the total number of time steps in the simulation is 42,000.

We find that the simulation replicates the classical elastic deflection $\Delta = \frac{PL}{AE}$, the classical compressive strength $P_c = F'_cA$, and the classical tensile strength $P_t = F_tA$ only approximately. Note that we can only represent the geometry – the diameter and length of the cylinder – to quantized lengths; that is to approximately the nearest lattice spacing of one centimeter. Thus, a lattice body that has 30 particles along its length simulates a cylinder that is between 29.5 and 30.5 lattice spacings long. Because the variation in this dimension is roughly one centimeter in 30 centimeters, or about 3.3%, we cannot expect to routinely match the classical results more exactly than 3.3%.

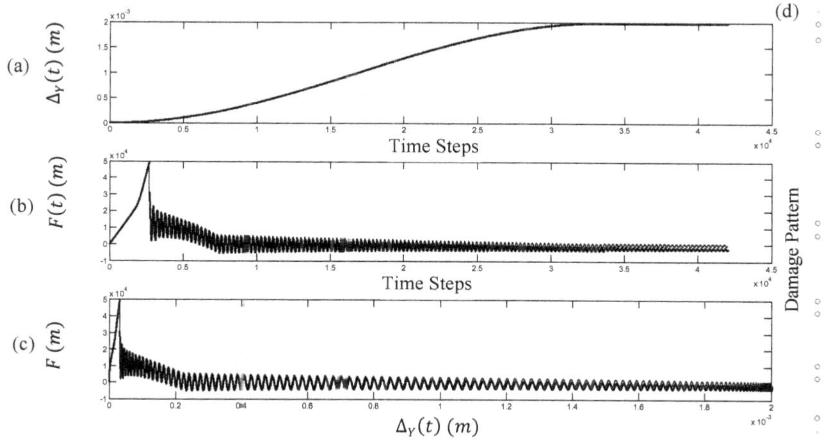

Figure 14.6. Load and displacement relations for 1D concrete cylinder loaded in tension. The image with 35 particles in (d) at right shows the final damage pattern, which is periodic, with complete damage occurring just below the top two loaded particles, and partial damage occurring in several other particle pairs along the length of the cylinder.

Fig. 14.6 shows the vertical displacement Δ_Y of the loaded region at the top and the vertical force F acting between the loaded region and the rest of the concrete body. The elastic behavior prior to tensile fracture is correct, but the simulation over-predicts the peak tensile stress by 1.33 percent. One notices that the cracking process releases the elastic energy stored prior to fracture as dynamic vibrational energy that gradually dies out due to damping.[g] The vibrational frequency of the top part is much

[g] This is the loud "bang" that one hears in the laboratory upon fracture of such a concrete cylinder loaded in tension. Depending upon where along the length of the cylinder the fracture occurs, the subsequent vibrational frequencies of the two remaining parts will

higher than the original vibrational frequency of the unbroken concrete cylinder.

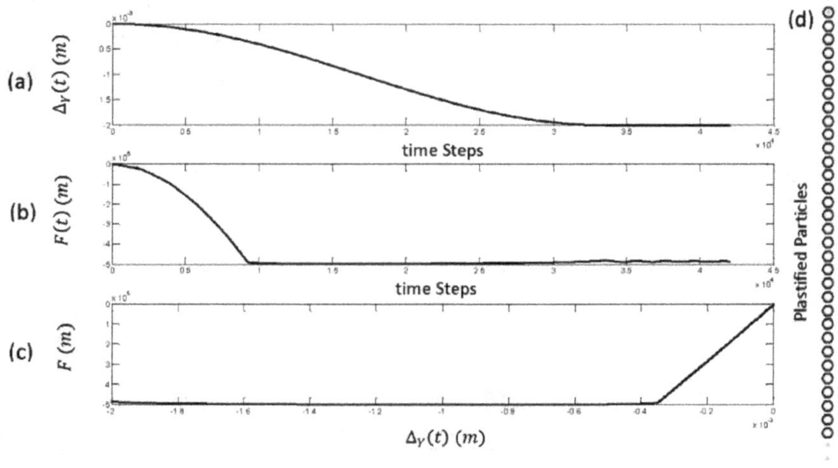

Figure 14.7. Load and displacement relations for 1D concrete cylinder loaded in compression. The 35 particles in (d) show that all particles in the cylinder plastify, except for three particles in the boundary regions.

Now, let us load the cylinder in compression. Fig. 14.7 shows the resulting displacement of the top loaded region and vertical force between the top region and the concrete cylinder. Fig. 14.7(d) shows that every particle (except for particles in the boundary regions at top and bottom) becomes plastic, in contrast to the tensile behavior shown in Fig. 14.6(d), in which damage is evident in only a few particles. Due to the rate-dependency of the yield load discussed in Section 14.1, the simulation over-predicts the yield stress, but by less than one percent.

The plastic model for compressive behavior of the concrete cylinder is of course a simplification; in reality, there would be post-peak reduction in compressive force as compressive deformation increases. Nonetheless, we conclude that the one-dimensional SPLM model is sufficiently realistic, for our present purposes, to model with reasonable fidelity, both the tensile and compressive behavior of a concrete test cylinder.

---

vary. Thus the simulation results subsequent to fracture are highly sensitive to initial conditions and numerical round-off errors, and are thus inherently unpredictable.

14.3.2 *Two-dimensional lattice*

We now simulate the concrete cylinder as a plane stress lattice of thickness 0.15 m in the Z-direction. Thus, we now approximate the cylinder as a cuboidal domain with a 0.15m by 0.15m square cross-section. Fig. 14.8 shows the three lattices, rotated counterclockwise by 0°, 10° and 30° about the Z-axis.

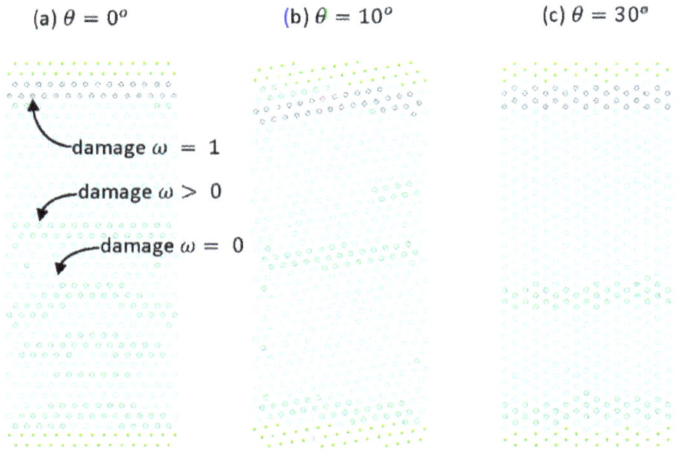

Figure 14.8. Damage patterns as a function of the rotated lattices used for two-dimensional plane-stress simulation of concrete cylinder.

Using the same applied compressive applied displacement as in the previous section, and as illustrated in Fig. 14.7(a), the yield stress is within one percent of the specified value for all three lattice rotations. However, the yield *force* F_{Yield} in the cylinder varies from 9% lower than the theoretical value (calculated as $F_{Yield} = \sigma_{Yield} \times Area$) for the lattice with 0° rotation, to 2% low for the lattice with 10° rotation, and finally 1% high for the lattice with 30° rotation. We ascribe these errors to variations in modeling the target cross-sectional width caused by the varying lattice rotations.[h] Nonetheless, the elastic and compressive yield behavior behaves essentially as expected – within the ability of the lattice to represent the specimen geometry.

---

[h] As we decrease the lattice spacing, or increase the cylinder diameter, these percent errors would be expected to decrease. Concrete is an inherently random material, and these errors are consistent with this randomness.

The damage and fracture behavior varies somewhat with lattice rotation, as shown in Fig. 14.8. The tensile strength $F_{tensile}$ in the cylinder varies from 11.5% lower than the theoretical value (calculated as $F_{Tensile} = \sigma_t \times Area$) for the lattice with 0° rotation, 18% low for the lattice with 10° rotation, and 12% low for the lattice with 30° rotation. We ascribe these errors to the artificial constraint of the loaded regions at the top and bottom of the cylinder, which are constrained to move as rigid bodies, and to boundary effects and to variations in modeling the target cross-sectional width caused by the lattice rotations. To save space, we do not show the load versus displacement behavior, but it is quite similar in character to that shown in Fig. 14.6 for the one-dimensional lattice model.

Each of the two-dimensional lattice simulations, employing approximately 570 particles and 42000 time steps, required approximately eight minutes running on a single processer computer.

14.3.3 *Three-dimensional lattice*

Next, we investigate the behavior of a three-dimensional lattice body simulating the concrete cylinder. Each problem contains approximately 8600 particles, and we again run each simulation for 42000 time steps, as in the previous two subsections. As in the previous subsection, we consider three lattice rotations: 0°, 10° and 30° about the Z-axis, as shown in Fig 14.9. Each problem required approximately one hour to run on two processors running in parallel.

For a three-dimensional FCC lattice representing a 6-inch-diameter cylinder, the variation in cross-sectional area is up to 14%. Therefore, with a 1-cm lattice spacing modeling our 6-inch-diameter cylinder, we should expect the axial stiffness results to vary by up to 15% due to varying lattice rotations and translations with respect to the classical analytical definition of the body.

The elastic load versus deformation relationship for the three lattice rotations is approximately correct, varying by up to 10% from the classical solution. As described in the previous subsection, the error in elastic load (for a given applied deformation of the top cap) is primarily due to the approximate nature of the geometric discretization.

The yield stress of the central particle is well within one percent of the specified yield stress.

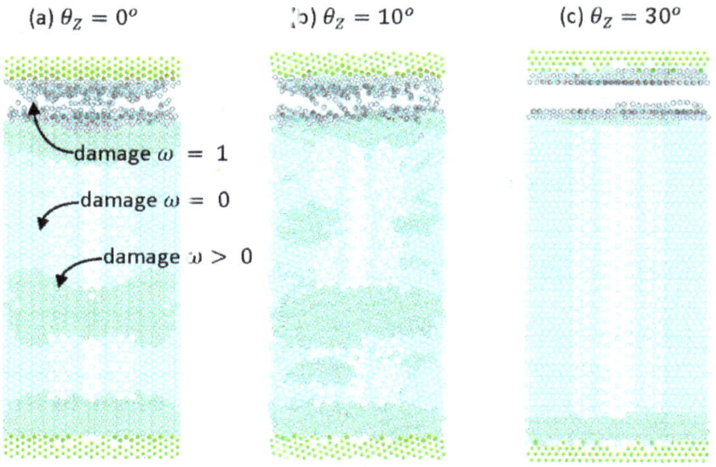

Figure 14.9. Damage of three-dimensional lattice model of concrete cylinder subject to three lattice rotations about the Z axis.

The yield force is 1.5% lower than the theoretical value for $\theta_Z = 0^o$, 1.5% lower than the theoretical value for $\theta_Z = 10^o$, and 6% above the theoretical value for $\theta_Z = 30^o$. Again these errors are ascribed to the approximate nature of the geometric discretization of the concrete cylinder.

When stretched, the tensile damage initiated at peak loads of 18%, 20%, and 15% below the theoretical value of tensile strength for the $\theta_Z = 0°$, $\theta_Z = 10°$, and $\theta_Z = 30°$ lattice rotations respectively. This under-prediction of the tensile strength is probably due to boundary effects causing increases in bond stretch and subsequent premature damage initiation. Just as in the laboratory, in the SPLM simulation it is difficult to avoid spurious effects of the method by which tensile loading is applied. Fig. 14.9 shows the damage patterns subsequent to complete fracture for three lattice rotations. Note that in each case some partial damage occurred in bands at various locations along the length of the specimen. Dynamic stress waves traveling along the axis of the cylinder likely caused this partial damage. However, in all cases, the fracture occurred within the particles nearest to the loaded end of the cylinder. The overall direction of the crack was unaffected by lattice rotation, but the morphology (width) of the crack band is somewhat affected by lattice rotation.

Without going into detail, in each case load-displacement behavior was similar to that shown in Fig. 14.6, and thus the fracture energy absorption seems to be reasonable.

14.3.4 *Brazilian split cylinder*

The Brazilian split cylinder test has been used over the years as an indirect (and notoriously poor) measure of the tensile strength of concrete. In this test, the standard $H = 12in.$ in height by $D = 6in.$ diameter concrete cylinder is loaded in compression along its diameter, as shown in Fig. 14.10.

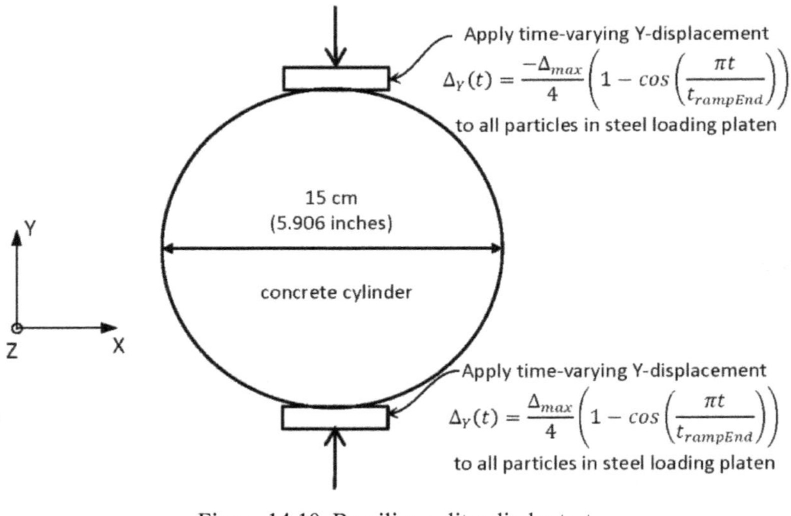
Apply time-varying Y-displacement
$\Delta_Y(t) = \frac{-\Delta_{max}}{4}\left(1 - cos\left(\frac{\pi t}{t_{rampEnd}}\right)\right)$
to all particles in steel loading platen
15 cm
(5.906 inches)
concrete cylinder
Y
Z
X
Apply time-varying Y-displacement
$\Delta_Y(t) = \frac{\Delta_{max}}{4}\left(1 - cos\left(\frac{\pi t}{t_{rampEnd}}\right)\right)$
to all particles in steel loading platen

Figure 14.10. Brazilian split cylinder test.

Assuming perfectly classical linear elastic behavior up to the point where a vertical crack through the center of the cylinder forms due to exceedance of the tensile strength σ_t, the peak load P_{max} that can be applied to the cylinder is given by

$$P_{max} = \frac{\pi HDF_t}{2} = \frac{\pi \times 0.30m \times 0.15m \times 2.758 \times 10^6 Pa}{2} = 195\ kN. \quad (14.6)$$

Fig 14.11 shows the SPLM simulation of the split cylinder at three different stages of compressive plastification and subsequent tensile cracking. The maximum applied relative displacement between the top and the bottom platen is 5.0 mm. A diametral tensile crack forms, but only after significant compressive plastification under the applied loads. The peak load of 513 kN, which is more than twice the theoretical value given by Eq. 14.6, occurs at approximately stage (a) in Fig. 14.11. This supports the well-known fact that the tensile strength obtained from the Brazilian split cylinder test produces approximately twice the tensile strength as that obtained from a direct tension test.

This peak load occurs well before the tensile crack, shown in Fig. 14.11(c), forms in the simulation, indicating that the proximate failure mechanism is plastification/crushing in the compressive regime near the applied loads, and not the formation of a tensile crack in the middle of the cylinder.

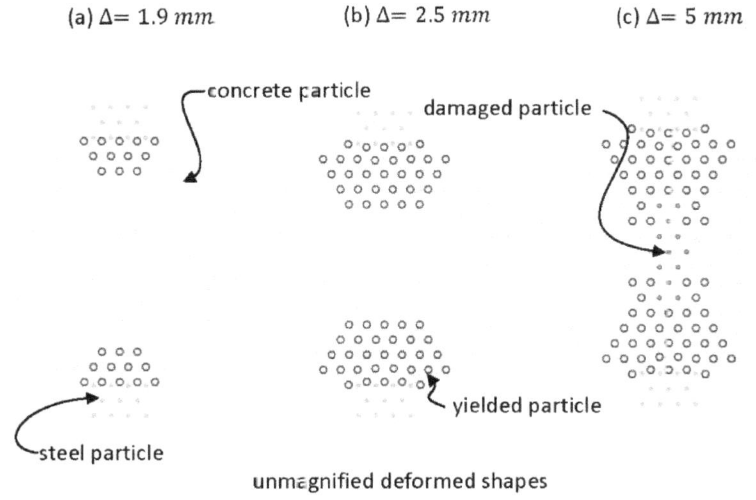

Figure 14.11. Brazilian split cylinder simulation results at three different stages of deformation.

14.4 Simulation of reinforced concrete beams

The most common element in concrete structures is the reinforced concrete beam. To exhibit the SPLM's capabilities in simulating reinforced concrete beams, we consider a centrally-loaded, simply-supported beam of width b, depth h, distance from centroid of tension reinforcement to top of beam d, and other dimensions shown in Fig. 14.12.

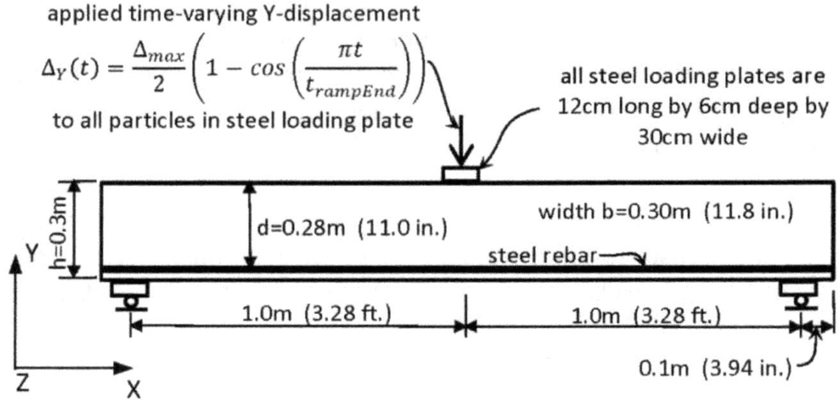

Figure 14.12. Geometry and loading of reinforced concrete beam.

To demonstrate various failure modes, we simulate the beam with varying flexural reinforcement ratios, defined as $\rho_s \equiv \frac{A_s}{bd}$, with $\rho_s = 0.2\%$, $\rho_s = 0.5\%$, $\rho_s = 1.0\%$, and $\rho_s = 2.0\%$.

We use the same SPLM simulation material models and parameters as were used in Section 14.3, with the material properties for concrete and for steel defined in Tables 14.1 and 14.2 respectively. We model the reinforced concrete beam and its steel supports as plane stress lattice bodies, and we model each reinforcing bar as a one-dimensional lattice body. The total number of lattice particles is 8147. The simulated duration of each simulation is eight fundamental periods, or $8 \times 0.020s = 0.16s$ (approximately 1.6 million time steps). We run the simulation on eight processors, and the simulation requires approximately 7.5 hours.

Although this duration would be sufficient to achieve a quasistatic elastoplastic solution, with damage and cracking, due to snapback behavior, the solution is dynamic at times of unstable cracking regardless of how slowly the beam is loaded.

Figs. 14.13 through 14.16 show the simulation results for each of the four steel ratios.

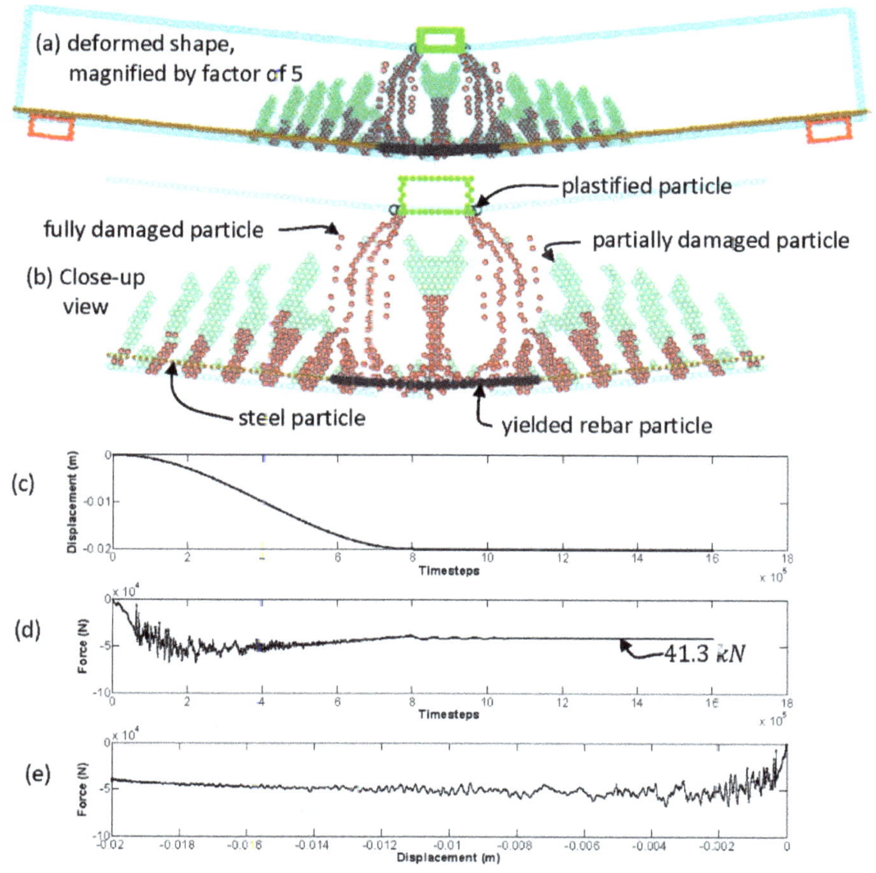

Figure 14.13. Simulation results of beam with flexural steel ratio $\rho_s = 0.2\%$. (a) Deformed shape at end of simulation; (b) Close-up view; (c) Applied load point displacement vs. time step; (d) Force between loading plate and concrete beam vs. time step; (e) Load vs. load point displacement.

The most lightly reinforced beam, with 0.2% flexural steel, failed primarily by yielding of the flexural steel, accompanied by plastification of concrete particles at the top of the beam – a typical flexural failure of an under-reinforced concrete beam. Fig. 14.13(d) shows that the force versus time graph is quite noisy, as would be expected as damage and consequent flexural cracks form. Defining the failure load $P_{N(SPLM)}$ as the steady-state force response at the end of the simulation we see the beam holds $P_{N(SPLM)} = 41.3\ kN$. This compares to the flexural strength according to the ACI Code flexural strength equations calculated as

$$\begin{gathered} A_s = A_s bd = 0.002 \times 0.3m \times 0.28m = 0.000168m^2 \\ a = \frac{A_s F_Y}{0.85F'_c b} = \frac{0.000168m^2 \times 413.7 \times 10^6 Pa}{0.85 \times 27.58 \times 10^6 Pa \times 0.3m} = 0.009882m \\ M_{N(ACI)} = A_s F_Y \left(d - \frac{a}{2}\right) \\ = 0.000168m^2 \times 413.7 \times 10^6 Pa \left(0.28m - \frac{0.009882m}{2}\right) \\ = 19.12\ kN - m \\ P_{N(ACIflexural)} = \frac{4M_{N(ACI)}}{span} = \frac{4 \times 19.12\ kN-m}{2m} = 38.2kN \end{gathered} \tag{14.7}$$

The SPLM simulation predicts a strength P_N that is 8.1% higher than the ACI code prediction. This difference is reasonable, as one can consider neither of these predictions highly precise. However, the SPLM simulation is far more rational and elucidates much more behavior (cracking spacing and width, deflections, dynamic effects, etc.) than do the ACI code formulas.

The ACI Code predicts the shear strength as

$$\begin{gathered} V_{N(ACI)} = 2\left[\sqrt{F'_c(in\ PSI)}(in\ PSI)\right] bd \\ = 2\sqrt{4000PSI} \times \frac{6894.8Pa}{PSI} \times 0.3m \times 0.28m = 73.3\text{kN}\ ; \\ P_{N(ACIshear)} = 2V_{N(ACI)} = 146.5kN\ , \end{gathered} \tag{14.8}$$

Therefore, shear failure clearly does not control the strength of this lightly reinforced beam.

Fig. 14.14 shows the results of the beam with 0.5% flexural reinforcement. The flexural steel yields in a roughly symmetrical fashion, but interestingly the rebar does not yield at the center of the beam. Apparently, the tension-stiffening behavior of the concrete prevents rebar yield at the beam center. Additionally, four tension cracks form near the top of the beam. This is surprising, and perhaps dynamic stress waves emanating from cracks forming on the tension side of the beam caused the secondary cracks at the top of the beam.

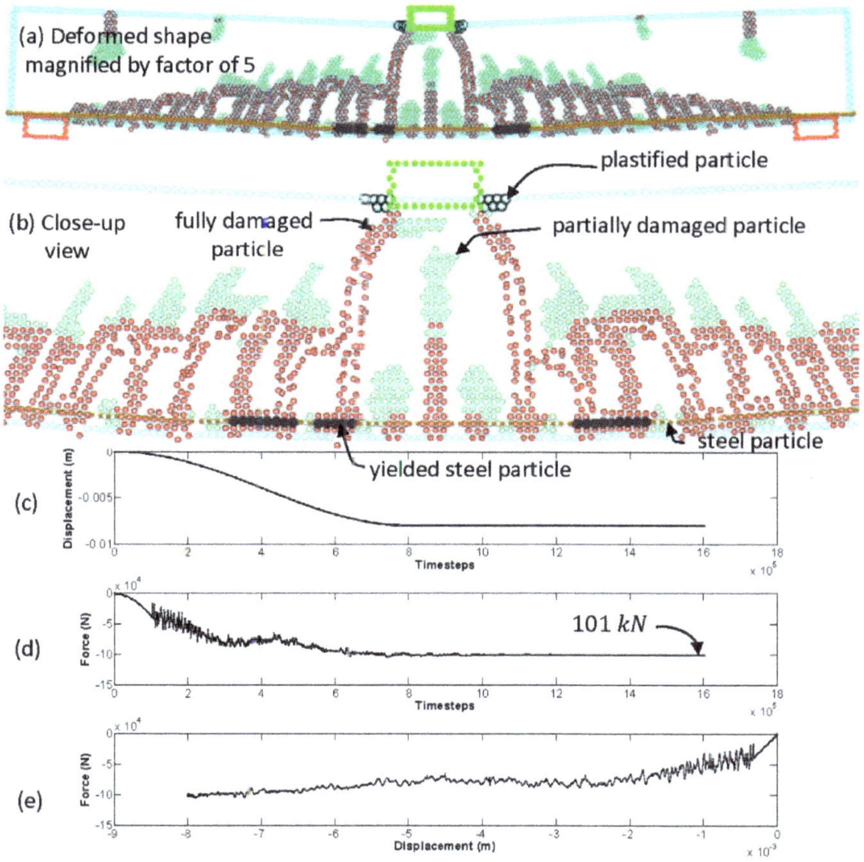

Figure 14.14. Simulation results of beam with flexural steel ratio ρ_s = 0.5%. (a) Magnified deformed shape at end of simulation; (b) Close-up view; (c) Applied load point displacement vs. time step; (d) Force between loading plate and concrete beam vs. time step; (e) Load vs. load point displacement.

Fig. 14.14(d) shows that the load point force plateaus at about 101 kN. The ACI code predicts a flexural strength of 93.0kN and a shear strength of 146.5 kN for this beam. The SPLM predicts a flexural strength 8.6% higher than is predicted by the ACI code. This difference is reasonable in light of the differing approximations made by both methods of analysis.

Fig. 14.15 shows the results of the beam reinforced with 1% flexural reinforcement. The ACI code predicts a flexural strength of $P_{N(flexuralACI)} = 177.4\ kN$ and shear strength of $P_{N(shearACI)} = 146.5\ kN$ for this beam; thus the ACI Code predicts a shear failure.

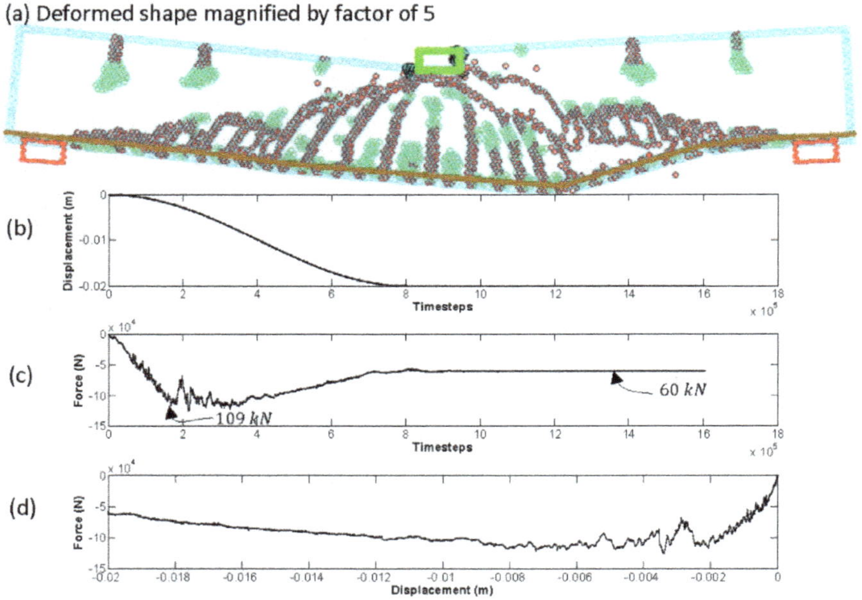

Figure 14.15. Simulation results of beam with flexural steel ratio $\rho_s = 1.0\%$. (a) Deformed shape at end of simulation; (b) Applied load point displacement vs. time step; (c) Force between loading plate and concrete beam vs. time step; (d) Load vs. load point displacement.

The SPLM simulation indeed also predicts a classical shear fracture mode of failure, at a load of $P_{N(SPLM)} = 109\ kN$, as is evident from the cracking pattern evident Fig. 14.15. Interestingly, after the main shear cracks form, the beam is still able to carry some residual load of $P_{N(SPLM)} = 60\ kN$, perhaps due to a combination of catenary action of the un-yielded rebar, and the plastically deforming particles at the top center of the beam.

The SPLM simulation predicts a shear strength that is 26% lower than the ACI prediction; this is probably because the SPLM model does not attempt to model post-cracking aggregate interlock along the main shear crack.

Fig. 14.16 shows the results of the beam reinforced with 2% flexural reinforcement – a very heavily reinforced beam. The ACI code predicts a flexural strength of $P_{N(flexuralACI)} = 320.5\ kN$ and predicts shear strength of $P_{N(shearACI)} = 146.5\ kN$; thus the ACI Code predicts a shear failure.

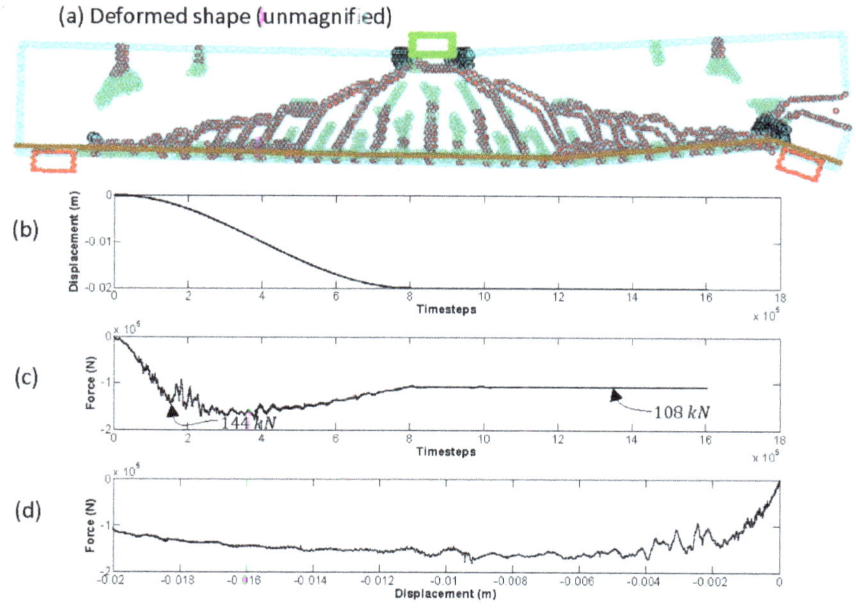

Figure 14.16. Simulation results of beam with flexural steel ratio $\rho_s = 2.0\%$. (a) Deformed shape at end of simulation; (b) Applied load point displacement vs. time step; (c) Force between loading plate and concrete beam vs. time step; (d) Load vs. load point displacement.

The SPLM simulation instead predicts a rebar anchorage-type failure at a load of $P_{N(SPLM)} = 144\ kN$, as is evident from the cracking pattern shown in Fig. 14.16. Even after the anchorage crack forms, the beam is still able to carry some residual load of $P_{N(SPLM)} = 108\ kN$.

The SPLM simulation predicts a shear strength that is 1.7% lower than the ACI prediction; the close correspondence is probably a coincidence, as the failure types are completely different. The ACI code does not provide specific provisions for anchorage failure; instead it requires a "development length", as a way to avoid having to predict rebar anchorage and splice failures.

Finally, let us add stirrups to the beam with 1% flexural reinforcement, as shown in Fig. 14.17. The stirrups are spaced at $s_V = \frac{h}{3} = 0.10\ m\ (3.94\ in.)$, and the area of each stirrup is $A_v = 0.005 \times h \times s_V = 0.00015\ m^2\ (0.23\ in.^2)$.

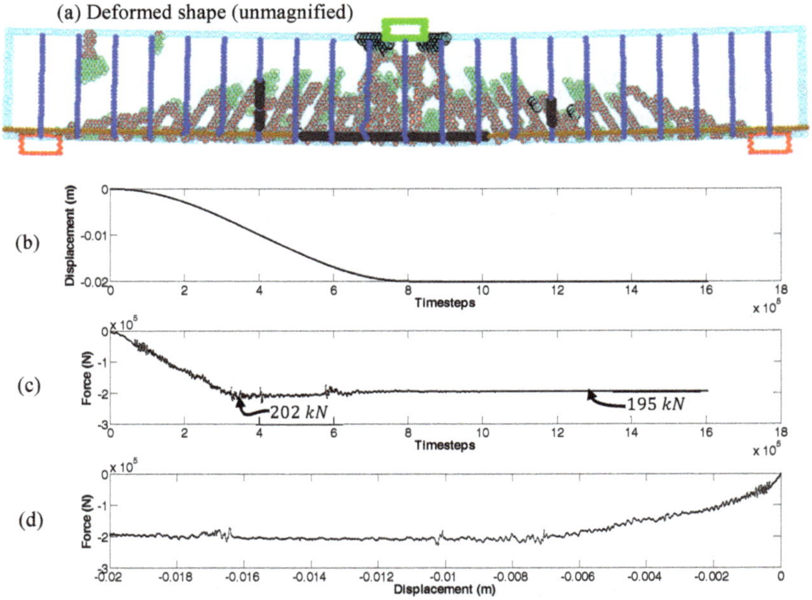

Figure 14.17. Simulation results of beam with stirrups spaced at $s_V = \frac{h}{3} = 0.10\ m$ (3.94 in.) with area $A_v = 0.005 \times h \times s_V = 0.00015\ m^2\ (0.23\ in.^2)$ and with flexural steel ratio $\rho_s = 1.0\%$. (a) Deformed shape at end of simulation; (b) Applied load point displacement vs. time step; (c) Force between loading plate and concrete beam vs. time; (d) Load vs. load point displacement.

Fig. 14.17 (c) shows that the strength of the beam is now approximately *202 kN* (or *195 kN*, depending upon how one defines strength). According to the ACI code, the shear strength is now increased due to the presence of the stirrups as

$$V_{N(ACI)} = V_C + V_S = 2\left[\sqrt{F'_c(in\ PSI)}(in\ PSI)\right] bd + \frac{A_v F_y d}{s_V}$$

$$= 73.3kN + \frac{0.00015m^2 \times 413.7\ \text{MPa} \times 0.28m}{0.10m} = 73.3kN + 173.8kN \quad (14.9)$$

$$247.1kN\ ;$$

$$= P_{N(ACIshear)} = 2V_{N(ACI)} = 494.2kN\ ,$$

Thus the shear failure mode no longer controls the strength of the beam, because the flexural mode of failure of $P_{N(flexuralACI)} = 177.4\ kN$ is now smaller. The SPLM prediction of 195 kN is 10 percent higher than the ACI code prediction. Perhaps the increased strength in the SPLM analysis is due to the funicular shape of the flexural steel helping to support the applied load.

14.5 Discussion

The examples included in this chapter have demonstrated the capabilities of the SPLM to simulate reinforced concrete beams, including elastic, plastic, damage, fracture, dynamic, and large deformation behaviors. The results are qualitatively convincing, although not highly precise.

In deterministic analysis of reinforced concrete structures, design engineers ought not to be too concerned with obtaining a high degree of precision – concrete is inherently random and thus imprecise. However, it is crucial to simulate all of the essential *features* of concrete behavior, even if only approximately. This, the SPLM does well.

Another area where the SPLM excels is in the simulation of dynamic behavior of structures – indeed the method is perhaps best suited to the simulation of short term dynamic events. As we have seen, however, even events normally thought of as being quasistatic, are not quasistatic if the dynamic nature of crack formation and propagation is included in the simulation. In this sense, the SPLM opens up new understanding about an important aspect of material behavior that researchers have so far understood only vaguely. See, for example (Gopalaratnam, V., Gerstle, W., Isenberg, J., and Mindess, S., 2004).

The method requires powerful computers. The reinforced beam examples presented in Section 14.4, each required 8147 lattice particles running for 1605000 time steps for approximately eight hours on an eight-processor parallel computer. We anticipate that as computer hardware and software becomes more capable in the coming years that the solution of increasingly large problems, with more particles and more time steps, will become ever more feasible.

One of the most important contributions of the SPLM is its clear statement about the smallest size scale of the model. By choosing the lattice spacing, this size scale is determined. Specifying this size scale is necessary if engineers are to develop understanding and confidence in the model. By fixing the lattice spacing, the SPLM achieves much computational, as well as conceptual, simplicity.

Although we have demonstrated that the method is reasonably objective with respect to lattice orientation and translation, there is still some lack of objectivity, especially in the post-peak damage regime. If we seek higher degrees of objectivity and isotropy, then we can extend the SPLM, using the techniques presented in this book, by including more bonds per particle, but at the cost of increased computing requirements. We believe that the SPLM presented in this book, with eighteen bonds per particle, is sufficiently effective for many purposes in predicting the behavior of reinforced concrete, and many other types of structures.

We have not yet addressed some aspects of the constitutive model, including creep, shrinkage, and thermal effects. Researchers can address these and other aspects of material behavior within the framework of the SPLM.

The main advantage of the SPLM over other contemporary computational simulation approaches is that the SPLM is a much simpler model, ideally suited to modern digital computers. The SPLM makes fewer, and different, assumptions about material behavior than the classical continuum mechanics approach, and is thus capable of predicting a wider range of behaviors than the more conventional models.

The conventional computational models will certainly continue to be important and necessary. The SPLM adds a valuable tool for the computational simulation of solid structures.

14.6 Exercises

14.1 Derive Eqs. 14.3 for the apparent Young's modulus $E_{apparent}$ of a bar with various levels of transverse constraint.

14.2 Derive Eqs. 14.4 for the apparent yield strength $\sigma_{Yield(apparent)}$ of a bar with various levels of transverse constraint.

14.3 Derive Eqs. 14.5 for the apparent tensile strength $\sigma_{t(apparent)}$ of a bar with various levels of transverse constraint.

14.4 Determine the fundamental period of axial vibration of a 6-inch long normal concrete cylinder. Develop an approximate analytical model of the cylinder. Compare your answer to the value of 0.00042 s given in Section 14.3.1. Also, compare your solution to a finite element modal analysis solution.

14.5 Perform a linear elastic plane stress finite element analysis of the Brazilian split cylinder discussed in Section 14.3.4. Discuss the results of your analysis, and compare these results to the simulation results of Section 14.3.4.

14.6 Verify the fundamental frequency of vibration of the beam shown in Fig. 14.12. (a) Develop an approximate analytical model; (b) Develop a plane stress finite element model; (c) run an SPLM simulation without damping, plasticity, and damage.

14.7 In Fig. 14.14, tensile cracks formed at the top of the beam. Using a time-history dynamical linear elastic finite element analysis, with the same applied displacement as in the example simulation, determine if you can explain why these tensile cracks might have occurred.

14.8 Repeat the analysis of the Brazilian split cylinder presented in Section 14.3.4, except decrease the lattice spacing by a factor of two, to 0.5 cm. How do the simulation results change?

14.9 Repeat the analysis of the Brazilian split cylinder presented in Section 14.3.4, except increase the specimen size by a factor of two, while keeping the lattice spacing at 1 cm. How do the simulation results change?

14.10 Repeat the analyses of the reinforced concrete beam presented in Section 14. 4, with flexural steel ratio $\rho_S = 0.2\%$, except decrease the simulation time by (a) a factor of two, to four fundamental periods; (b) a factor of four, to two fundamental periods. How do the simulation results depend upon the duration of load application?

Bibliography

ACI318. (2014). *ACI 318 Building Code Requirements for Structural Concrete.* Farmington Hills Michigan: American Concrete Institute.

ACI446. (2009). *ACI 446.3R-97: Finite Element Analysis of Fracture in Concrete Structures: State of the Art.* Farmington Hills, Michigan: American Concrete Institute.

Anderson, T. L. (2005). *Fracture Mechanics.* Boca Raton, Florida: Taylor and Francis.

Barenblatt, G. I. (1962). The mathematical theory of equilibrium cracks in brittle fracture. *Advances in Applied Mechanics*, 55-129.

Bazant, Z. P. and Jirasek, M. (2002). Nonlocal Integral Formulations of Plasticity and Damage: Survey of Progress. *Journal of Engineering Mechanics ASCE*, 1119-1149.

Bazant, Z. P. and Oh, B. -H. (1983). Crack band theory for fracture of concrete. *Materials and Structures*, 155-177.

Bazant, Z. P. and Planas, J. P. (1998). *Fracture and size effect in concrete and other quasibrittle materials.* Boca Raton, Florida: CRC Press.

Beazley, D.M. and Lomdahl, P. S. (1994). Message-Passing Multi-Cell Molecular Dynamics on the Connection Machine 5. *Parallel Computing 20 (2)* (pp. 173-195). Amsterdam: Elsevier.

Bessuelle P. and Rudnicki, J. W. (2004). Localization: Shear bands and compaction bands. In M. Gueguen Y. and Bouteca, *Mechanics of Fluid-Saturated Rock* (pp. 219-321). London: Academic Press.

Bittencourt, T. N., Wawrzynek, P. A., Ingraffea, A. R. and Sousa, J. L. (1996). Quasi-automatic simulation of crack propagation for 2D LEFM problems. *Engineering Fracture Mechanics*, 321-334.

Bolander, J. E. and Hong G. S. (2002). Rigid-body-spring network modelling of prestressed concrete members. *Structural Journal of the ACI*, 595-604.

Boresi, A.P., Schmidt, R.J., and Sidebottom, O.M. (1993). *Advanced Mechanics of Materials.* New York: John Wiley and Sons.

Cauchy, A. L. (1823). Recherches sur l'equilibre et le mouvement interieur des corps solides ou fluides, elastiques ou non elastiques. *Bulletin de la Societe Philomatique*, 9-13.

Cauchy, A. L. (1827). De la pression ou tension dans un corps solide. *Exercices de mathematiques*, 42-56.

Cauchy, A. L. (1827). Sur la condensation et la dilitation des corps solides. *Exercises de Mathematiques, Volume 2*, 60-69.

Cosserat, E. and Cosserat, F. (1909). *Theorie des Corps Deformables.* Paris: Hermann et Fils.

Cusatis, G. (2001). *Tridimensional Random Particle Model for Concrete.* Milano: Politecnico di Milano.

Dugdale, D. S. (1960). Yielding of steel sheets containing slits. *Journal of the Mechanics and Physics of Solids*, 100-104.

Erdogan, F., and Sih, G. C. (1963). On the crack extension in plates under plane loading and transverse shear. *ASME Journal of Basic Engineering*, 519-527.

Eringen, A. C. (2002). *Nonlocal Continuum Field Theories.* New York: Springer.

Felippa, C. A. (1994). 50 Year Classic Reprint: An Appreciation of R. Courant's 'Variational Methods for the Solution of Problems of Equilibrium and Vibrations,' 1943. *International Journal for Numerical Methods in Engineering*, 2159-2187.

Gerstle, K. H. (1974). *Basic Structural Analysis.* Englewood Cliffs, N.J.: Prentice Hall.

Gerstle, W. H. and Ingraffea, A.R. (1991). Does Bond-Slip Exist? *Concrete International, ACI*, 44-48.

Gerstle, W. H. and Sau, N. (2004). Peridynamic modeling of Concrete Structures. *Proceedings of the 5th Intl. Conf. on Fracture Mechanics of Concrete Structures* (pp. 949-956). IA-FRAMCOS.

Gerstle, W. H., Sau, N., and Aguilera, E. (2007). Micropolar modeling of concrete structures. *Proc. 6th Intl. Conf. on Fracture Mechanics of Reinforced Concrete Structures.* Catania, Italy: Ia-FRAMCOS.

Gerstle, W., and Sau, N., and Silling, S. (2007). Peridynamic Modeling of Concrete Structures. *Journal of Nuclear Engineering and Design*, 1250-1258.

Gerstle, W., Honarvar Geitanbaf, H., and Asadollahi, A. (2013). Computational simulation of reinforced concrete using the micropolar state-based peridynamic hexagonal lattice model. *FRAMCOS 8* (pp. 261-270). Barcelona, Spain: CIMNE.

Gerstle, W., Ingraffea, A. R., and Perucchio, R. (1988). Three-Dimensional Fatigue Crack Propagation Analysis Using the Boundary Element Method. *International Journal of Fatigue*, 189-192.

Gerstle, W., Sakhavand, N., Rahman, A., and Tuniki, B. K. (2011). Simulation of Concrete using Micropolar Peridynamic Hexagonal Lattice Model. *11th U.S. Nat. Congress on Computational Mechanics.* Minneapolis.

Gerstle, W., Sau, N., and Silling, S. (2005). Peridynamic modeling of plain and reinforced concrete structures. *18th Intl. Conf. on Structural Mechanics in Reactor Technology (SMiRT18).* Beijing, China: Atomic Energy Press.

Gerstle, W., Silling, S., Read, D., Tewary, V., and Lehoucq, R. (2008). Peridynamic Simulation of Electromigration. *Computers, Materials and Continua*, 75-92.

Gopalaratnam, V., Gerstle, W., Isenberg, J., and Mindess, S. (2004). *Dynamic Fracture: State of the Art.* Dearborn MI.: American Concrete Institute.

Grattan-Guinness, I. (1990). *Convolutions in French Mathematics, 1800-1840.* Basel-Boston-Berlin: Birkhauser Verlag.

Griffith, A. A. (1921). *The Phenomena of Rupture and Flow in Solids.* London: Phil. Trans. Roy. Soc. of London.

Hibbeler, R. C. (2005). *Mechanics of Materials.* London: Pearson Education.

Hill, R. (1962). Acceleration waves in solids. *Journal of the Mechanics and Physics of Solids*, 1-16.

Hillerborg, A., Modeer, M., and Petersson, P. E. (1976). Analysis of crack formation and crack growth in concrete by means of fracture mechanics and finite elements. *Cement and Concrete Research*, 773-782.

Honarvar Gheitanbaf, H. (2013). *MS Thesis: Parallel Simulation of Particle Dyanamics with Application to Micropolar Peridynamic Lattice Modeling of Reinforced Concrete Structures.* Albuquerque: University of New Mexico.

Hrennikoff, A. (1941). Solution of Problems in Elasticity by Framework Method. *ASME Journal of Applied Mechanics*, A169-A175.

Hughes, T. J. R., Cottrell, J. A., and Bazilevs, Y. (2005). Isogeometric analysis: CAD, finite elements, NURBS, exact geometry and mesh refinement. *Computer Methods in Applied Mechanics and Engineering*, 4135-4195.

Hui, C. Y. and Ruina, A. (1995). Why K? High order singularities and small scale yielding. *International Journal of Fracture, 72*, 97-120.

Hussain, M. A., Pu, S. I., and Underwood, J. (1974). Strain energy release rate for a crack under combined mode I and mode II. *Fracture Analysis, ASTM STP 560* (pp. 2-28). Philadelphia: American Society for Testing and Materials.

Inglis, C. E. (1913). *Stresses in a Plate due to the Presence of Cracks and Sharp Corners.* Institution of Naval Architects.

Irwin, G. R. (1957). *Analysis of stresses and strains near the end of a crack traversing a plate* (Vol. 24). Journal of Applied Mechanics.

Jirásek, M. (1998). Nonlocal models for damage and fracture: comparison and approaches. *International Journal of Solids and Structures*, 4133-4145.

Kachanov, M. (1958). Time of rupture processes under creep conditions. *Izv. Akad. Nauk. SSR, Otd. Nauk Tekh. Nauk.*, 26-31.

Krajcinovic, D. (1983). Creep of structure - a continuous damage mechanics approach. *Journal of Structural Mechanics*, 1-11.

Kroner, E. (1981). *Continuum Theory of Defects.* Amsterdam: North-Holland.

Krumhansl, J. A. (1968). Some considerations of the relation between solid state physics and generalized continuum mechanics. In E. Kroner, *Mechanics of Generalized Continua* (pp. 298-311). New York: Springer-Verlag.

Kunin, I. A. (1983). *Elastic Media with Microstructure II.* Berlin: Springer-Verlag.

Kutta, M. W. (1901). Beitrag zur näherungsweisen Integration totaler Differentialgleichungen. *Zeitschrift für Mathematik und Physik*, 435-453.

Lehoucq, R. B., and Silling, S. A. (2007). Force Flux and the Peridynamic Stress Tensor. *Journal of Mechanics and Physics of Solids.*

Lemaitre, J. and Chaboche, J.-L. (1974). A nonlinear model of creep-fatigue damage cumulation and interaction. *Proceedings of IUTAM symposium on mechanics of viscoelastic media and bodies.* Gothenburg: Springer Verlag.

Madenci, E. and Oterkus, E. (2014). *Peridynamic Theory and its Applications.* New York: Springer.

Man H.-K and Van Mier, J. G. M. (2007). Size and shape effects in fracture strength of concrete. *FramCos 6* (pp. 39-44). London: Taylor and Francis.

Maugin, G. A. (2013). *Continuum Mechanics Through the Twentieth Century.* Dordrecht: Springer.

Mazars, J., and Lemaitre, J. (1984). application of constitutuve damage mechanics to the strain and fracture behavior of concrete. In S. Shah, *Proceedings of the NATO advanced research workshop on application of fracture mechanics on cementitious composites* (pp. 507-520). Dordrecht, Netherlands: Evanston University.

Navier, C.-L. (1827). Memoire sur les lois de l'equilibre et du mouvement des corps solides elastiques. *Memoires de l'Institut*, 375-384.

Newmark, N. M. (1959). A method of computation for structural dynamics. *ASCE Journal of Engineering Mechanics*, 67-94.

Newton, I. (1999). *The Principia.* Berkeley: University of California Press.

Osakada, K. (2008). History of Plasticity and Metal Forming Analysis. *ICTP 2008 - 9th International Conference on Technology of Plasticity* (pp. 22-43). Gyeongju, Korea: KSTP, Korean Soc. for Technology of Plasticity.

Plimpton, S. J. (1995). Fast Parallel Algorithms for Short-Range Molecular Dynamics. *Computational Physics, 117*, 1-19.

Rabotnov, Y. N. (1971). Creep rupture under stress concentration. In A. A. Smith, *Advances in Creep Design* (pp. 3-19).

Radjaï, F. and Dubois, F. (2011). *Discrete Element Modeling of Granular Materials.* New York: John Wiley and Sons.

Rahman, A. S. (2011). *Lattice-Based Peridynamic Modeling of Linear Elastic Solids.* Albuquerque, NM: University of New Mexico.

Rankine, W. J. (1842). *On the causes of the unexpected breakage of the journals of railway axles, and on the means of preventing such accidents by observing the law of continuity in their construction.* Institution of Civil Engineers, Minutes of Proceedings.

Rashid, Y. R. (1968). Analysis of prestressed concrete pressure vessels. *Nuclear Engineering and Design*, 334-344.

Reidel, W. (1927). Beitrage zur Losung des ebenen Problems eines elastischen Korpers mittles der Airyschen Spannungsfunktion. *Zeitschrift fur angewandte Mathematik und Mechanik*, 169-188.

Rudnicki J. W. and Rice J. R. (1975). Conditions for the localization of the deformation in pressure sensitive dilatant materials. *Journal of Mechanics and Physics of Solids*, 371-394.

Sakhavand, N. (2011). *MS Thesis: Parallel Simulation Of Reinforced Concrete Structures Using Peridynamics.* Albuquerque: University of New Mexico.

Schlangen, E. and Van Mier, J. G. M. (1992). Experimental and numerical analysis of micromechanisms of fracture of cement-based composites. *Journal of Cement and Concrete Composites*, 105-118.

Sih, G. C. (1974). Strain-energy-density factor applied to mixed-mode crack problems. *International Journal of Fracture*, 305-321.

Silling, S. A. (2000). Reformulation of Elasticity Theory for Discontinuities and Long-Range Forces. *Journal of the Mechanics and Physics of Solids*, 175-209.

Silling, S.A., Epton, M., Weckner, O., Xu, J., and Askari, E. (2007). Peridynamic States and Constitutive Modeling. *Journal of Elasticity*, 151-184.

Tada, H., Paris, P. C., and Irwin, G. R. (2000). *The Stress Analysis of Cracks Handbook.* ASME Press.

Tadmor, E. B.; Miller, R. E.; and Elliot, R. S. (2012). *Continuum Mechanics and Thermodynamics.* Cambridge: Cambridge University Press.

Thomas, T. Y. (1961). *Plastic Flow and Fracture in Solids.* New York: Academic Press.

Timoshenko, S. P. (1953). *History of Strength of Materials.* New York: McGraw-Hill.

Truesdell, C. (1960). A Program toward Rediscovering the Rational Mechanics of the Age of Reason. *Archive of History of Exact Sciences*, 3-36.

Truesdell, C. (1965). *The Elements of Continuum Mechanics.* New York: Springer-Verlag.

Turner, M. J., Clough, R. W., Martin, H. C., and Topp, L. J. (1956). Stiffness and Deflection Analysis of Complex Structures. *Journal of the Aeronautical Sciences*, 805-823.

Verlet, L. (1967). Computer Experiments on Classical Fluids. I. Thermodynamical Properties of Lennard-Jones Molecules. *Physical Review Letters*, 98-103.

Warren, T. L., Silling, S. A., Askari, A., Weckner, O., Epton, M. A., and Xu, J. (2009). A non-ordinary state-based peridynamic method to model solid material deformation and fracture. *International Journal of Solids and Structures*, 1186-1195.

Weighardt, K. (1905). Uber einen Grenzubergang der Elastizitatslehre und Seine Anwendung auf die Statik hochgradig statisch unbestimmter Fachwerke. *Verhandlunge des Vereins zur Beforderung des Gewehrbefleisses*, 139-176.

Westergaard, H. M. (1939). *Bearing Pressures and Cracks* (Vol. 61). J. Appl. Mech.

Williams, M. L. (1957). *On the Stress Distribution at the Base of a Stationary Crack* (Vol. 24). J. Appl. Mech. Trans. ASME.

Xie, M. and Gerstle, W. (1995). Energy-Based Cohesive Crack Propagation Modeling. *ASCE J. Engng. Mechanics*, Vol. 121, No. 12.

Index

www.ingramcontent.com/pod-product-compliance
Lightning Source LLC
Chambersburg PA
CBHW050535190326
41458CB00007B/1787
9789814699549